Optical Holography

Optical Holography

Materials, Theory and Applications

Edited by

PIERRE-ALEXANDRE BLANCHE, PHD
Research Professor
College of Optical Sciences
The University of Arizona
Tucson, Arizona, United States

ELSEVIER

OPTICAL HOLOGRAPHY-MATERIALS, THEORY AND APPLICATIONS ISBN: 978-0-12-815467-0

Publisher: Matthew Deans
Acquisition Editor: Kayla Dos Santos
Editorial Project Manager: Fernanda Oliveira
Production Project Manager: Poulouse Joseph
Cover Designer: Alan Studholme

3251 Riverport Lane
St. Louis, Missouri 63043

List of Contributors

Pierre-Alexandre Blanche, PhD
Research Professor
College of Optical Sciences
The University of Arizona
Tucson, AZ, United States

V. Michael Bove, Jr., SB, SM, PhD
Principal Research Scientist
Media Lab
Massachusetts Institute of Technology
Cambridge, MA, United States

Marc Georges, PhD
Doctor
Centre Spatial de Liège – STAR Research Unit
Liège Université
Angleur, Belgium

Tom D. Milster, BSEE, PhD
Professor
College of Optical Sciences
University of Arizona
Tucson, AZ, United States

Professor
Electrical and Computer Engineering
University of Arizona
Tucson, AZ, United States

Silvio Montresor, PhD
Le Mans Université
LAUM CNRS 6613
Le Mans, France

Izabela Naydenova, PhD, MSc
Professor
School of Physics and Clinical and Optometric Sciences
College of Sciences and Health
TU Dublin
Dublin, Ireland

Pascal Picart, PhD
Professor
Le Mans Université
LAUM CNRS 6613
Le Mans, France

Ecole Nationale Superieure d'Ingenieurs du Mans
Le Mans, France

Vincent Toal, BSc, MSc, PhD
Centre for Industrial and Engineering Optics
Dublin Technological University
Dublin, Ireland

Director for Research
Optrace Ltd.
Dublin, Ireland

Preface

More than 70 years after its discovery, holography is still mesmerizing the public with its ability to display 3D images with crisp depth rendering and shimmering colors. Today, holograms are more than a curiosity, and they have found applications in a large variety of products ranging from security tags to head-up displays and gun sights. In addition to mirrors and lenses, holograms have become an essential tool that enables scientists to control light in novel ways.

However, one application eludes our quest: the highly anticipated holographic television. The reason holographic televisions are not available at your local electronics store, explained in detail in this book, is the extraordinarily large amount of information that must be processed and displayed in order to generate dynamic holograms. Fortunately, the emergence of new display technologies such as spatial light modulators and micromirror devices are helping engineers develop prototypes that are becoming more convincing. It is my belief that holographic television will emerge very soon.

Working in the field of holography is extremely gratifying because the research is at the forefront of some very exciting new techniques and developments. In recent years, we have seen the appearance of the holographic microscope, the holographic optical tweezers, and holographic sensors.

In this book, seven accomplished scientists explain where in their own field holography occupies a center stage. They guide the reader from the essential concepts to the latest discoveries.

The first chapter by Pierre-A Blanche is an introduction to the world of holography. It starts with a short history and takes the approach of describing holography using diffraction gratings, which can easily be generalized. This chapter explains the basic concepts such as thick vs. thin holograms or transmission vs. reflection geometries. The scalar theory of diffraction with its rigorous mathematical expressions is developed next. This chapter concludes with a section describing the major optical configurations that have been developed for recording holograms and how they produce holograms with different characteristics.

The second chapter, also by Pierre-A Blanche, describes holographic recording material and their processing. To understand the different material characteristics and metrics, this chapter starts by explaining the terminologies used in this field. Permanent materials that can only record the hologram once are introduced first, followed by refreshable materials where the hologram can be recorded, erased, and recorded again. This chapter also reviews electronic devices that can dynamically record or display holograms.

Chapter 3 by Tom D. Milster details algorithms that can compute holographic patterns, such as the Gerchberg–Saxton iterative Fourier transform algorithm. Starting from this seminal work, Milster discusses its convergence property and then expands to more modern variations that are now used to reduce noise and improve computational speed.

Michael Bove authored Chapter 4 about holographic television. After a brief overview of the different techniques that have been developed, the chapter discusses the limitations due to the very large spatiotemporal bandwidth required to generate dynamic holograms. As a way to overcome this limitation, different technologies of light modulators and microdisplays are introduced and their performance compared in the prospect of their use for the future holographic television. This chapter concludes with a very interesting take on holographic augmented and virtual reality.

In Chapter 5, Marc Georges presents the holographic interferometry technique. This technique allows the measurement of the phase of an object or a scene, which evolves over time, and is used to detect defects in laminated material. It can also be used for measuring the vibration modes of industrial components such as turbine blades. After defining the characteristics of an ideal system, Georges reviews the different implementations that have been proposed, moving from analog systems to the more modern electronic speckle pattern interferometry. Because the sensor resolution keeps improving, it is now possible to detect the interference fringes directly+, which leads to the most recent digital holographic interferometry techniques, which are described at the end of the chapter.

Chapter 6, written by Pascal Picart and Silvio Montresor, is dedicated to digital holography. Digital holography is the inverse problem of a computer-generated hologram and is about digitally reconstructing the optical wavefront from a recorded interference pattern. Picart and Montresor start by introducing the fundamentals of Fourier optics and then move to the different configurations for the recording of digital holograms, followed by the description of different algorithms for the numerical reconstruction of digital holograms. Finally, the noise in digital holographic images is discussed, and different techniques for its reduction are compared.

Holographic sensors are introduced in Chapter 7, where Izabela Naydenova describes this unique and fascinating aspect of holograms. Starting with a brief historical overview, the chapter describes holograms as a sensor platform, the fabrication of the photonic structures, and the different approaches to functionalize the holographic materials. The chapter ends by listing the challenges facing the future development of holographic sensors.

Chapter 8 is dedicated to the use of holography for security. In this chapter, Vincent Toal explains the problem of counterfeit products and its prevention using security tags such as holograms. This application is enabled by the mass production of holograms as well as their serialization, which are both described. What makes holograms so interesting for security is that they can be used in a nonimaging way such as match filtering and joint transform correlation. Toal also explains how encryption methods can be used to make the security even more unbreakable. Finally, holographic techniques for the imaging of concealed objects are presented.

Contents

CHAPTER 1

Introduction to Holographic Principles

PIERRE-ALEXANDRE BLANCHE, PHD

A SHORT HISTORY

Welcome to the beautiful world of holography. With their shimmering color and ghostlike appearance, holograms have taken a hold in the popular imagination, and buzz marketing alike. This is a rare accomplishment for a scientific technique, that worth to be noted. Together with this general appreciation, comes the misinterpretation. The word "hologram" is sometimes associated to the phenomena that have nothing to do with the scientific usage of the term. It is not problematic in everyday life, but it can become conflicting when the technology penetrates the market. We have all heard about holographic glass, holographic how from deceased artists, holographic television, princess Leia hologram, etc. Some are holograms indeed, some are not. This book will help demystify holography, and I hope it will help you gain a new appreciation for the technique that can be applied in a lot of different circumstances.

There exist three possible ways to alter or change the trajectory of light: reflection, refraction, and diffraction. In our everyday experiences, we mostly encounter reflections from mirrors and flat surfaces, and refraction when we look through water, or wear prescription glasses. Scientists have used reflection and refraction for over 400 years to engineer powerful instruments such as telescopes and microscopes. Isaac Newton [1] championed the classical theory of light propagation as particles, which accurately described reflection and refraction. Diffraction, on the other hand, could not be explained by this corpuscular theory, and was only understood much later with the concept of wave propagation of light, first described by Huygens [2], and extensively developed later by Young [3] and Fresnel [4].

Wave propagation theory predicts that when the light encounters an obstacle such as a slit, the edges do not"cut" a sharp border into the light beam, as the particle theory predicted, but rather there is formation of wavelets that propagate on the side in new directions. This is the diffraction phenomenon. Eventually, the particle and wave points of view will be reconciled by the quantum theory, and the duality of wave particle was developed by Schrödinger [5] and de Broglie. [6,7].

While the light propagation from mirrors and lenses can be explained with the thorough understanding of reflection and refraction, holography can only be explained by recognizing diffraction. A hologram is nothing but a collection of precisely positioned apertures that diffracts the light and forms a complex wave front such as a three-dimensional (3D) image. In addition because the light is considered as a wave in these circumstances, both the amplitude and the phase can be modulated to form the hologram. Amplitude modulation means local variation of absorption, and phase modulation means a change in the index of refraction or thickness of the material. In the latter case of phase modulation, the holographic media can be totally transparent, which account for a potentially much efficient diffraction of the incident light. We will describe the different properties of holograms in Section 2.

Holograms are very well known for the awe-inspiring 3D images they can recreate. But they can also be used to generate arbitrary wavefronts. Examples of such wavefronts are focalization exactly like a lens, or reflection exactly like a mirror. The difference of the hologram from the original element (lens or mirror) is that, in both cases, diffraction is involved, not reflection or refraction. That type of hologram, called holographic optical element, is found in optical setups where for reason of space, weight, size, complexity, or when it is not possible to use classical optical elements. Some examples include combiner in head-up display, dispersion grating in spectrometers, or spot array generators for cameras and laser pointers.

There are two very different techniques for manufacturing holograms. One can either compute it or record it optically. Computing a hologram involves the calculation of the position of the apertures and/or

Optical Holography-Materials, Theory and Applications. https://doi.org/10.1016/B978-0-12-815467-0.00001-3

phase shifters, according to the laws of light propagation derived by Maxwell [8]. This calculation can be fairly easy for simple wavefronts such as a lens, for extremely complicated for high-resolution 3D images. On the other hand, optically recording a hologram implies the registration of both the amplitude and the phase of the wavefront. Capturing the light intensity was first achieved with the invention of photography by Niépce in 1822. But recording the phase eluded scientists until 1948. Although the concept of optical interference was known for ages, it is only when Dennis Gabor introduced the concept of making an object beam interfere with a reference beam that recording the phase became possible [9,10]. Indeed, when two coherent beams intersect, constructive and destructive interferences occurs according to the phase difference, this transforms the phase information into intensity information that can be recorded the same way photographs are taken. In some sense, the reference beam is used to generate a wave carrier that is modulated by the information provided by the object wave (similarly to AM radio).

Gabor coined the term holographic from the Greek words *holos*: "whole" and *graphe*: "drawing" because the technique recorded for the first time the entire light field information: amplitude and phase. Gabor used the technique to increase the resolution in electron microscopy and received the Nobel Prize in Physics in 1971 for this discovery.

Owing to the very short coherence length of the light sources available to Gabor at the time, the object and reference beams required to be colinear. Unfortunately, this configuration yield to very poor imaging quality because the transmitted beam and ± 1 diffracted orders were superimposed, leading to high noise and a "twin-image" problem.

Holographic imaging will have to wait for the invention of the visible light laser in 1960 by Maiman [11], and for Leith and Upatnieks to resolve the twin-image problem [12,13]. Using a long coherence length laser source, one may divide a beam into two parts—one to illuminate the object (the object beam) and the other (the reference beam) is collimated and incident at an angle to the hologram recording material. As a result of the high degree of coherence, the object and reference beams will still interfere to form the complex interference pattern that we call the hologram. On reconstruction, a monochromatic beam is incident to the recorded hologram and the different diffracted waves are angularly separated. This way, the 0, $+1$, and -1 orders can be observed independently, solving the problem of both noise and twin images observed in in-line holograms. Section 6 will describe the different configuration to record holograms.

In parallel, and independently to Leith and Upatnieks, Denisyuk worked on holograms where the object and reference beams are incident the hologram plane from opposite directions [14–16]. Such holograms are formed by placing the photosensitive medium between the light source and a diffusely reflecting object. In addition of being much simpler and more stable to record, these reflection holograms can be viewed by a white light source because only a narrow wavelength region is reflected back in the reconstruction process. We will see the fundamental reason for this selectivity in Section 2.3 about the characteristics of thick holograms.

Once high-quality imaging and computer-generated holograms (CGHs) were demonstrated [17,18], the research on holography experienced a phenomenal growth, expanding to encompass a large variety of applications such as data storage [19], information processing [20], interferometry [21], and dynamic holography [22] to cite only a few. Today, with the widespread access to active LCoS and MEMS devices, there is a rejuvenation of the holographic field where a new generation of researchers is applying the discoveries of the past decades to electronic-controlled spatial light modulators. New applications are only limited by the imagination of scientists and engineers, and developments are continuously being reported in the scientific literature.

This chapter will continue by developing the theory of thick and thin diffraction gratings. Once these bases have been established, we will move to the scalar theory of diffraction that shows how to calculate the light field from a diffractive element, and vice versa. We will finish by describing several important experimental setup used to record holograms.

DIFFRACTION GRATINGS

Waves and Interference

A great deal can be understood about holography without the complication of imaging, and by simply looking at the properties of diffraction gratings. Diffraction gratings are particular holograms where the interferences fringes, or Bragg's planes, are parallel. As such, they transform one plane wave into another plane wave with a different direction. This simple action on the light beam makes the mathematical formalism much easier to understand.

After the analysis of simple gratings, holographic images can simply be viewed as the superposition of several planar wavefronts, and the hologram itself can be viewed as the superposition of several gratings, much like Fresnel and Fourier decompositions.

Maxwell's equation defines the properties of the electromagnetic field. In addition, for most holographic applications, the magnetic field can be neglected without loss of generality. In that case, only the Helmholtz equation remains to define the electric field **E**:

$$\left(\frac{1}{c^2}\right)\left(\frac{\partial^2 \mathbf{E}}{\partial t^2}\right) - \nabla^2 \mathbf{E} = 0 \tag{1.1}$$

with c being the speed of light and bold face font used to represent vectors.

A solution of this differential equation has the form of a plane wave:

$$\mathbf{E}(\mathbf{r}, t) = \mathbf{A}\cos(\mathbf{k}\cdot\mathbf{r} - \omega t + \phi) \tag{1.2}$$

where **A** is an imaginary vector describing the direction of the electric field oscillation, and contain the polarization information, **k** is the wave vector pointing in the direction of light propagation which magnitude is related to the wavelength $|\mathbf{k}| = 2\pi/\lambda$. **r** is the position vector defining the position at which the field is calculated, ω is the frequency, and ϕ the phase of the wave. Two equivalent representation of a plane wave are illustrated in Fig. 1.1. It has to be noted that a spherical wavefront is also solution of the Helmholtz equation.

Using Euler's formula $\exp(ix) = cosx + i\ sinx$, the plane wave solution can be rewritten as:

$$\mathbf{U}(\mathbf{r}, t) = A \quad \exp[i(\mathbf{k}\cdot\mathbf{r} - \omega t + \phi]\ \hat{\mathbf{a}} \tag{1.3}$$

where the polarization vector $\hat{\mathbf{a}}$ has been extracted from the amplitude vector **A** which is now the scalar A.

One need to keep in mind that the actual electric field **E** is the real part of the complex notation **U** in Eq. (1.3):

$$\mathbf{E}(\mathbf{r}, t) = \Re[\mathbf{U}(\mathbf{r}, t)] = \frac{1}{2}U(\mathbf{r}, t) + \frac{1}{2}U^*(\mathbf{r}, t) \tag{1.4}$$

where the * denotes the complex conjugate.

When two plane waves of the form of Eq. (1.3) cross each other, interference occurs. The total field can be described as:

$$\begin{aligned}\mathbf{U}_{total}(\mathbf{r}, t) = {} & A_1 \exp[i(\mathbf{k}_1\cdot\mathbf{r} - \omega_1 t + \phi_1]\hat{\mathbf{a}}_1 \\ & + A_2 \exp[i(\mathbf{k}_2\cdot\mathbf{r} - \omega_2 t + \phi_2]\hat{\mathbf{a}}_2\end{aligned} \tag{1.5}$$

where the subscripts number 1 and 2 describing the two waves.

In this formulation, we can see that the pattern is not necessarily static but change as a function of time. It is only in the special case where $\omega_1 = \omega_2$ that Eq. (1.5) becomes time invariant and can be expressed in a simpler form, where the total intensity is

$$\begin{aligned} I(\mathbf{r}) = {} & \int \mathbf{U}_{total}(\mathbf{r}, t)\mathbf{U}^*_{total}(\mathbf{r}, t) \\ & \propto A_1^2 + A_2^2 + 2A_1A_2|\hat{\mathbf{a}}_1\cdot\hat{\mathbf{a}}_2|\cos[(\mathbf{k}_1 - \mathbf{k}_2)\cdot\mathbf{r} + \arg(\hat{\mathbf{a}}_1\cdot\hat{\mathbf{a}}_2)] \\ & = I_1 + I_2 + 2\sqrt{I_1 I_2}|\hat{\mathbf{a}}_1\cdot\hat{\mathbf{a}}_2|\cos[(\mathbf{k}_1 - \mathbf{k}_2)\cdot\mathbf{r} + \arg(\hat{\mathbf{a}}_1\cdot\hat{\mathbf{a}}_2)]\end{aligned} \tag{1.6}$$

To maximize the contrast between dark (destructive interference) and bright (constructive interference) regions of the interference pattern, the polarization of the wave should be identical $\hat{\mathbf{a}}_1 = \hat{\mathbf{a}}_2$, and the equation reduces to the familiar form:

$$I(\mathbf{r}) = I_1 + I_2 + 2\sqrt{I_1 I_2}\cos[(\mathbf{k}_1 - \mathbf{k}_2)\cdot\mathbf{r}] \tag{1.7}$$

This intensity modulation can be recorded inside a material as an index modulation or absorption modulation pattern to form a diffraction grating.

Eq. (1.7) describing the intensity modulation in space, can be recast as a static plane wave with a wave vector defined as:

$$\mathbf{K} = \mathbf{k}_1 - \mathbf{k}_2 \tag{1.8}$$

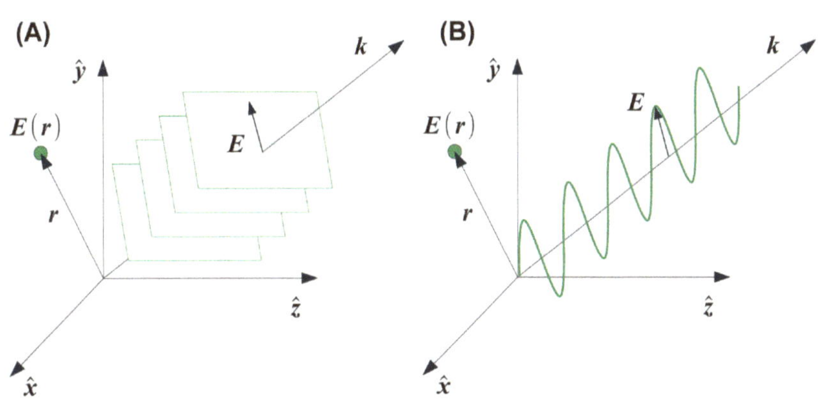

FIG. 1.1 Plane wave representation as **(A)**: plane of equal field intensity or **(B)**: oscillation of the amplitude of the field along the wave vector **k**.

Similar to the definition of $\mathbf{k}_1$ and $\mathbf{k}_2$ that are the wave vectors of the light beams, $\mathbf{K}$ is the grating vector of the interference pattern. The magnitude of the grating vector $\mathbf{k}_2$ is related to the spacing Λ between two planes of equal magnitude, also called the Bragg's planes:

$$|\mathbf{K}| = \frac{2\pi}{\Lambda} \tag{1.9}$$

During the reading of a diffraction grating, an incident plane wave defined by the wave vector $\mathbf{k_i}$ encounters the recorded grating $\mathbf{K}$ and is diffracted in the direction $\mathbf{k_d}$ according to the K-vector closing condition:

$$|\mathbf{K}| = \mathbf{k_d} - \mathbf{k_i} \tag{1.10}$$

The K-vector closing condition is identical to the grating equation devised from crystallographic measurements where the modulation planes were actually rows of atoms:

$$\sin(\theta_d) - sin(-\theta_i) = m\frac{\lambda}{\Lambda} \tag{1.11}$$

where θ_d is the angle of diffraction, θ_i is the angle of incidence, and m is an integer number that defines the diffraction order.

Both geometries are shown in Fig. 1.2 with the definition of the angles.

The angular dispersion as a function of the frequency can be directly derived from the grating Eq. (1.11)

$$\frac{d\theta_d}{d\lambda} = \frac{m}{\Lambda\cos\theta_d} \tag{1.12}$$

From Eq. (1.12), it can be seen that the lower frequencies (red) are diffracted at a larger angle than the higher frequencies (blue). This is the reverse from what is observed with a refractive prism (with normal index dispersion), where higher frequencies exit at a larger angle. This opposition can be used to make an optical system achromatic, with the diffractive element compensating the dispersion of the refractive lens.

The direction of the diffraction maximums are given by the Bragg's law that can be understood as the condition for constructive interference for the light interacting with two successive diffraction planes:

$$\sin(\theta_B + \varphi) = m\frac{\lambda}{2n\Lambda} \tag{1.13}$$

where θ_B is the angle of incidence for which there is a maximum of diffracted intensity, also called Bragg's angle, φ is the slant angle of the grating, that is, the angle between the grating vector and the normal to the grating surface (see Fig. 1.2).

It should be noted that the grating equations expressed in Eqs. (1.10)–(1.13) do not give any indication on the intensity of the wave being diffracted, only the *direction and frequency*. The calculation of the wave intensity, or diffraction efficiency, according to the grating parameters will be derived in Section 2.3 for thick volume grating, and in Section 2.4 for thin or surface relief grating with different format modulation.

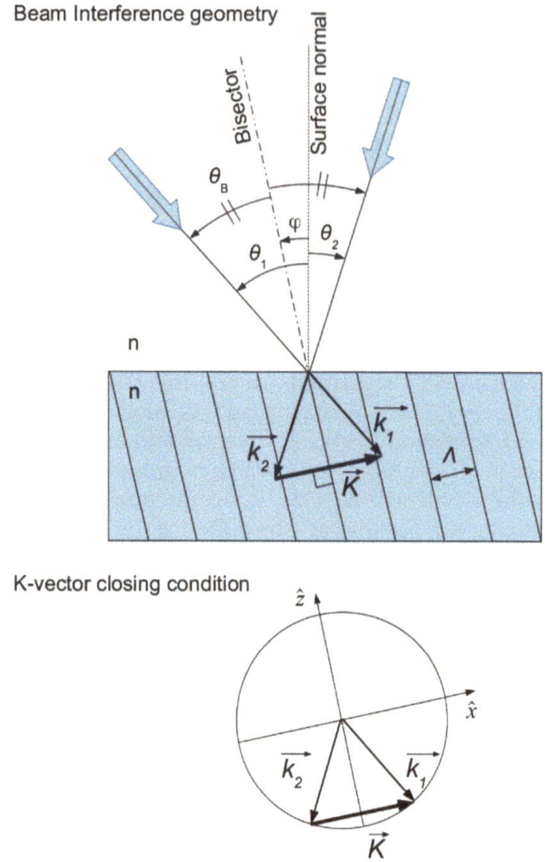

FIG. 1.2 Definition of the angles and vectors for the beam interference geometry, and K-vector closing condition for the same geometry.

Point Source Interference

Armed with the general equation for the interference between two waves (Eq. 1.7), let us derive some specific cases and observe the pattern formed by the fringes.

Two-plane waves

In the case of two-plane wave that have different incidence angles, the phase and intensity are respectively given by:

$$\phi_i = \frac{2\pi}{\lambda}x\sin\theta_i \qquad I_i = 1 \tag{1.14}$$

Inserted into Eq. (1.7), we found for the interference pattern:

$$I(x) = 2 + 2\cos\left[\frac{2\pi}{\lambda}x(\sin\theta_1 - \sin\theta_2)\right] \tag{1.15}$$

which is identical to the grating Eq. (1.11) with a pattern frequency of:

$$\Lambda = \frac{\sin\theta_1 - \sin\theta_2}{\lambda} \tag{1.16}$$

The geometry of this configuration along with the interference pattern formed is shown in Fig. 1.3.

We can see that the interference pattern only varies along one dimension (x). The recording of this pattern inside a material forms a diffraction grating.

Arbitrary point sources

Instead of using plane waves, we can use spherical waves with arbitrary origins (x_i, y_i, z_i). Their phase and intensity is now described by:

$$\phi_i = \frac{2\pi}{\lambda}\left[\sqrt{(x-x_i)^2 + (y-y_i)^2 + (z-z_i)^2}\right]$$
$$I_i = \left(\frac{A_i}{r_i}\right)^2 \tag{1.17}$$

In all generalities, the interference pattern is

$$I(P) = \left(\frac{A_1}{r_1}\right)^2 + \left(\frac{A_2}{r_2}\right)^2 + 2\left(\frac{A_1}{r_1}\right)^2\left(\frac{A_2}{r_2}\right)^2$$
$$\times\cos\left[\frac{2\pi}{\lambda}\left(\sqrt{\left(x-\frac{\Delta x}{2}\right)^2 + \left(y-\frac{\Delta y}{2}\right)^2 + \left(z-\frac{\Delta z}{2}\right)^2}\right.\right.$$
$$\left.\left.-\sqrt{\left(x+\frac{\Delta x}{2}\right)^2 + \left(y+\frac{\Delta y}{2}\right)^2 + \left(z+\frac{\Delta z}{2}\right)^2}\right)\right] \tag{1.18}$$

with $\Delta x = x_1 - x_2$.

This expression only became interesting by looking at particular cases such as the two that follow.

Side-by-side point sources

For point sources that are located side by side along the x axis, with a separation distance of Δx, and if we consider a constant intensity, we can describe the phase and intensity as:

$$\phi_i(x,y) = \frac{2\pi}{\lambda}\left[\sqrt{\left(x \pm \frac{\Delta x}{2}\right)^2 + y^2 + z^2}\right]$$
$$I_i = 1 \tag{1.19}$$

So the interference pattern becomes a relatively simple expression:

$$I(P) = 2 + 2\cos\left[\frac{2\pi}{\lambda}\left(\sqrt{\left(x-\frac{\Delta x}{2}\right) + y^2 + z^2}\right.\right.$$
$$\left.\left.-\sqrt{\left(x+\frac{\Delta x}{2}\right) + y^2 + z^2}\right)\right] \tag{1.20}$$

The geometry of this configuration along with the interference pattern formed is shown in Fig. 1.4.

Collinear point sources

For point sources that are located on the z axis but at different distances, and considering a constant intensity, the phase and intensity are given by:

$$\phi_i(x,y) = \frac{2\pi}{\lambda}z_i + \frac{\pi}{\lambda z_i}(x^2 + y^2)$$
$$I_i = 1 \tag{1.21}$$

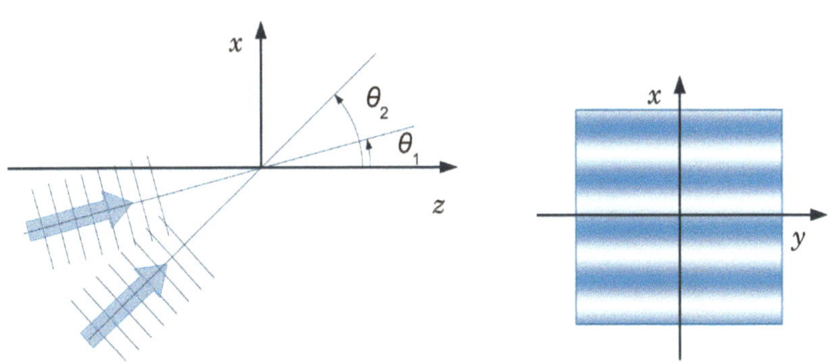

FIG. 1.3 Geometry and interference pattern produced by two-plane waves incident at different angles.

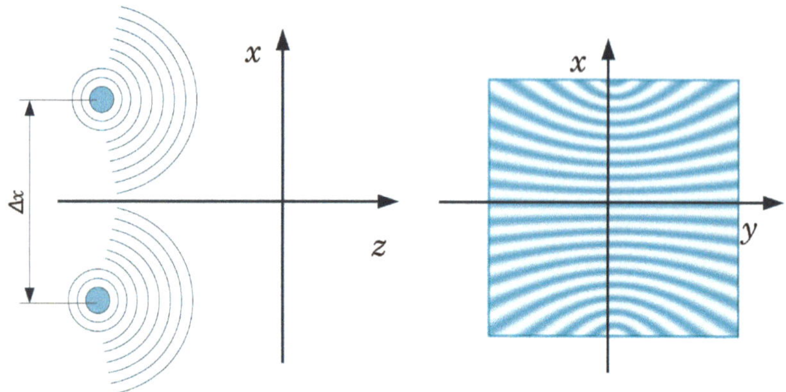

FIG. 1.4 Geometry and interference pattern formed by two point sources located side by side.

The interference pattern is

$$I(P) = 2 + 2\cos\left[\frac{\pi}{\lambda}\left(\frac{1}{z_1} - \frac{1}{z_2}(y^2 + z^2)\right)\right] \tag{1.22}$$

The geometry of this configuration along with the interference pattern formed is shown in Fig. 1.5. We will see an identical pattern when we will study the Gabor zone plate in Section 3.4.4.

It has to be noted that from the perspective of the point sources, the two cases we just developed are identical. The two interference patterns are formed either on the side, or along the axis of separation of the two sources.

Thick Grating's Characteristics

The interference pattern formed by two-plane waves was introduced in Fig. 1.3. Once this pattern is recorded inside a material, it forms a diffraction grating. The direction of the diffracted wave can be determined using the grating Eq. (1.11). Now we are going to study the distribution of energy around the Bragg's angle and the Bragg's wavelength for thick diffraction gratings.

These distributions have first been derived by Kogelnik in his coupled wave theory [23]. Another derivation that also give very good results is called the parallel stacked mirror model and has been introduced by Brotherton-Ratcliffe [24,25]. These two models give analytical solutions in the case when the grating satisfies the Bragg's condition for "thick" gratings.

Thick grating criteria

This thick grating condition is somewhat misnamed because it is not based on the physical thickness of the material, but on the premise that most of the energy is concentrated in the first diffraction order. This condition of operation is also called the Bragg regime and is indeed observed in gratings and holograms for which the recoding media are rather thick. This is because, in this condition, the incident beam interacts several times

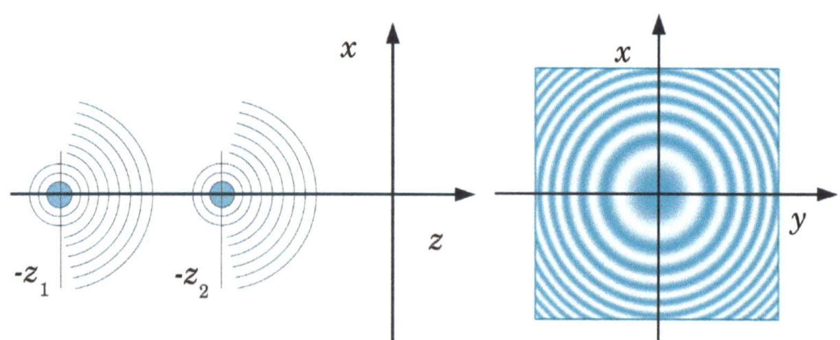

FIG. 1.5 Geometry and interference pattern formed by two point sources located along the axis of light propagation.

with the grating structure, and there is a progressive transfer of energy into the diffracted beam.

By contrast, "thin" gratings operated in the Raman-Nath regime of diffraction where an appreciable amount of energy can be found in higher orders of diffraction. Owing to their smaller thickness, the transfer of energy is not restricted to the first orders. The energy distribution diffracted by thin grating cannot be calculated using the Kogelnik theory or the parallel stacked mirror model and requires the more extensive and laborious rigorous coupled wave analysis developed by Moharam and Gaylord [26].

There is not a clear dividing boundary between thin and thick gratings. Instead, several criteria have been devised according to the approximations used in solving the coupled wave equation, and according to the results observed experimentally.

Two of the most used criteria to distinguish between thick and thin gratings are the Klein and Cook criteria [27], and the Moharam and Young criteria [28].

Klein and Cook criteria:

$$Q' = \frac{2\pi\lambda d}{n\Lambda^2 \cos\theta} \tag{1.23}$$

with $Q' < 1$ for thin gratings, and $Q' > 10$ for thick gratings.

The Moharam and Young criteria:

$$\rho = \frac{\lambda^2}{n\Delta n\Lambda^2 \cos\theta} \tag{1.24}$$

where Δn is the material index modulation, and $\rho < 1$ defines thin gratings, when $\rho \geq 1$ defines thick gratings. We can see that this criterion does not even take the physical thickness (d) of the grating into account.

To satisfy the thick grating criterion, the Bragg's planes need to extend to a certain volume inside the material (thus the name). Such a diffraction structure cannot be just overlaid on the surface. The advantage of thick grating is that most of the diffracted energy is found in the first order. For that reason, thick grating are of particular interest in holographic imaging and engineering because one does not have to deal with light present in higher diffraction orders, which induce noise and reduce efficiency in the desired first-order image.

Efficiency of thick gratings

The manufacturing of thick grating generally involves the recording of an interference pattern using an optical setup, techniques that will be detailed in Section 6. The reason for optical recording is that the diffractive structures need to be embedded inside the volume of the material, which is difficult to access otherwise. It is also possible to produce diffractive structures that satisfy the thick grating conditions using multilayer coating of dielectric layers. Such structures are better known as dichroic mirrors or interference filters.

The rigorous derivation of thick grating diffraction efficiency, expressed as the ratio between the wave intensity in the first order and the incident intensity, can be found in the original publications [23,24]. Here, we will summarize the principal results in the special cases of unslanted ($\varphi = 0$ or $\pi/2$), phase (Δn), and amplitude ($\Delta\alpha$) sinusoidal modulation. For these specific conditions, the mathematical expressions simplify dramatically, and it is helpful to keep in mind the general trend as they give a good intuition for other cases.

In addition to the modulation format (phase and amplitude), two different configurations of the grating will be discussed: transmission and reflection. Illustration of these two geometries are shown in Fig. 1.6, where a slant angle φ has been introduced for the sake of generality.

In transmission geometry, the diffracted light exits the grating by the opposite side of the incident light: the light goes through the grating. To do so, the Bragg's planes are oriented more or less orthogonal to the grating surface. The grating frequency in transmission geometry range from 300 to 3000 line pairs per mm (lp/mm) for visible light.

In the reflection geometry, the diffracted light exits the grating by the same side of the incident light, the light reflected back from the grating. In this geometry, the Bragg's planes are oriented more or less parallel to the surface, and the grating frequency is over 4000 lp/mm.

Phase gratings, where the index of refraction is modulated, can reach 100% efficiency either in transmission or in reflection. The expression for TE (transverse electrical) mode is respectively given by:

$$\eta_{TE} = \sin^2\left[\left(\frac{\pi\Delta n}{\lambda}\right)\left(\frac{d}{\cos\theta_i}\right)\right] \tag{1.25}$$

For transmission configuration and:

$$\eta_{TE} = \tanh^2\left[\left(\frac{\pi\Delta n}{\lambda}\right)\left(\frac{d}{\cos\theta_i}\right)\right] \tag{1.26}$$

For reflection configuration.

Plots of these functions according to the index modulation amplitude (Δn) are shown in Fig. 1.7.

It should be observed that in the case of a transmission grating, the efficiency is a periodic function (sin) that reaches its maximum when the phase modulation equal $m\pi/2$ (m being an integer). When the phase modulation extends past the first maximum, the grating is said to be overmodulated, and the light starts to be

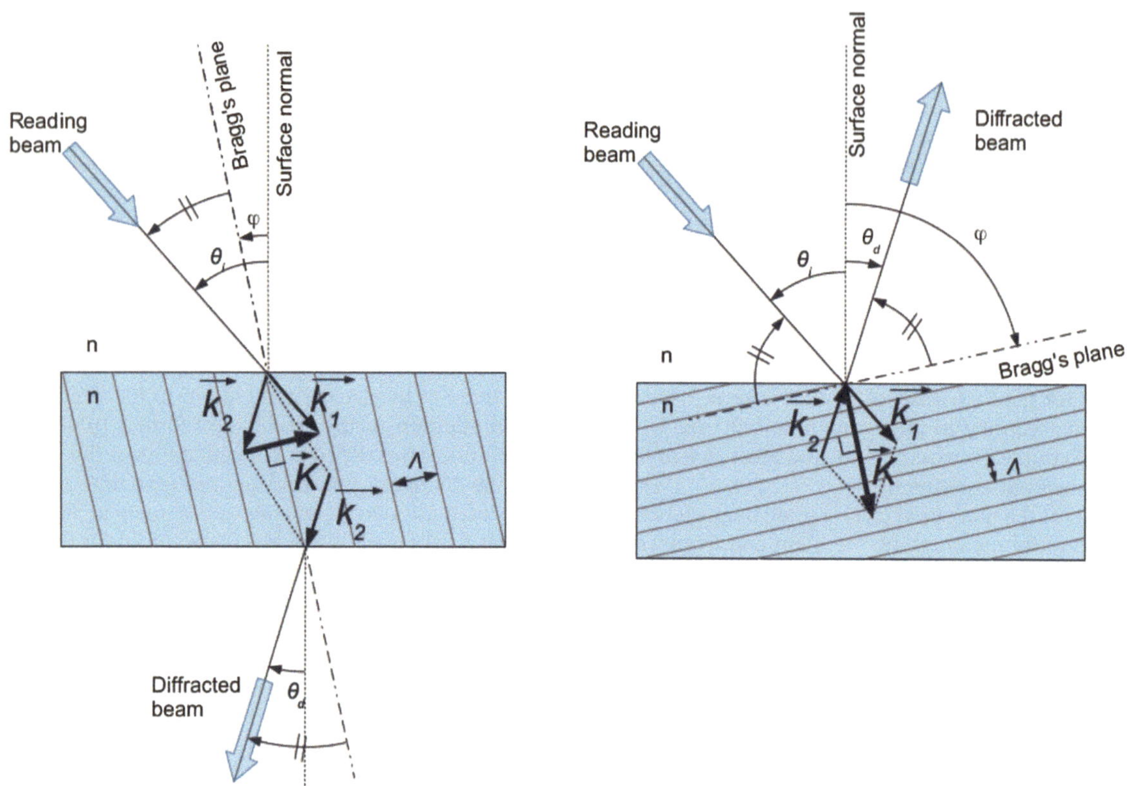

FIG. 1.6 Definition of the angles and vectors for transmission (left) and reflection (right) gratings. The K-vector closure condition is also drawn for each geometry.

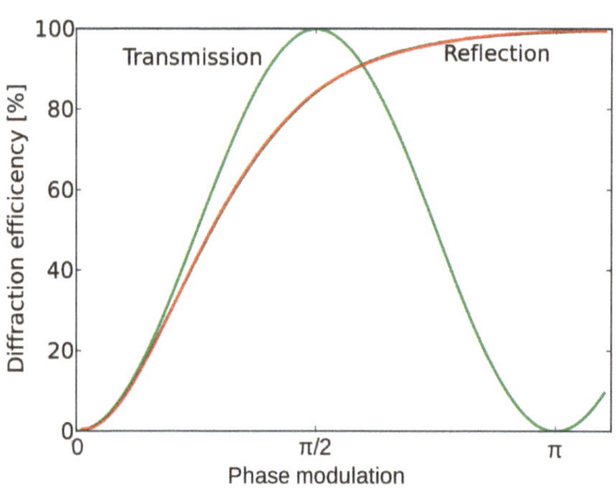

FIG. 1.7 Diffraction efficiency of a phase Bragg's grating in transmission or reflection according to the index modulation amplitude Δn.

coupled back to the zero order, reducing the efficiency. The minimum is reached when $\Delta n = m\pi$.

In the reflection configuration, the diffraction efficiency monotonically increases, and there is no overmodulation possible. The reason is that the light transferred into the first order exits the media and does not propagate further into the volume where it would have had a chance to be coupled back into the zero order.

Amplitude gratings are less efficient than phase gratings because the modulation is based on the absorption of a portion of the incident light. Volume amplitude gratings do exceed 7.2% efficiency. The expression for TE for a transmission grating is

$$\eta_{TE} = \exp\left(\frac{-2\alpha d}{\cos\theta_i}\right)\sinh^2\left[\left(\frac{\Delta\alpha}{2}\right)\left(\frac{d}{\cos\theta_i}\right)\right] \quad (1.27)$$

where α is the absorption coefficient and can reach values higher than 1, $\Delta\alpha$ is the modulation of that coefficient.

Eq. (1.27) exhibits a maximum efficiency of 3.7% when $\alpha = \Delta\alpha = \ln 3$.

For any value of the absorption α, the maximum efficiency is achieved when the modulation is maximum: $\Delta\alpha = \alpha$.

For the reflection case, the TE efficiency is given by:

$$\eta_{TE} = \frac{-\Delta\alpha^2}{4}\left[A + \sqrt{A^2 - \left(\frac{\Delta\alpha^2}{4}\right)}\coth\sqrt{A^2 - \left(\frac{\Delta\alpha^2}{4}\right)}\right]^{-2} \quad (1.28)$$

where the parameter A denotes the absorption term: $A = \alpha d/\cos\theta_i$.

The efficiency for the reflection amplitude grating monotonously increases with the modulation, asymptotically approaching a maximum of 7.2%.

Plots of the diffraction efficiency according to the modulation when $\alpha = \Delta\alpha$ for transmission (Eq. 1.27) and reflection (Eq. 1.28) Bragg's gratings are shown in Fig. 1.8.

The solutions for the TM (transverse magnetic) mode should be adjusted from the TE solutions with the coupling factor $\kappa_{\|}$ added to the modulation factor, either Δn or $\Delta\alpha$:

$$\kappa_{\|} = cos(2\theta_i) \quad (1.29)$$

Dispersion of thick gratings

The coupled wave theory also allows the derivation of dispersion for each configuration according to wavelength and the angular incidence [23]. More importantly, one needs to recognize that even with the same characteristics of thickness and modulation amplitude, transmission and reflection gratings have very different behaviors when it comes to dispersion and selectivity.

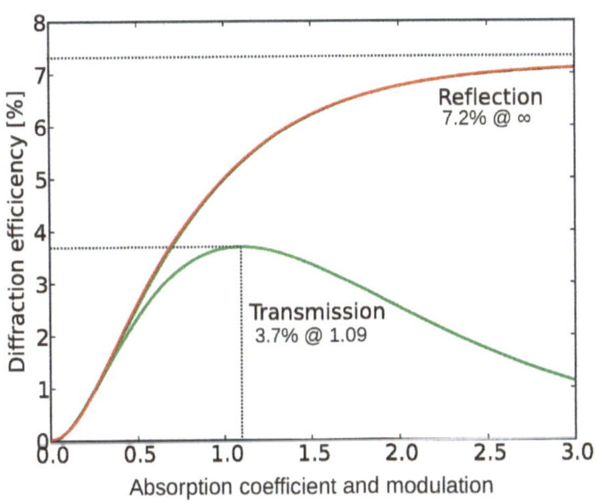

FIG. 1.8 Diffraction efficiency of an amplitude Bragg's grating in transmission or reflection according to the absorption coefficient, and assuming maximum modulation amplitude. $\alpha = \Delta\alpha$.

This is due to the very different Bragg's plane frequencies (Λ) observed for these geometries. In the transmission geometry, the grating frequency ranges from 300 to 3000 lp/mm, whereas in the reflection geometry the grating frequency is more than 4000 lp/mm.

The expressions for phase gratings in TE mode for the different cases are the following:

TRANSMISSION CONFIGURATION

Angular Dispersion

$$\eta(\theta_i) = \frac{\sin^2\left(\sqrt{\Delta\theta_i}\,\frac{\pi\Delta nd}{\lambda_B\cos\theta_i}\right)}{\Delta\theta_i} \quad (1.30)$$

with the detuning parameter $\Delta\theta_i$ given by:

$$\Delta\theta_i = 1 + \left(\frac{2n\sin\theta_B(\sin\theta_i - \sin\theta_B)}{\Delta n}\right)^2 \quad (1.31)$$

Spectral dispersion

$$\eta(\lambda_i) = \frac{\sin^2\left(\sqrt{\Delta\lambda_i}\,\frac{\pi\Delta nd}{\lambda_i\cos\theta_B}\right)}{\Delta\lambda_i} \quad (1.32)$$

with the detuning parameter $\Delta\lambda_i$ given by:

$$\Delta\lambda_i = 1 + \left[\frac{2n\frac{\lambda_i}{\lambda_B}\left[1 - \cos(2\theta_B)\left(1 - \frac{\lambda_i}{\lambda_B}\right)\right]}{\Delta n}\right]^2 \quad (1.33)$$

REFLECTION CONFIGURATION

Angular Dispersion

$$\eta(\theta_i) = \frac{\sinh^2(\Delta\theta_i A_i)}{\sinh^2(\Delta\theta_i A_i) + \Delta\theta_i^2} \quad (1.34)$$

With the parameter A_i given by:

$$A_i = \frac{\pi\Delta nd}{\sqrt{|\cos\theta_i - \cos\theta_B|\cos\theta_i}\,\lambda_B} \quad (1.35)$$

And the detuning parameter $\Delta\theta_i$ is given by:

$$\Delta\theta_i = \left[1 - \frac{4n^2(\cos\theta_i - \cos\theta_B)^2\cos\theta_i cos^2\theta_B}{|\cos\theta_i - 2\cos\theta_B|\Delta n^2}\right]^{1/2} \quad (1.36)$$

Spectral dispersion

$$\eta(\lambda_i) = \frac{\sinh^2(\Delta\lambda_i B_i)}{\sinh^2(\Delta\lambda_i B_i) + \Delta\theta_i^2} \quad (1.37)$$

With the parameter B_i given by:

$$B_i = \frac{\pi \Delta n d}{\cos\theta_B \lambda_i} \tag{1.38}$$

And the detuning parameter $\Delta\lambda_i$ given by:

$$\Delta\lambda_i = \left[1 - \frac{4n^2(\lambda_i/\lambda_B)^2\cos^4\theta_B[1-(\lambda_i/\lambda_B)]^2}{|1-2(\lambda_i/\lambda_B)|\Delta n^2}\right]^{1/2} \tag{1.39}$$

Note that these dispersion Eqs. (1.30–1.34), (1.37) can be further approximated by sinc^2 functions with the appropriate detuning coefficients.

Typical angular and spectral dispersion characteristics of reflection Bragg's gratings are illustrated in Figs. 1.9 and 1.10. Generally speaking, transmission grating are diffracting a large bandwidth of wavelengths each at a very specific angles, they produce a rainbow. They are wavelength tolerant and angularly selective. On the other hand, reflection gratings diffract a very narrow bandwidth at any angle, they act as a notch filter, reflecting one single color. Reflection Bragg's gratings are wavelength selective and angularly tolerant.

Although these rules of thumb behavior for the different type of gratings are very useful to keep in mind, they can be proved wrong for particular cases such as edge-lit gratings ($\varphi = \pi/4$) that fall in between the two categories, or for very thick gratings ($d > 100$ μm) that are extremely selective in both angle and wavelength independently of their configuration.

Edge-lit gratings have a slant angle close to 45° (Fig. 1.11). Their name comes from the fact that to achieve this angle, one of the beams needs to be incident from the side (edge) of the material. This type of grating is useful for injecting or extracting the light to and from a waveguide. This type of configuration, using a waveguide, has recently gained popularity for solar concentration application [29], augmented reality see-through displays [30,31], and head-up display [32].

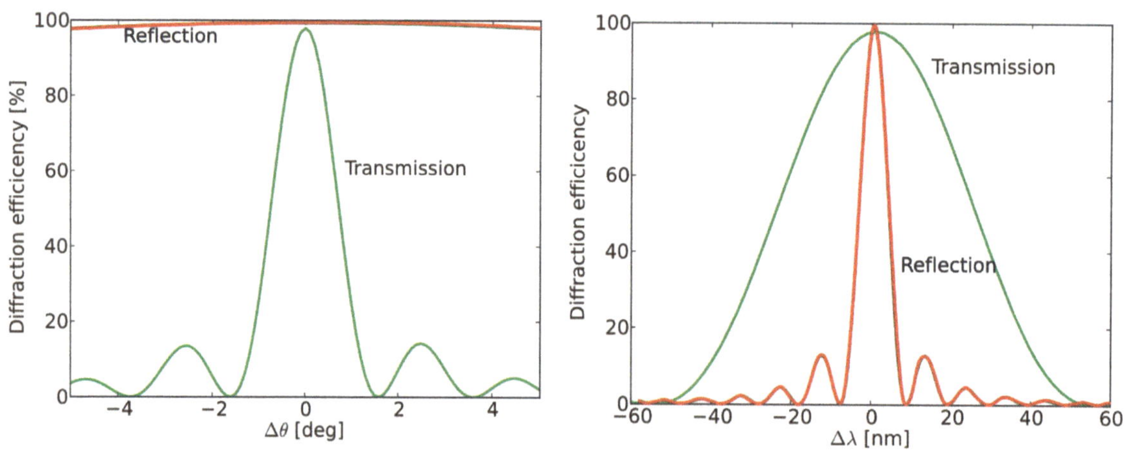

FIG. 1.9 Typical angular (left) and spectral (right) dispersion of a transmission or reflection Bragg's gratings. Generally speaking, transmission gratings are angularly selective, when reflection gratings are wavelength selective.

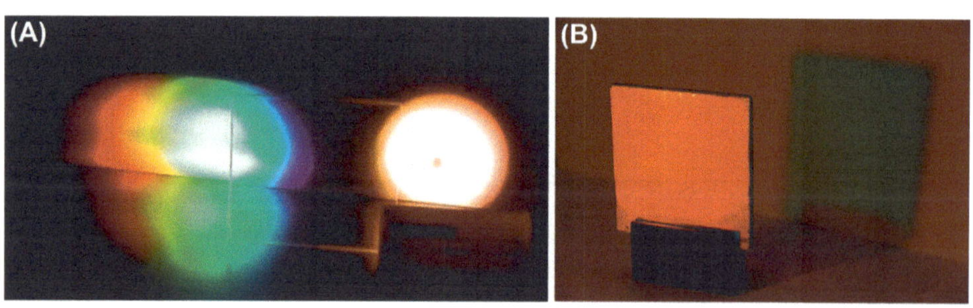

FIG. 1.10 Picture of volume phase holographic gratings. **(A)** Transmission grating that disperses the incident light into a rainbow. **(B)** Reflection grating that selectively diffracts the red portion of the spectrum. Both holograms are made from the same material (dichromated gelatin) and are illuminated by a halogen white light.

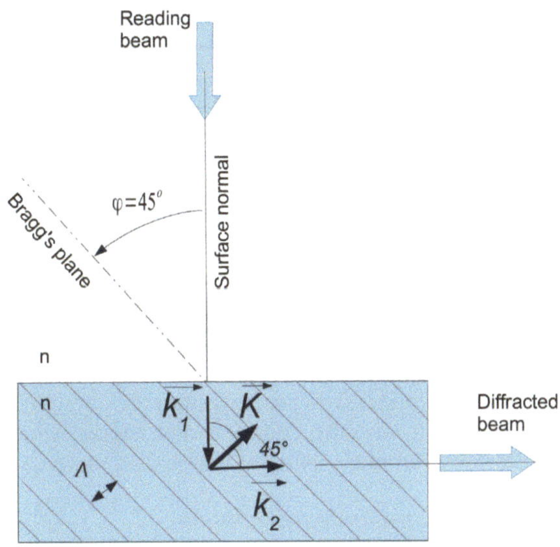

FIG. 1.11 Geometry of an edge-lit hologram. The diffracted beam is evanescent, that is, directed parallel to the surface of the material.

The angle and frequency selectivity properties of edge-lit holograms are in between those observed for transmission and reflection gratings (Fig. 1.9). These properties are identical either if the edge-lit hologram is used in reflection (hologram placed at the bottom of the waveguide), or transmission (hologram placed at the top of the waveguide). This is because the hologram parameters, angle and frequency, change only very slightly between the two orientations.

Multiplexing

Because thick gratings can be made highly selective according to the reading angle or wavelength, it is possible to record multiple holograms at the same location, and in the same material, which do not interfere with each other. This means that one hologram can be read without having any light diffracted by the others holograms. This technique, known as multiplexing, is particularly useful for data storage where the memory capacity can be increased thousands of times [33]. It is also used for creating color holograms from three holograms diffracting individually the red, green, and blue colors [34]. A particular case of wavelength-multiplexed hologram is the Lippmann photography that will be introduced in Section 6.10.1 [35].

The grating vector **K** can be modified in two aspects: magnitude and direction. So, two types of multiplexing are possible: angular and wavelength. In angular multiplexing, the direction of the grating vector is changed by using different incidence angles for each hologram. In wavelength multiplexing, the magnitude of the grating vector is changed for each hologram by using different wavelengths to record them.

As a general rule, the efficiency of each hologram during multiplexing follows a $1/N_H^2$ law where N_H is the number of holograms. Indeed, if the maximum dynamic range (amplitude or phase) of the material is ΔM each multiplexed hologram is using a portion of this range, so the modulation per hologram is $\Delta M/N_H$. As the efficiency is proportional to the square of the modulation (Eqs. 1.25 and 1.26), we obtain that $\eta \propto 1/N_H^2$.

This relationship is only valid for the cases where the hologram cannot be overmodulated. Overmodulation means that the optical path difference ($\Delta n \cdot d$) that can be achieved in the material is larger than the $\pi/2$ necessary to obtain maximum efficiency: $\eta = 100\%$. When the material is extremely thick, or when the modulation can be made extremely large, it is possible to record several multiplexed holograms with each one having 100% efficiency for different incident angles or wavelengths, of course.

An important metric in multiplexed holograms is the cross talk. This is the ratio between the sum of the energies diffracted by the hologram that are not interrogated, and the energy diffracted by the holograms that is being read. The cross talk is part to of signal-to-noise ratio (SNR) for the system. As such, it is often expressed in decibel (dB).

Fig. 1.12 shows the angular selectivity of two holograms with the same parameters, but with different slant angles to shift the diffraction peaks by 3.5°. Although the angular separation of the main lobe is larger than their full-width half max (1°), the cross talk is increased due to the presence of secondary lobes in the diffraction profile. At zero degree, the cross talk is 4.2% or −27.5 dB. A lower cross talk could be achieved by using a shift of either 2.5° or 4°, where the main lobes would be aligned with a minimum from the other hologram.

Thin Grating's Characteristics

Thin gratings operate in the Raman-Nath regime where the incident wave interacts only a few times with the modulation. This can be only one single time as for the case of surface relief gratings. Eqs. (1.23) and (1.24) mathematically describe the condition for the thin grating regime. In this mode of operation, a substantial amount of energy can be coupled in higher diffraction orders ($m > 1$). However, this is not always the case and a thin surface relief grating can be made highly efficient as we will see in this section.

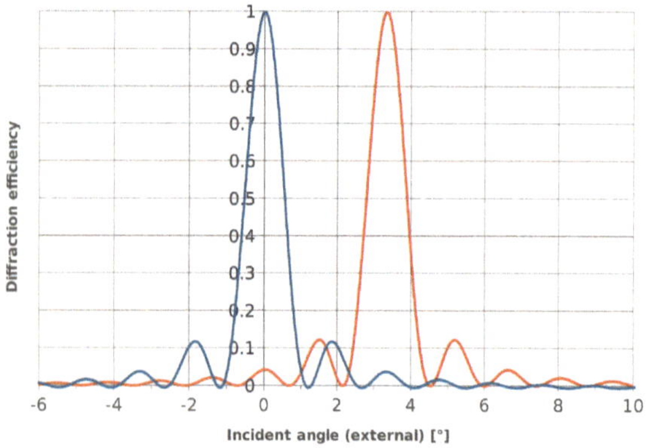

FIG. 1.12 Angular selectivity of two transmission holograms with 634 lp/mm, in 100 μm thick material, and at 800 nm wavelength. Although the angular separation of the holograms is larger than the main lobe full-width half max of 1°, the cross talk is increased due to the presence of secondary lobes in the diffraction.

Thin gratings are extremely important because they can easily be manufactured by printing a structure obtained by computer calculation. Most of the holograms encountered in daily life such as security tag on banknotes and luxury goods fall in that category and are made by the embossing technique (see Chapter 2 on holographic materials). Thin gratings can also be dynamically displayed using electronically controlled spatial light modulators such as LCoS (liquid crystal on silicon) and DLP (digital light processor).

The efficiency of thin gratings depends on the shape of the modulation [36]. As for thick gratings, one can distinguish between amplitude or phase modulation, but for thin gratings, it is also important to recognize the geometrical format of the profile such as square, sinusoidal, or sawtooth pattern.

The rigorous calculation of the efficiency and number of orders is based on Fourier decomposition of the complex amplitude of the transmitted wave function $t(x)$ according to the grating modulation $M(x)$. By finding an expression of the form:

$$t(x) = \sum_{m=-\infty}^{\infty} A_m exp(i\, m\mathbf{K}x) \tag{1.40}$$

The portion of the intensity in the *mth* order is $\eta_m = |A_m^2|$, and the direction of propagation for that order is given by the vector $m\mathbf{K}$.

We are going to analyze six cases that are relevant to today's holographic manufacturing and displays. These include the three modulation profiles shown in Fig. 1.13: sinusoidal, binary (or square), and sawtooth (or blazed). For each of these shapes, the two possible modulation format will be investigated: amplitude and phase. We will also consider what happen when the sawtooth profile is digitized into m discrete levels.

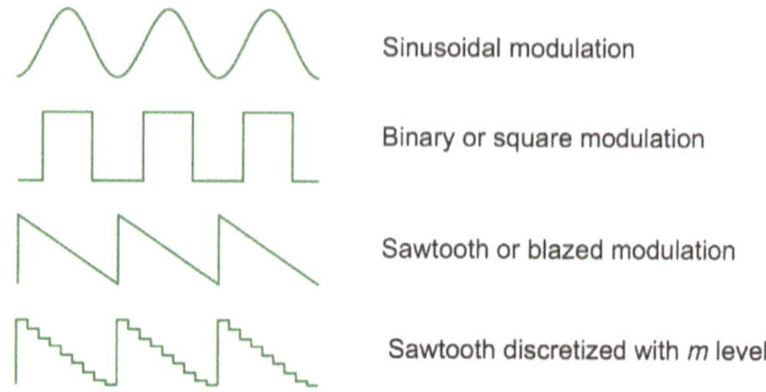

FIG. 1.13 Shape of the modulation format for thin gratings that are mathematically analyzed in the text.

Sinusoidal amplitude modulation

For a sinusoidal amplitude grating, the modulation format is given by:

$$\begin{aligned} |t(x)| = M(x) &= M_0 + \frac{\Delta M}{2}\sin(\mathbf{K}x) \\ &= M_0 + \frac{\Delta M}{4}\exp(i\,\mathbf{K}x) + \frac{\Delta M}{4}\exp(-i\,\mathbf{K}x) \end{aligned} \tag{1.41}$$

where $M_0 \in [0,1]$ is the average transmittance, $\Delta M \in [0,1]$ is the transmittance peak to valley modulation, and $|\mathbf{K}| = 2\pi/\Lambda$ is the wave vector.

The three different terms on the right side of Eq. (1.41) are associated with the amplitude of the different diffraction orders: 0, +1, −1 respectively. There are no higher orders for such grating. The diffraction efficiency ($\eta = |t_{\pm 1}|^2$) found in the ±1 order for this modulation is given by:

$$\eta_{\pm 1} = \left(\frac{\Delta M}{4}\right)^2 \leq 6.25\% \tag{1.42}$$

which is maximum when $M_0 = \Delta M/2 = 1/2$.

The behavior of the diffraction efficiency as a function of the amplitude modulation ΔM is plotted in Fig. 1.14.

Sinusoidal amplitude gratings can be fabricated by recording an interference pattern into thin layer of silver halide emulsion and then chemically processed to reveal the latent image.

Sinusoidal phase modulation

For a sinusoidal phase grating, there is no absorption: $t(x) = 1$, but the complex amplitude transmittance is given by:

$$t(x) = \exp(-iM_0)\exp\left(-i\frac{\pi\delta M}{2}\cos(\mathbf{K}x)\right) \tag{1.43}$$

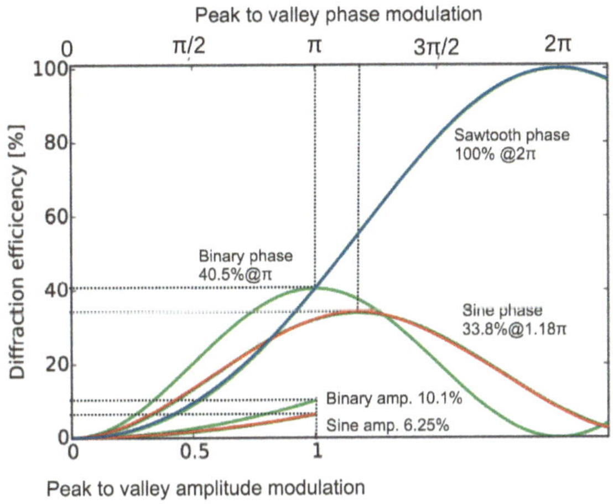

FIG. 1.14 Diffraction efficiency of thin gratings according to the modulation shape (sine, square, sawtooth), format (amplitude or phase), and modulation amplitude ΔM.

where M_0 is a constant phase shift, and ΔM is the peak to valley phase modulation.

Ignoring the constant phase shift, the right-hand side of Eq. (1.43) can be expanded in a Fourier series as:

$$t(x) = \sum_{m=-\infty}^{\infty} J_m\left(\frac{\pi \Delta M}{2}\right)\exp(i\, m\mathbf{K}x) \tag{1.44}$$

where J_m is the Bessel function of the first kind with the *mth* order representing the amplitude of the waves, when the exponential terms represent plane waves direction, that is, the diffracted orders.

From the decomposition given in Eq. (1.44), it can be seen that there is an infinite number of diffraction orders (one for each term of the sum).

The diffraction efficiency in the first orders is given by:

$$\eta_{\pm 1} = J_1^2\left(\frac{\pi \Delta M}{2}\right) \leq 33.8\% \tag{1.45}$$

which is maximum when $\delta M = 1.18$. Note here that this value of δM means that the peak to valley phase modulation should be slightly larger than π to maximize the efficiency.

The behavior of the diffraction efficiency according to the peak to valley phase modulation is shown in Fig. 1.14.

Sinusoidal phase gratings can be fabricated by recording an interference pattern into a very thin layer of photopolymer or dichromated gelatin and processing the emulsion to boost the index modulation.

Binary amplitude modulation

For a binary amplitude grating, the modulation is a square function, and the Fourier decomposition is expressed as:

$$\begin{aligned} M(x) &= M_0 + \frac{2\Delta M}{\pi}\sum_{m=1}^{\infty}\frac{\sin[(2m-1)\mathbf{K}x]}{2m-1} \\ &= M_0 + \frac{\Delta M}{\pi}\sum_{m=1}^{\infty}\frac{\exp[i(2m-1)\mathbf{K}x] + \exp[-i(2m-1)\mathbf{K}x]}{2m-1} \end{aligned} \tag{1.46}$$

The terms of this decomposition are all odds due to the $2m-1$ expression in the exponential functions. In consequence, there are no even diffraction orders for this type of modulation. The diffraction efficiency for the ± 1 orders is given by:

$$\eta_{\pm 1} = \left(\frac{\Delta M}{\pi}\right)^2 \leq 10.1\% \tag{1.47}$$

Maximum efficiency is achieved when $M_0 = \Delta M/2 = 1/2$.

The behavior of the diffraction efficiency according to the peak to valley amplitude modulation is shown in Fig. 1.14.

Binary amplitude gratings have been historically manufactured by using office printers on transparent films. Nowadays, this type of modulation is found when a holographic pattern is displayed on a DLP light modulator. The DLP pixels are composed of mirror that can be flipped left or right. For the incident light beam, the mirrors act as nearly perfect reflector or absorber depending of the direction they are oriented.

Binary phase modulation

For a binary phase grating, the complex amplitude transmittance is given by the following expression:

$$t(x) = \exp(-iM_0)\exp\left[-i\pi\frac{\Delta M}{2}\sum_{m=1}^{\infty}\frac{\sin[(2m-1)\mathbf{K}x]}{2m-1}\right] \tag{1.48}$$

The terms of the decomposition are all odds due to the $2m-1$ expression in the exponential functions, so there are no even diffraction orders as we have seen in the case for the amplitude binary grating. However, conversely to the amplitude case, the phase modulation term is now contained in the exponential and needs to be expanded to find the value of the efficiency. For the $\pm m$ orders, the efficiency is

$$\eta_{\pm m} = \left[\operatorname{sinc}\left(\frac{m}{2}\right)\sin\left(\frac{\pi\Delta M}{2}\right)\right]^2 \tag{1.49}$$

with $\operatorname{sinc}(x) = \sin(\pi x)/(\pi x)$.

For the first orders efficiency, we have

$$\eta_{\pm 1} = \left[\frac{2}{\pi}\sin\left(\frac{\pi\Delta M}{2}\right)\right]^2 \leq 40.5\% \tag{1.50}$$

which is maximum for $\Delta M = 1$, that is, a peak to valley phase modulation of $\pi/2$.

The behavior of the diffraction efficiency according to the phase modulation is shown in Fig. 1.14.

Binary phase grating can be manufactured by using single-layer photolithographic process where a photoresin is selectively exposed and removed. The pattern can be used as it, in this case, the phase modulation is given by thickness of the resin layer times its index modulation minus 1: $\Delta M = d(n-1)$. Alternatively, the resin can be covered by a layer of metal that make the structure reflective. In this case, the modulation is given by twice the thickness of the resin layer due to the double pass of the light in the grooves: $\Delta M = 2d$.

A counter-intuitive, but nonetheless important, result from this decomposition exercise is that the maximum diffraction efficiency in the first orders is larger for square gratings (40.5% for phase, 10.1% for amplitude) than for sinusoidal gratings (33.8% for phase, 6.25% for amplitude).

Sawtooth phase modulation

For a sawtooth phase grating, which is also called blazed grating, the complex amplitude transmittance is

$$t(x) = \exp(-iM_0)\exp\left[-i\pi\frac{\Delta M}{2}\sum_{m=1}^{\infty}\frac{(-1)^m}{m}\sin(mKx)\right] \tag{1.51}$$

The diffraction efficiency for the +1 or −1 orders is

$$\eta_1 = \operatorname{sinc}^2\left(1-\frac{\Delta M}{2}\right) \leq 100\% \tag{1.52}$$

To maximize the efficiency, the amplitude of the modulation should be $\Delta M = 2$, which is equal to a peak to valley phase modulation of 2π (see Fig. 1.14).

Note that when phase patterns are used in a reflection configuration, the modulation is half of the one obtained for a transmission configuration because the path length difference is twice as large due to the double pass of the light.

Discretized sawtooth phase modulation

For a discretized sawtooth phase grating, the ramp is composed of m levels spaced apart at equal amplitude (see Fig. 1.13). This configuration is important to derive because, for many manufacturing processes, it is not possible to reproduce a perfectly smooth sawtooth profile. Instead, the slope is composed of multiple discrete steps. For example, it is possible to expose and etch photo-resin several times to make such a stepped sawtooth profile. It is also the case of LCoS modulators that generate that type of modulation where the ramp is approximated by the digital dynamic range of the pixels.

For this type of modulation, the diffraction efficiency for the +1 or −1 orders given by [37,38]:

$$\eta_1 = \left[\frac{1}{\pi}\sin\left(\pi\left(1-\frac{\Delta M}{2}\right)\right)\frac{\sin\left(\frac{\pi}{m}\right)}{\sin\left(\frac{\pi}{m}\left(1-\frac{\Delta M}{2}\right)\right)}\right]^2 \leq 100\% \tag{1.53}$$

Expression 1.53 yields the same result as Eq. (1.52) for the limit where $m \to \infty$.

Similarly than for the blazed profile, the maximum efficiency is achieved when the phase modulation is 2π: $\Delta M = 2$, but it varies with m, the number of levels:

$$\eta_1 = \operatorname{sinc}^2\left(\frac{1}{m}\right) \leq 100\% \tag{1.54}$$

It also has to be noted that, because the lateral spacing between the steps is fixed by the resolution of the process or by the pixel pitch in the case of an LCoS LSM, the maximum grating spacing achievable (Λ) is divided by the number of levels used to define the ramp. This reduction of the grating spacing limits the maximum diffraction angle achievable by the diffraction pattern according to the Bragg Eq. (1.13). In Fig. 1.15, we plotted both the behavior of the efficiency, which increases with the number of levels, and the diffraction angle, which decreases with the same number. Thus, the user is often confronted to a choice in selecting either high efficiency or larger diffraction angle.

Although the function of Eq. (1.54) is continuous, in the real world, m can only takes discrete values, starting at 2. $m = 2$ is the case of a binary grating, for which Eq. (1.54) logically gives the same value of efficiency (40.5%) as when computed directly by Eq. (1.49), describing binary phase grating.

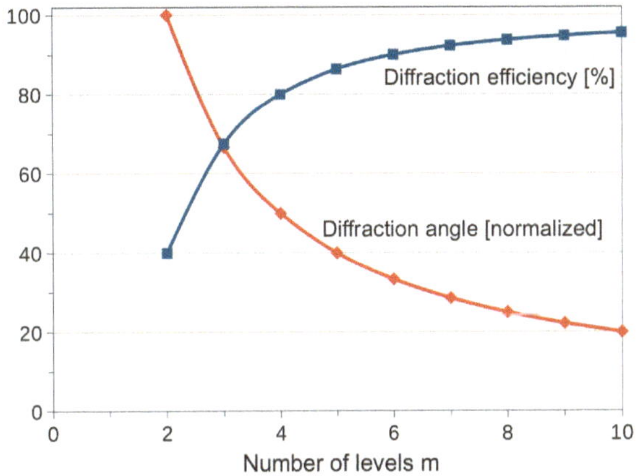

FIG. 1.15 Diffraction efficiency and angle of a discretized sawtooth grating structure with a 2π phase modulation, according to the number of levels defining the sawtooth function.

It has to be noted that the modulation amplitude for maximum efficiency is $\Delta M = 2$ for Eq. (1.54), and $\Delta M = 1$ for Eq. (1.49), because during the digitization of the modulation, the average level is multiplied by a factor of two.

SCALAR THEORY OF DIFFRACTION

Now that we have seen the diffraction by various periodic gratings, we are going to generalize the formalism for any structure. Finding the mathematical formulation for the transformation between the aperture geometry and the diffracted field will allow us not only to determine the form of the wave diffracted by a specific structure but also to calculate the pattern to generate a particular field, the so-called CGH.

Computing the diffraction pattern from an object or the retrieval of the object from the observed diffraction pattern is the field of digital holography. Chapter 5 of this book is dedicated to these computations, and the present section is meant as an introduction to the field.

Digital holography offers many advantages compared with regular digital photography. For example, by capturing on both phase and amplitude of the diffracted field, we can compute the 3D structure of the original object, change the focus of the instrument after the data has been captured (postfocusing), or observe wavelength scale deformation of an object [39,40].

Kirchhoff Diffraction Integral

To start, we would like to determine the propagation of the field after going through an arbitrary aperture as shown in Fig. 1.16.

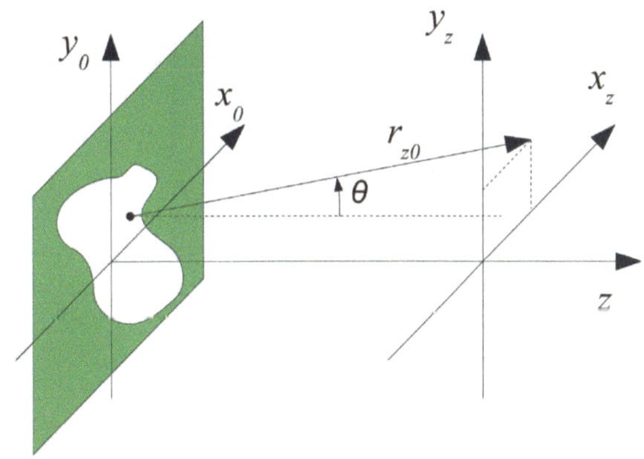

FIG. 1.16 Geometry and definition of the coordinate systems for the propagation of the electromagnetic field through an aperture.

The energy carried by the magnetic field is usually much weaker than the energy in the electrical field ($|\mathbf{B}| = |\mathbf{E}|/c$), so we are going to simplify the calculation by limiting ourselves to the electric field.

According to the Huygen's principle, the aperture acts as a homogeneous light source, and the field is null in the opaque portions of the aperture. So, at distance z, the field is given by the summation over all the points of the aperture multiplied by the wave propagation function to that distance:

$$E(x_z, y_z) = \sum_{aperture} [\text{incident field at}(x_0, y_0)] \times [\text{wave propagation to } z\text{: } r_{z0}] \tag{1.55}$$

The wave propagation is solution of the Helmholtz equation introduced in Eq. (1.1), and we will choose the spherical wave solution:

$$\mathbf{E}(\mathbf{r}, t) = \frac{\mathbf{A}}{r}\cos(\mathbf{k}\cdot\mathbf{r} - \omega t + \phi) \tag{1.56}$$

Inserting Eq. (1.56) Into expression 1.55, we obtain the *Kirchhoff diffraction integral*:

$$\boxed{E(x_z, y_z) = \frac{1}{i\lambda}\int_{aperture} E(x_0, y_0)\frac{\exp(ikr_{z0})}{r_{z0}}\cos\theta\, ds} \tag{1.57}$$

where, for Cartesian coordinates: $r_{z0} = \sqrt{z^2 + (x_z - x_0)^2 + (y_z - y_0)^2}$.

Fresnel Diffraction Integral

As elegant as the Kirchhoff diffraction integral (Eq. 1.57) is, it is very hard to compute and some simplifications are necessary to obtain a manageable expression.

Let us consider the expansion of the z term of the r_{z0} expression in Taylor series: sqrt $1+\varepsilon = 1 + \frac{\varepsilon}{2} - \frac{\varepsilon^2}{8} + \ldots$, such that:

$$r_{z0} = z + \frac{1}{2}\left[\left(\frac{x_z - x_0}{z}\right)^2 + \left(\frac{x_z - x_0}{z}\right)^2\right] + \ldots \quad (1.58)$$

Our first set of approximations will be to neglect the third term inside the complex exponential ($\exp(ikr_{z0})$), and the second term in the denominator (r_{z0}) of Eq. (1.57). This set of simplifications is referred as the paraxial approximation as it can be applied for a small aperture in regard to the distance z:$z \gg x_z - x_0$ and $z \gg y_z - y_0$.

This leads to the *Fresnel diffraction integral*:

$$\boxed{E(x_z, y_z) = \frac{\exp(ikz)}{i\lambda z}\int_{aperture} E(x_0, y_0)\exp\left[\frac{ik}{2z}\left((x_z - x_0)^2 + (y_z - y_0)^2\right)\right] ds} \quad (1.59)$$

The paraxial approximation validity criteria, used to truncate the Taylor series, can also be expressed as the Fresnel number F:

$$F = \frac{(D/2)^2}{z\lambda} \leq 1 \quad (1.60)$$

where D is the aperture diameter.

The Fresnel number inequality (Eq. 1.60) expresses the fact that the distance z should be larger than the wavelength λ, but not necessarily much larger than the aperture D. So, the Fresnel approximation is valid in the so-called "near field."

Fraunhofer Diffraction Integral

If we are interested by the solution for an observation plane farther away from the aperture, that is, in the "far field," where $z \ll k(x_0^2 + y_0^2)_{max}$, further approximation can be used.

If we expend the quadratic terms of the Fresnel diffraction integral (Eq. 1.59) as $(a-b)^2 = a^2 + b^2 - 2\,ab$:

$$E(x_z, y_z) = \frac{\exp(ikz)}{i\lambda z}\exp\left[\frac{ik}{2z}(x_z^2 + y_z^2)\right] \times \int_{aperture} E(x_0, y_0)\exp\left[\frac{ik}{2z}\left((-2x_z x_0 - 2y_z y_0) + (x_0^2 + y_0^2)\right)\right] ds \quad (1.61)$$

The quadratic phase factor can be set to unity over the entire aperture:

$$\int_{aperture} E(x_0, y_0)\exp\left[\frac{ik}{2z}(x_0^2 + y_0^2)\right] ds = 1 \quad (1.62)$$

Therefore, Eq. (1.61) can be written as:

$$\boxed{E(x_z, y_z) = \frac{\exp(ikz)}{i\lambda z}\exp\left[\frac{ik}{2z}(x_z^2 + y_z^2)\right] \times \int_{aperture} E(x_0, y_0)\exp\left[\frac{ik}{z}(x_z x_0 + y_z y_0)\right] ds} \quad (1.63)$$

which is known as the *Fraunhofer diffraction integral.*

This result is particularly important once it is recognized that the integration term is simply the *Fourier transform of the aperture*. Furthermore, because it is the optical intensity that is relevant for most applications: $I = |E^2|$, the phase factor in front of the integral can be neglected.

Ultimately, this long mathematical development leads to the very convenient and elegant formulation:

$$\boxed{E(x_z, y_z) = \mathscr{F}\left(aperture(x_0, y_0)\right)} \quad (1.64)$$

The criteria for the Fraunhofer diffraction integral to be valid is that the observation distance z must be much larger than the aperture size and wavelength:

$$F = \frac{(D/2)^2}{z\lambda} \ll 1 \quad (1.65)$$

This condition is known in optics as the "far field" approximation.

Diffraction by Simple Apertures

Considering the relative simplicity of the Fraunhofer diffraction integral (Eq. 1.63), it is possible to find analytical solutions for simple apertures illuminated by a plane wave:

$$U(x, y) = A\exp[i(2\pi ct/\lambda)] \quad (1.66)$$

We are going to develop the cases of the following apertures:

- a slit
- a circular pinhole
- multiple slits
- a Fresnel zone plate

Diffraction by a slit

The slit is a rectangular function located at $z = 0$ of width W along the x dimension:

$$f(x) = \mathrm{rect}\left(\frac{x}{W}\right) \quad (1.67)$$

The integration of the field over the slit is given by:

$$U(x, z) = A\int_{-W/2}^{W/2}\exp\left(-i\frac{2\pi}{\lambda z}xx'\right)dx = \frac{-Az}{ikx}\left|\exp\left(-i\frac{2\pi}{\lambda z}xx'\right)\right|_{-W/2}^{W/2} \quad (1.68)$$

Using Euler's formula, we have

$$U(x,z) = AW\frac{\sin\left(\frac{W\pi x}{\lambda z}\right)}{\frac{W\pi x}{\lambda z}} = AW\mathrm{sinc}\left(\frac{W\pi x}{\lambda z}\right) \tag{1.69}$$

Or, in cylindrical coordinates ($\sin\theta = x/z$):

$$U(\theta) = AW\ \mathrm{sinc}\left(\frac{W\pi}{\lambda}\sin\theta\right) \tag{1.70}$$

Because the intensity can be expressed as: $I=UU^*$

$$I(\theta) = I_0 W\ \mathrm{sinc}^2\left(\frac{W\pi}{\lambda}\sin\theta\right) \tag{1.71}$$

The intensity distribution of Eq. (1.71) is shown in Fig. 1.17.

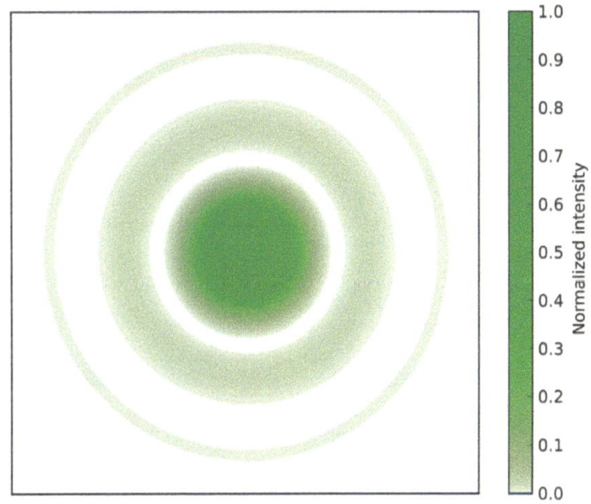

FIG. 1.18 Diffraction pattern formed by a circular aperture: the Airy disk.

Diffraction by a circular pinhole

The diffraction by a circular aperture of diameter D is a two-dimensional (2D) generalization of the case of a slit with a rotational symmetry applied to it. The intensity distribution in the far field becomes

$$I(\theta) = I_0 D\left[\frac{2J_1\left(\frac{D\pi}{\lambda}\sin\theta\right)}{\frac{D\pi}{\lambda}\sin\theta}\right]^2 \tag{1.72}$$

The diffraction pattern formed by the pinhole aperture is called the Airy disk and is shown in Fig. 1.18. A cross section of this Airy disk pattern is a sinc^2 function plotted in Fig. 1.17.

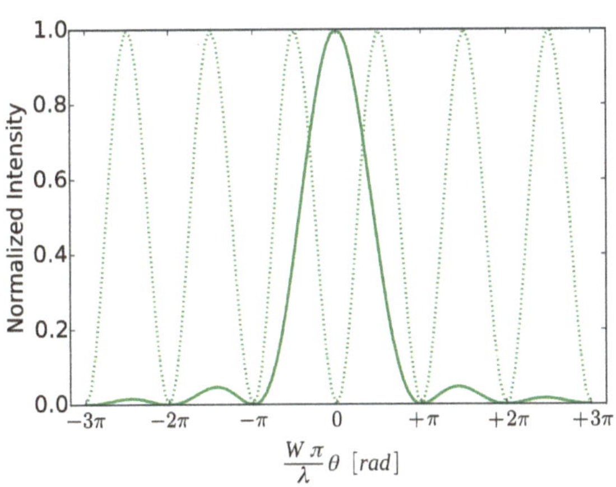

FIG. 1.17 Interference by a slit of width W. The doted line is a $\sin^2$ function for which secondary minima are collocated.

Diffraction by multiple slits

Under the Fraunhofer condition explained in Section 3.3, the intensity distribution diffracted by m slits of width W each separated by a distance Δx is

$$I(\theta) = I_0 W\ \mathrm{sinc}^2\left(\frac{W\pi}{\lambda}\sin\theta\right) \times \left[\frac{\sin\left(m\frac{\Delta x\pi}{\lambda}\sin\theta\right)}{\frac{\Delta x\pi}{\lambda}\sin\theta}\right]^2 \tag{1.73}$$

This equation can be obtained by the Fourier transform of the aperture as expressed in Eqs. (1.63) and (1.64). Alternatively, it can be derived by multiplying the expressions for the diffraction by a single slit (which gives the first term of Eq. 1.73) with the interference of m slits (which gives the second term of Eq. 1.73).

Fig. 1.19 shows an example of diffracted intensity obtained for three slits, with the different terms of Eq. (1.73) plotted independently.

Fresnel zone plate

The Fresnel zone plate is a binary structure that acts as a lens. The zone plate diffracts the incident plane wave into a focal spot. The pattern of a Fresnel zone plate is comprised of alternating opaque and transparent rings that act like slits. The radii R_m of these rings is such that the interference is constructive along the axis at the focal distance f:

$$R_m = \sqrt{m\lambda f + \frac{m^2\lambda^2}{4}} \tag{1.74}$$

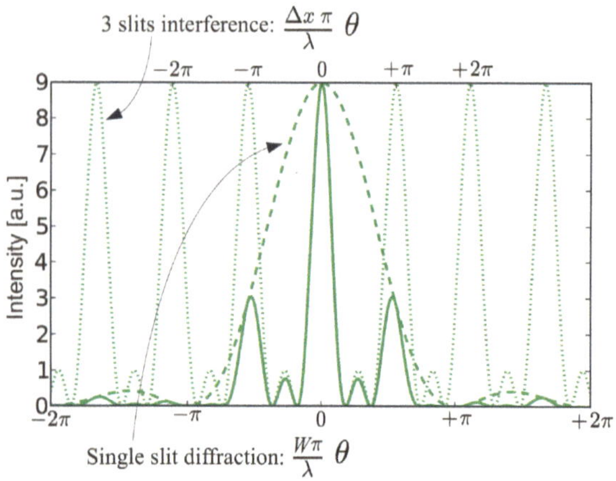

FIG. 1.19 Far field intensity pattern formed by the diffraction from 3 slits (plain line). The pattern is obtained by multiplying the interference of 3 slits (doted line) by the diffraction by a single slit (dashed line).

Fig. 1.20 shows a Fresnel zone plate structure and the condition on the radii R_m of the rings to obtain constructive interference at a distance f, which is that the distance from the radii to the focal point must be a multiple of half wavelength ($m\lambda/2$).

It has to be noted that the black and transparent rings composing the Fresnel zone plate can be inverted without any alteration in the diffraction properties.

The Fresnel zone plate is a binary amplitude modulation diffractive element. In that regard, we have seen in Section 2.4 on thin grating characteristics that such a structure diffracts multiple odd orders. In the case of the zone plate, the positive higher orders $(2m+1)$ will form multiple focal points at $f/(2m+1)$, when the negative orders $(-2m+1)$ will act as *negative lenses* with focal lengths $-f/(2m+1)$.

To reduce the number of higher orders and concentrate the energy into fewer focal spots, it is possible to replace the Fresnel binary modulation pattern by a sinusoidal gray scale modulation. Such a gray scale zone plate will only diffract into the $+1$ and -1 orders (see Section 2.4). To increase the throughput efficiency of the zone plate, it is also possible to replace the amplitude modulation with a phase modulation.

The gray scale zone plate is called the Gabor zone plate and is shown in Fig. 1.21. The amplitude of the modulation M according to the distance to the center r is given by:

$$M(r) = \frac{1}{2}\left[1 \pm \cos\left(\frac{kr^2}{f}\right)\right] \tag{1.75}$$

with $k = 2\pi/\lambda$.

The Gabor zone plate modulation is the same as the one obtained by making interfere a plane wave with a collinear point source located at a distance f from the plane of the interferogram. We will see this geometry when discussing the holographic recording setup in Section 6.

Zone plates are particularly interesting to obtain thin and light optical elements that can replace bulky refractive lenses. More complex diffractive structures can also be computed (i.e., CGH), such that they perform more elaborate optical functions. These CGHs can replace aspherical lenses that are difficult and costly to manufacture. Zone plates and CGHs can also be used to manipulate electromagnetic radiations for which there is no refractive material, such as for X-rays, or when the refractive materials are too expensive (chalcogenide glasses for thermal infrared radiation).

Deriving the shape of an aperture to obtain a specific wavefront involves finding the inverse function of the diffraction integral. We can use either the Kirchhoff

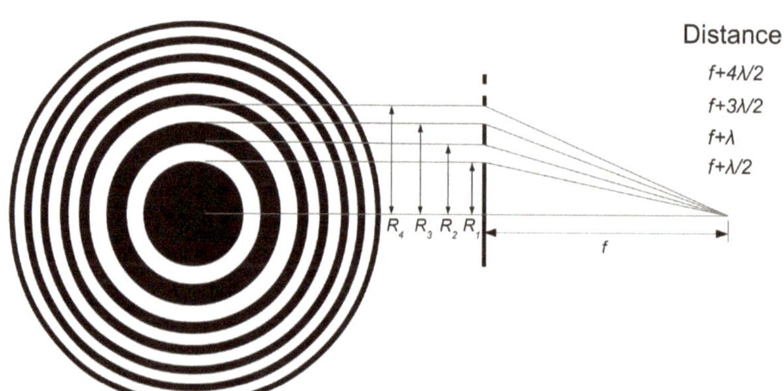

FIG. 1.20 Fresnel zone plate diffractive structure and radii of the successive rings to obtain constructive interference at the distance *f*.

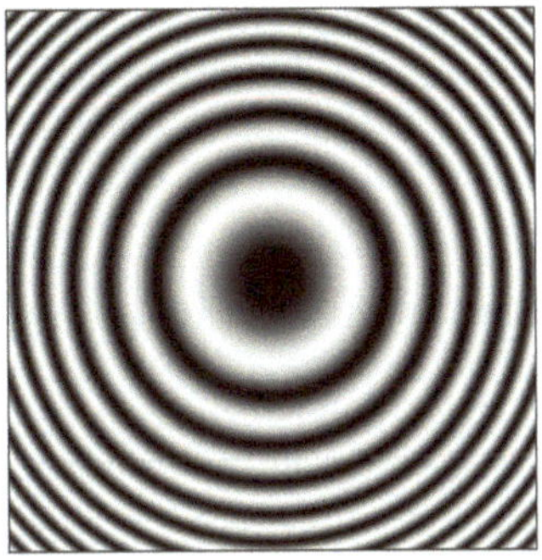

FIG. 1.21 Gabor zone plate pattern: sinusoidal phase or amplitude modulation that focuses an incident plane wave into a focal spot.

(Eq. 1.57), the Fresnel (Eq. 1.59), or the Fraunhofer (Eq. 1.63) diffraction integrals, to calculate the specific aperture that will generate the desired wavefront.

For the most general case, an exact solution for these equations cannot be found, and the diffraction pattern (i.e., the shape of the aperture) must be calculated using a computer. The field of CGHs started in the late 1960s when scientists gained greater access to computers. The field expands rapidly with the implementation of the fast Fourier transform (FFT) algorithm that made possible to compute Fourier transform over 22D images of significant size [17,18,41].

However, even with today's computers, it is not yet possible to compute CGHs in their most rigorous form, using the Kirchhoff diffraction integral, in real time, for complex 3D images. This problem, known as the computational bottle neck, can be get around using simpler expressions (Fresnel, Fraunhofer) and by using lookup tables that contain precalculated values for the aperture to generate specific shapes and wavefronts [42–44]. The optimization of the algorithms to compute CGH is still an active topic of research today.

COMPUTER-GENERATED HOLOGRAMS

Fourier Hologram

The easiest path for the computation of the diffraction pattern is to use the Fraunhofer diffraction integral (Eq. 1.63) and take the inverse Fourier transform of each sides:

$$\text{aperture}(x_0, y_0) \propto \mathcal{F}^{-1}[E(x_z, y_z)] \tag{1.76}$$

Of course, the same far field conditions regarding the image distance being much greater than the aperture size and wavelength apply to both this expression and the diffraction integral. The common shorthand for these conditions to be respected is that the holographic image will be formed at infinity: $z \to \infty$. To observe the image at a more convenient distance, one can use a lens which will form the image at its focal length. This configuration is shown in Fig. 1.22.

Considering the Fourier transform of 2D function is always a 2D function, and that the solution is independent of the distance z, there is only one image plane for the Fourier hologram and the image will be two dimensional.

Example of a Fourier binary amplitude hologram is shown in Fig. 1.23.

Fresnel Hologram

Finding the inverse function for the Fresnel diffraction integral (Eq. 1.59) is somewhat more complicated than for the Fraunhofer equation because the field E is now a function of the propagation distance z. Eq. (1.59) can be rewritten by creating a parabolic wavelet function such as:

$$h(z) = \exp\left[\frac{ik}{2z}\left((x_z - x_0)^2 + (y_z - y_0)^2\right)\right] \tag{1.77}$$

By substituting $h(z)$ in Eq. (1.59), it now possible to find an expression for the aperture:

$$\text{aperture}(x_0, y_0) \propto \frac{1}{E(x_0, y_0)} \times \mathcal{F}^{-1}\left[\frac{\mathcal{F}(E(x_z, y_z))}{h(z)}\right] \tag{1.78}$$

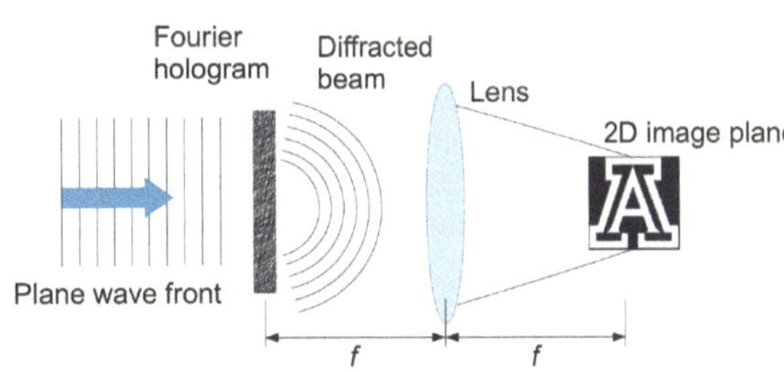

FIG. 1.22 Formation of the image at the focal of a lens with a Fourier hologram.

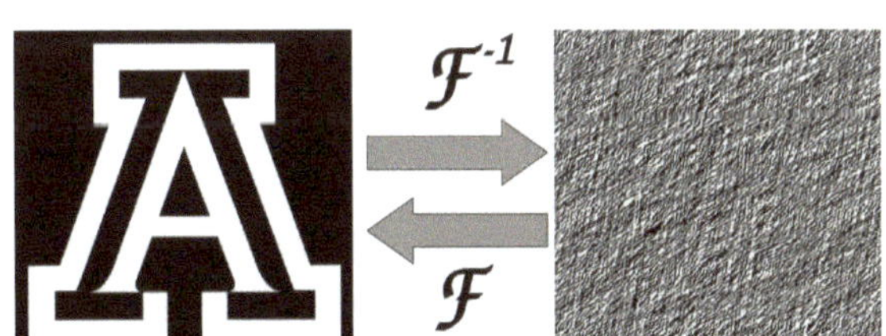

FIG. 1.23 Example of a Fourier holographic pattern (right) computed from a 2D field distribution (right).

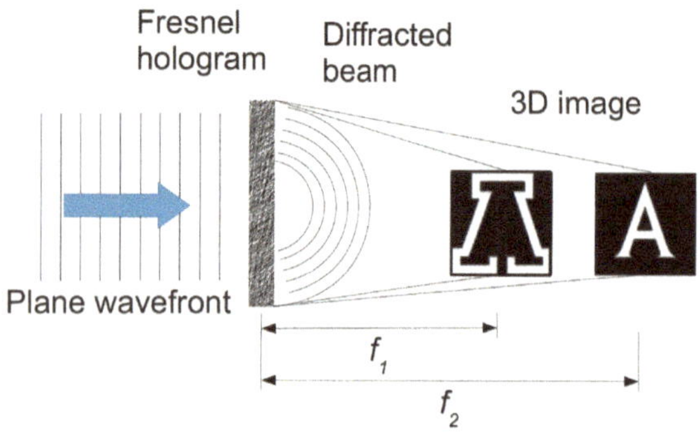

FIG. 1.24 Image formation with a Fresnel computer-generated hologram.

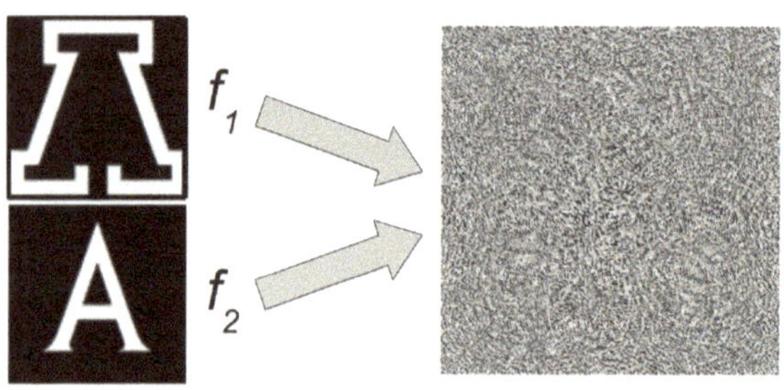

FIG. 1.25 Example of a Fresnel holographic pattern (right) computed from a two-field distribution located at different distances (right).

Eq. (1.78) is certainly more daunting than the simpler Fourier expression 1.76. However, the benefit of Fresnel hologram is that there is no need to include a lens in the setup in order to bring the image to a focus, the diffraction pattern does the focusing by itself (see Fig. 1.24). The inclusion of the wavelet propagation generates Fresnel zone plate-like structures in the hologram that acts as diffraction lenses to focus the image at finite locations.

In addition, the image generated by the Fresnel hologram can be three dimensional, that is, composed of several focal planes. An example of a computer-generated Fresnel hologram is shown in Fig. 1.25 where the two sections of the image will be formed at different distances. A close observation of the diffractive pattern will reveal some centrosymmetric structures that are due to the Fresnel zone plates.

The Fresnel diffraction integral equation can also be used to reconstruct an object when the hologram (or interferogram) is captured as an image. Today CMOS and CCD sensors have pixels small enough to resolve the interference produced by small objects located near the sensor plane. The advantages of not imaging the object of interest are that, first, no lens is needed, and second, the 3D information can be reconstructed. This technique has been used to demonstrate very high−resolution holographic microscopes capable of resolving single cells such as red blood cells and lymphocytes [45].

Iterative Computation of Hologram, the Gerchberg-Saxton Algorithm

When computing a diffraction pattern from an image, or when reconstructing an image from a diffraction pattern (inverse transformation), the Fourier transform operation generates two terms: a real part, which is the transmittance (amplitude); and an imaginary part, which is the phase modulation. Quite often one or the other is not captured during the measurement, or not reproduced during the display of the hologram. This is due to the properties of the image sensor (amplitude only), or because of the characteristics of the display element: LCoS SLM are phase-only elements, and DLPs are amplitude-only elements. Left unaddressed, this problem reduces the efficiency of the hologram and increases the noise in the image.

An effective way to minimize the degradation is to use an iterative computation such as the Gerchberg-Saxton algorithm [46]. The principle of this algorithm is that the phase (intensity) of the *mth* iteration can be used along the source intensity (phase) distribution to calculate the $m + 1$st function via Fourier transform and its inverse. A schematic diagram of the iteration is shown in Fig. 1.26 where a phase hologram is computed from the image intensity distribution. Chapter 3 of this book develops the different variations of the Gerchberg-Saxton algorithm and their use for different applications.

For imaging purpose, the Gerchberg-Saxton algorithm converges quite rapidly as it can be seen in Fig. 1.27, where only three iterations are necessary for the hologram to eliminate most of the noise in the image it reproduces.

However, for applications where the noise needs to be reduced to a minimum, it could take several tens of iterations to optimize the SNR [47]. For a more detailed discussion about the Gerchberg-Saxton algorithm as well as more advanced computational techniques, see Chapter 3 by Tom D. Milster.

Resolution of Computer-Generated Holograms

The limitation in the resolution of CGH comes from several factors. The first one being the computation of the Fourier transform, which usually uses a FFT algorithm that samples the function and limits the number of frequencies. Because of this sampling, the result is

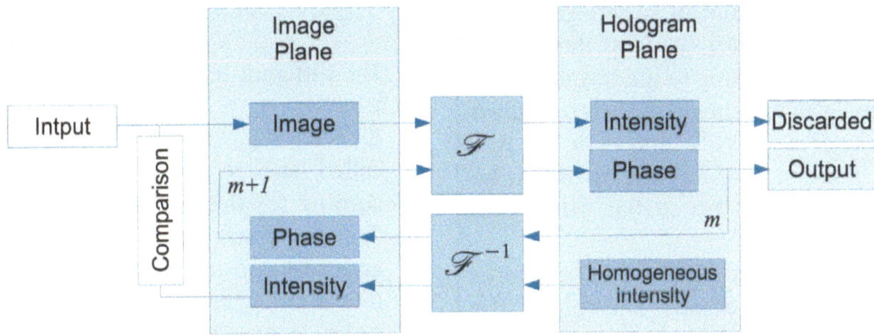

FIG. 1.26 Flow diagram of the iterative Gerchberg-Saxton algorithm.

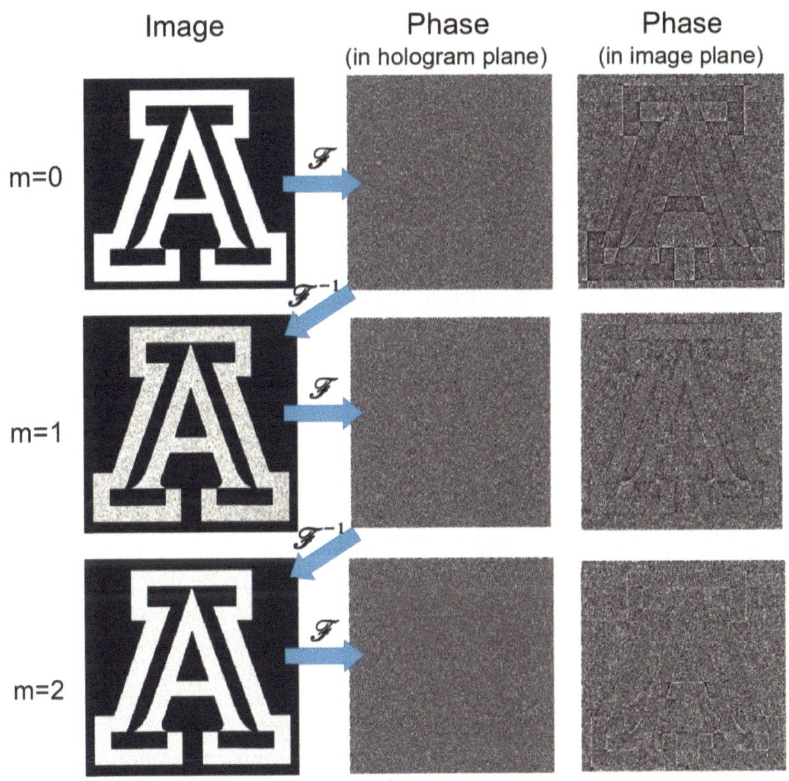

FIG. 1.27 Convergence of a Gerchberg-Saxton algorithm. The top-left image is the input, subsequent images are computed back from the phase hologram. Note the noise in the second image that is dramatically reduced in the second iteration.

not continuous but discretized, which generates some high-frequency noise. However, it can be argued that this resolution can be increased arbitrarily by using a finer sampling mesh, even though this lengthens the computation time.

More noise is generated during the physical reproduction of the hologram. This is due to the finite pixel size and pitch of the modulator. Here too, quantization is introduced, which limits the frequency band that can be reproduced. Whether it is a printer, lithography, or an electronic spatial light modulator, the technique has a limited space-bandwidth product (SBP) and is not able to replicate the entire spectrum of frequencies that are contained in the holographic pattern.

More rigorously, the SBP is a measure of the information contained in a signal or the rendering capacity of a device. For an optical system, it is defined as the product of the spatial frequency ($\Delta\nu$) by the spatial extent of the image (Δx). According to the Nyquist sampling theorem, a signal can only be perfectly reproduced by a system only if the area of its SBP fits inside the area of the system SBP. The shape of the SBP itself can be modified by lenses, reducing the spatial extent but increasing the frequency.

When computing a hologram, the Fourier transform rotates the SBP by 90 degrees because the role of the space and frequency are inverted. For a 2D image, and an image sensor, both the bandwidth and spatial extent are two dimensional. However, in holography, the image can be three dimensional, but the holographic pattern is only two dimensional, which imposes a very high burden on the system SBP:

$$SBP_{signal} = \Delta x^3 \cdot \Delta\nu^3 \leq SBP_{systel} = \Delta x'^2 \cdot \Delta\nu'^2 \tag{1.79}$$

To satisfy the Nyquist theorem, we see that the number of "pixels" composing the system (i.e., the hologram) should be larger by a power of 6/4 to fully reconstruct the 3D image (i.e., the signal). This is the fundamental reason why CGHs are still not able to reproduce small details such as object textures, even with today high-resolution SLM and computer capacity. New devices, such as leaky mode waveguides, with very high SBP might help in that regard in the near future [34].

HOLOGRAPHIC RECORDING AND READING FORMALISM

General Case of Hologram Recording and Reading

In the most general terms, the intensity modulation pattern created when a reference beam R and an object beam O interfere is

$$|O+R|^2 = |O|^2 + |R|^2 + OR^* + O^*R \tag{1.80}$$

When this intensity profile is recorded inside a material (silver halide, dichromated gelatin, photopolymer, etc.), the response of the material itself (β) should be included to obtain the physical modulation pattern:

$$T(x,y) = \beta|O|^2 + \beta|R|^2 + \beta OR^* + \beta O^*R \tag{1.81}$$

Once recorded, this modulation pattern can be interrogated with a reading beam R, which for now is assumed to be identical to the reference beam. In this case, the output field can be expressed as:

$$E_{out}(x,y) = \beta|O|^2R + \beta|R|^2R + \beta O|R|^2 + \beta O^*R^2 \tag{1.82}$$

The different terms of Eq. (1.82) can be interpreted as follows:

- $\beta|O|^2R = E_{scat}$ is an intermodulation term, also called halo, resulting from the interference of wave coming for the different points of the object. This "information" is contained in the term $|O^2|$ and is generally considered as noise.
- $\beta|R|^2R = E_{trans}$ is the transmitted beam or the zero order. It does not contain any object information, only R terms.
- $\beta O|R|^2 = E_{+1}$ is the +1 diffraction order. It is the reconstructed object beam because it contains an O term, which is the exact wavefront as the one scattered by the object. This diffracted beam will produce a *virtual image* of the object located at the object position.
- $\beta O^*R^2 = E_{-1}$ is the −1 diffraction order. It contains a O^* term reconstructing a conjugate image of the object. This diffracted beam will produce a *real image* of the object that will appear pseudoscopic: the relief is inverted as the front part is seen on the back and the background on the front (like a molding cast seen from the inside out).

A graphical representation of the different terms of Eq. (1.82) is shown in Fig. 1.28.

Aberrations in Holograms

To write Eq. (1.82), we assumed that the reading beam R was identical, in shape, direction, and wavelength, to the reference beam used to record the hologram. If it is not the case, and $\mathbf{k}_{ref} \neq \mathbf{k}_{read}$, the diffracted beam will not reconstruct the exact same object beam.

We are going to calculate the difference that occurs for two different cases that are often seen in the laboratory: difference in wavelength and difference in source spatial extend.

In addition, aberration in the reconstructed hologram can come from distortion in the material, changing the Bragg plane orientation between recording and reading. We are going to see the impact of material swelling or shrinkage on the diffracted beam.

The general geometry for the different point sources used for the recording and reading of a hologram is shown in Fig. 1.29.

The difference between the position of a point source at the object location and the reconstruction of this point can be calculated using the grating equation

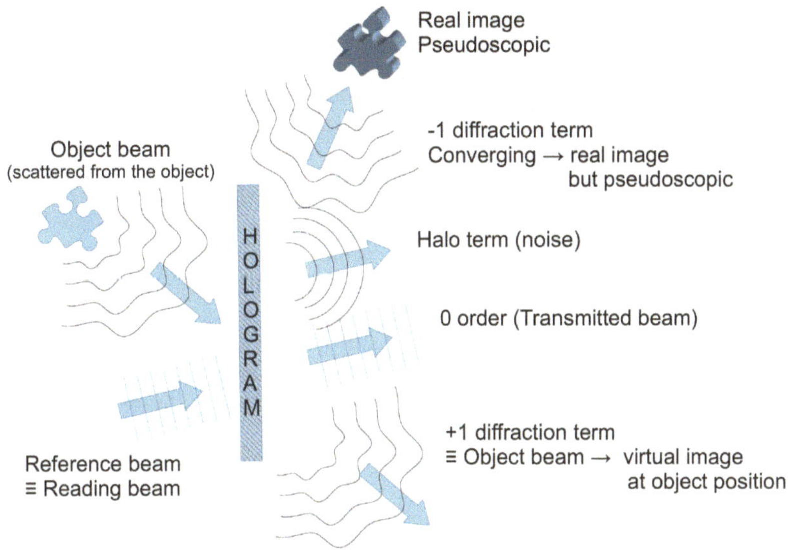

FIG. 1.28 Recording beams and diffraction terms produced by a hologram.

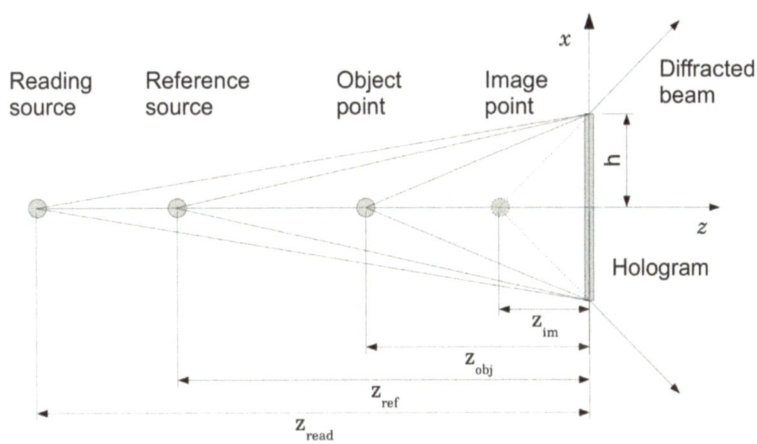

FIG. 1.29 Geometry of the different point sources for the recording and reading of a hologram.

(Eq. 1.11) and trigonometric relations. The general expression for the position for the image point is

$$\frac{1}{z_{im}}=\frac{1}{h}\tan\left[\sin^{-1}\left[\frac{\lambda_{read}}{\lambda_{write}}\left(\sin\left(\tan^{-1}\frac{h}{z_{obj}}\right)-\sin\left(\tan^{-1}\frac{h}{z_{ref}}\right)\right)+\sin\left(\tan^{-1}\frac{h}{z_{read}}\right)\right]\right] \tag{1.83}$$

It can be seen that this relationship includes the distance from the optical axis (h), so any variation in the reading source will induce some off-axis-type aberration.

Using the paraxial approximation $h \ll z$, expression 83 becomes more manageable:

$$\frac{1}{z_{im}}\approx\frac{\lambda_{read}}{\lambda_{write}}\left(\frac{1}{z_{obj}}-\frac{1}{z_{ref}}\right)+\frac{1}{z_{read}} \tag{1.84}$$

which indeed gives back $z_{im}=z_{obj}$ in the ideal case of $\lambda_{read}=\lambda_{write}$ and $z_{read}=z_{ref}$.

Writing and reading wavelength difference

It often happens that the hologram recorded with a laser source is replayed with another source that does not have the exact same wavelength: $\lambda_{read}\neq\lambda_{write}$.

If the chromatic variation is the only one we consider and the reading source is correctly located: $z_{read} = z_{ref}$, expression 83 becomes

$$\frac{1}{z_{im}} = \frac{1}{h}\tan\left[\sin^{-1}\left(\frac{\lambda_{read}}{\lambda_{write}}\sin\left(\tan^{-1}\frac{h}{z_{obj}}\right)\right)\right] \tag{1.85}$$

which simplifies in the paraxial approximation ($h \ll z$) into:

$$\frac{z_{obj}}{z_{im}} \approx \frac{\lambda_{read}}{\lambda_{write}} \tag{1.86}$$

We note that longer wavelengths are more deflected, and the image position is closer as shown in Fig. 1.30. This can also be observed in Figs. 1.10A, and 1.39, where the holograms are illuminated with a white light source.

Source spatial extent

The spatial extent of the reading source induces some blurring in the holographic image:

$$\frac{D_{read}}{D_{im}} \approx \frac{z_{read}}{z_{im}} \tag{1.87}$$

This relationship is shown in Fig. 1.31.

This blurring is the reason why point-like sources are preferred for the sharp reproduction of hologram. An example of this effect can be seen in Fig. 1.37, where a reflection hologram is illuminated with a point-like source on the left and a source with large spatial extend (fluorescent tube) on the right.

Media thickness variation

When the material in which the hologram is recorded swell or contract, the orientation and spacing of the Bragg plane (direction and magnitude of the grating vector **K**) changes as shown in Fig. 1.32.

The distance of the image point according to the thickness variation ($d\,d'^{-1}$) is given by:

$$z_{im} = \frac{d}{d'}(z_{ref} + z_{obj}) - z_{ref} \tag{1.88}$$

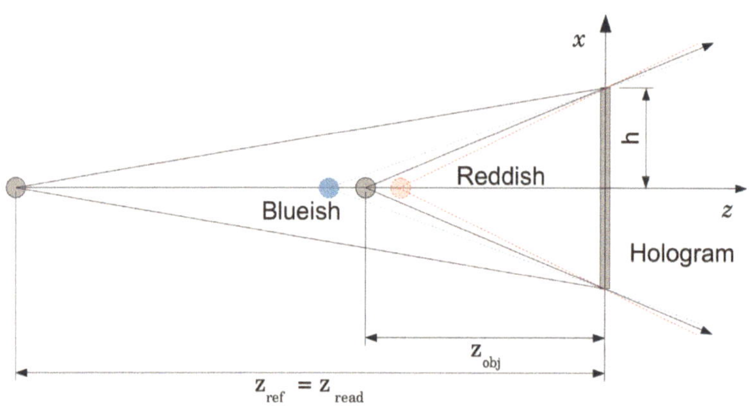

FIG. 1.30 Chromatic aberration in hologram when replayed with a source with a different wavelength than the recoding laser.

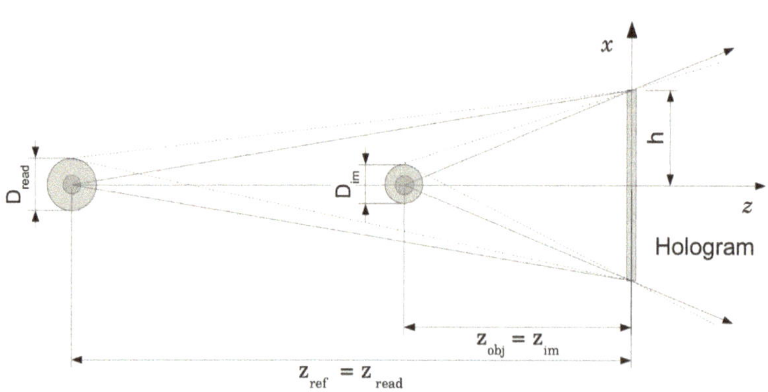

FIG. 1.31 Blurring of the image due to the spatial extend of the reading source: $D_r ead$.

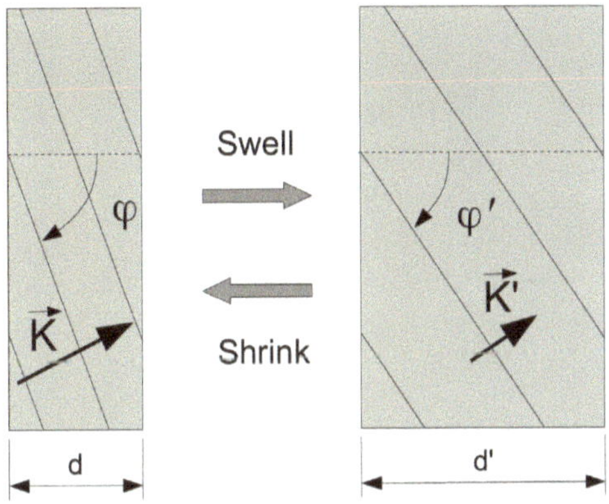

FIG. 1.32 Variation of the Bragg plane spacing and orientation with holographic material thickness.

Because the Bragg plane separation changes with the thickness, the Bragg's wavelength also shifts from the recording wavelength:

$$\lambda_{Bragg} = \lambda_{ref}\left(\cos^2\varphi + \frac{d'}{d}\sin^2\varphi\right) \tag{1.89}$$

where φ is the slant angle of the Bragg planes according to the surface normal.

This shift in wavelength is particularly noticeable for reflection holograms that are highly wavelength selective, and for which the slant angle is large ($\varphi \approx 90^\circ$).

Phase Conjugate Mirror

If a hologram is read with a beam that is the conjugate of the reference beam (R^*), the diffraction contains a term that is the exact conjugate of the object beam and propagates in a backward direction. If the equation of the diffracted terms introduced earlier (Eq. 1.82) is written with a conjugated reading beam, we obtain

$$E_{out}(x,y) = \beta|O|^2R^* + \beta|R|^2R + \beta OR^{*2} + \beta O^*|R|^2 \tag{1.90}$$

where we see appear the object conjugate term: $\beta O^*|R|^2$

This means that the diffracted beam has the same amplitude and phase as the object beam but is propagating in the opposite direction. It is as though the wave propagates back on the exact same track. This is the reason why phase conjugation is also called optical time reversal or wavefront reversal.

With a regular mirror, when the incident beam is reflected, the direction of propagation is just bent, but the phase keeps propagating unaffected: a diverging beam continues to diverge. With a phase conjugate mirror, the direction of propagation is reversed and the phase is reversed: a diverging beam converges back to its source.

The phase conjugate mirror process is particularly useful with dynamic holographic recording materials where the hologram and diffracted beam constantly adjust to the incoming object beam. Therefore, if the initial object beam is deformed by going through a perturbing media, the diffracted beam goes back to the same path and emerges undistorted (see Fig. 1.33). This technique of dynamic phase conjugation is used for imaging through scattering media [48] and internal cavity beam cleanup in a laser [49].

Self-phase conjugation happens when there is no external reference beam. Instead, the interference pattern is formed between the object beam and some of its own scattered energy or the back reflection at the material interface. This creates a reflection hologram that reflects part of the incident energy. Self-phase conjugation can be used to make saturate absorber and can cause Brillouin scattering in optical fibers [50].

For more information about optical phase conjugation, see Ref. [51], or [52], and more recently [53].

HOLOGRAPHIC SETUPS

In this section, we will describe various optical setups for the recording and replaying of holograms. As importantly, we will also discuss the respective diffraction properties of the holograms recorded in these particular setups. Because of the historical role that the different geometries have played in the development of holography, these configurations are associated with the name of their inventor.

Researchers have developed many adaptations of these basic configurations to fulfill the specific needs of targeted applications. Some of these more elaborate designs will be presented in the subsequent chapters of this book dedicated to specific utilization.

Fig. 1.34 shows an example of a holographic recording setup for the production of high-quality diffraction gratings. The mirrors on the left are used to collimate the laser beams. The mirrors on the right are used to adjust the angle between the recording beams. Air laminar fluxes on the back of the optical table help to prevent dust particle in the setup (clear room of class 100). The room is lightened with orange light to prevent the sensitization of the material before exposure.

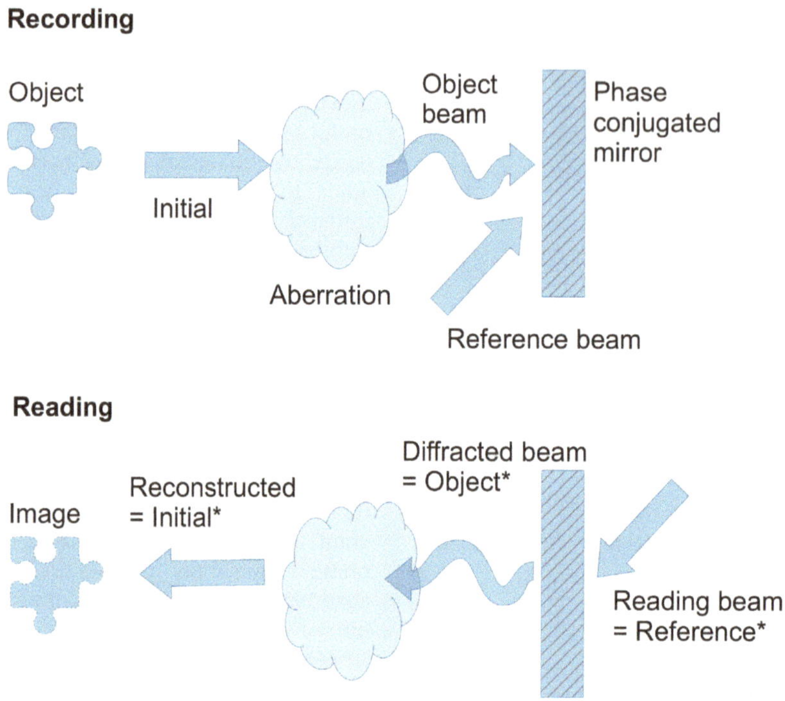

FIG. 1.33 Illustration of the phase conjugation process where a hologram is read with the conjugate of the reference beam and produce a diffracted beam, which is the conjugate of the original object beam. If an aberration medium has deformed the original image, the conjugated object goes through the aberration in reverse and is restored, forming an unaberrated image.

FIG. 1.34 Example of a holographic recording setup for the production of high-quality diffraction gratings.

Inline Transmission Hologram (Gabor)

Introduced by Denis Gabor to improve the resolution of electron microscopes [9,10], this particular recording geometry is depicted in Fig. 1.35. The object is positioned in front of the recording media and the interference occurs between the wavefront transmitted, but not perturbed through the object, and the light transmitted and scattered by the object. Obviously, the object must to be transparent for this configuration to work.

The advantages of this configuration are that the coherence of the light source can be minimal because the path difference between the object and reference beam is very small. Keeping the path length difference as small as possible was critical in Gabor's original work, which occurred before the invention of the laser, and with electron beam anyway.

Another advantage of this type of holograms is that they can be read with a polychromatic light source because the chromatic dispersion only occurs when moving off-axis.

The major problem observed when using this configuration is that all the terms of Eq. (1.82): transmission,

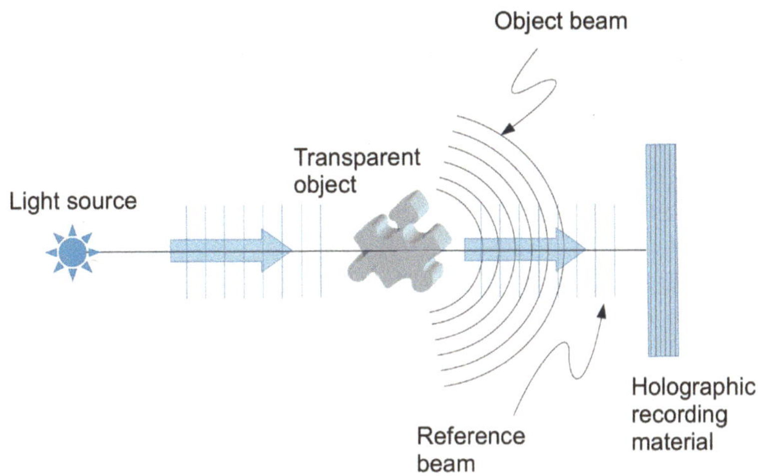

FIG. 1.35 Inline transmission hologram recording geometry: Gabor.

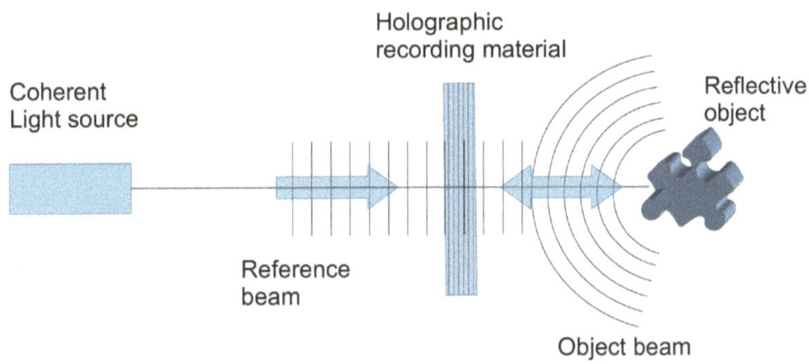

FIG. 1.36 Inline reflection hologram recording geometry: Denisyuk.

diffraction orders and halo, superimpose in the same direction, which reduces the visibility of the information.

Inline Reflection Hologram (Denisyuk)

Introduced by Yuri Denisyuk for 3D imaging [14–16], the object to be recorded is located behind the holographic recording material, as shown in Fig. 1.36. The interference is produced between the original beam going through the material, and the light scattered back from the object.

The advantages of this geometry are that the setup is quite stable because the optical path difference between the object and reference beams could be kept to a minimum if the object positioned very close to the recording medium. In addition, because reflection holograms are wavelength selective (see Section 2.3), the hologram can be read with a polychromatic light source, reproducing the color at which the hologram was recorded.

The disadvantages of this geometry are that the beam reflected by the front face of the hologram is directed in the same direction as the reconstructed beam and can superimpose, showing an annoying glare. Second is that the color of the holographic image is dictated by the wavelength of the recording light source. So, to produce color 3D images with this geometry, three different light sources centered on red, green, and blue are required (see section/refsec:Color). Finally, because the hologram has a very large acceptance angle, it can diffract the light from different points of an extended source, smearing the reproduced image. This especially happen in planes that are further away from the recording material, which limits the depth of field of the image (see Fig. 1.37). To limit this effect, point light sources such as halogen lamps or diodes are preferred for the display of Denisyuk holograms.

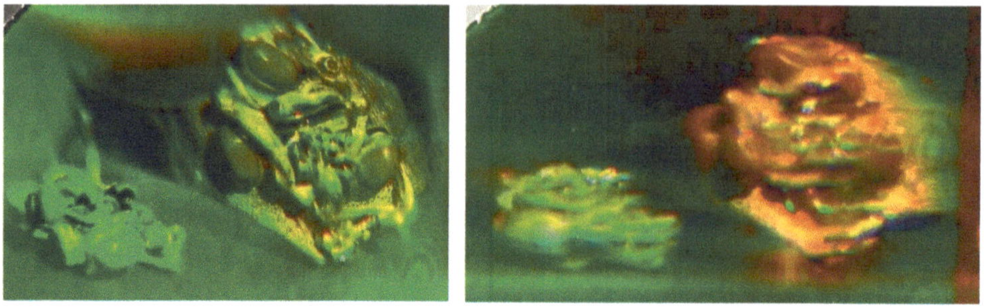

FIG. 1.37 Pictures of an inline reflection hologram that has been recorded with a green laser on bleached silver halide. The hologram is replayed by, left: a halogen (polychromatic, point like) light source, right: fluorescent tube (large étendue) light source (Hologram courtesy of Arkady Bablumian).

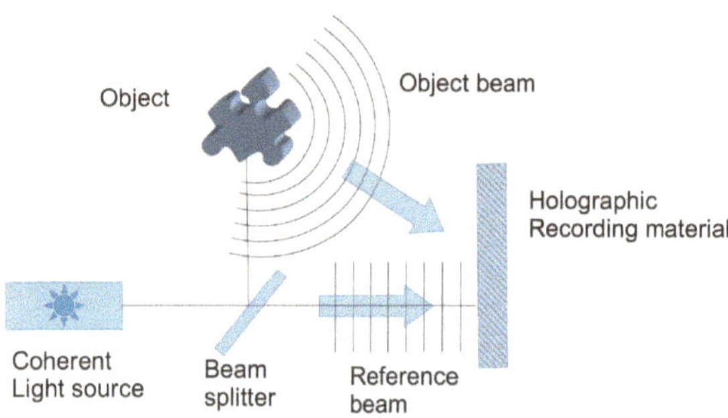

FIG. 1.38 Off-axis transmission hologram recording geometry: Leith and Upatnieks.

Off-Axis Transmission Hologram (Leith and Upatnieks)

To separate the image (+1 order) from the transmitted (0 order) and halo beams, Emmett Leith and Juris Upatnieks used an off-axis geometry where the object and reference recording beams are incident to the material at different angles [12,13,54]. The geometry is shown in Fig. 1.38.

When reading such a hologram, a monochromatic point-like source is needed. If a polychromatic light source is used instead, the wavelength dispersion of the hologram is such that multiple copies of the image superimpose with different colors, and the object cannot be observed (see Fig. 1.39).

For transmission holograms, the color of the reproduced image is given by the color of the source used to illuminate the hologram. The wavelength of the recording source is irrelevant, contrary as it is the case for reflection holograms.

Although light coherence is not needed to replay holograms, oftentimes a laser is used to read transmission hologram because these sources are monochromatic. The narrower the bandwidth is, the sharper the image appears. However, with very narrow spectral line, speckle pattern becomes visible, which gives a grainy aspect to the image (see Fig. 1.39 left). The speckle is produced by the interference between different points of the image on the observer's retina or on the detector. The speckle pattern is subjective as it changes with the observer position and pupil size.

To avoid this disruptive speckle texture, a source with a larger coherence length is preferred. Thus, a trade-off should be made between a source with narrow bandwidth, which produces speckle and a source with large bandwidth which produces a less sharper image.

Color can be reproduced with transmission holograms by recording three different holograms at three different angles, and reproducing each one with a different monochromatic light source (red, green, and blue). To make sure the three different diffracted images superimpose in the same direction during the replay, the recording angles should be calculated using the Bragg's law (Eq. 1.13) to correct for the wavelength difference between recording and reading sources. Setups to record color holograms are discussed in more detail in Section 6.7.

Transfer Holograms: H2

In both the Denisyuk and Leith and Upatnieks configuration, the reconstructed image appears on the back side of the recording medium. For a more dramatic effect, it is often wished that the image appears floating *in front* of the plate. To do so, a second hologram called transfer hologram (H2) can be recorded from the first one, called the master hologram (H1).

The recording geometry for a transfer hologram is shown in Fig. 1.40, for a transmission master hologram. The master hologram is recorded as described in Fig. 1.38. When replaying it with a reading beam that is the conjugate (*) of the initial reference beam, it forms a real image that is pseudoscopic (inside-out

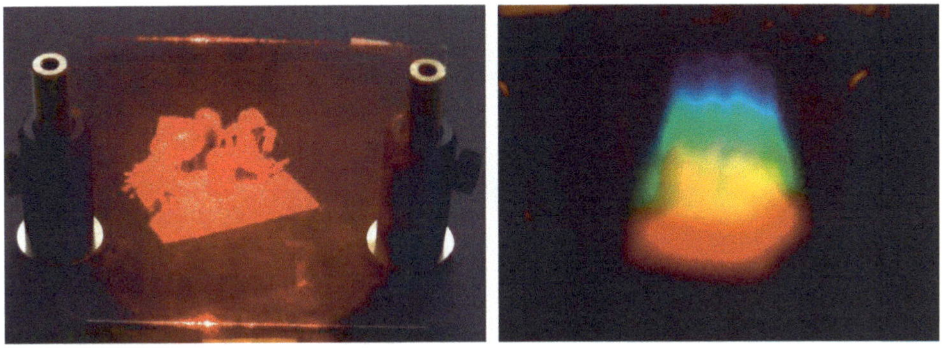

FIG. 1.39 Picture of an off-axis transmission hologram recorded on bleached silver halide and replayed by left: a red monochromatic laser diode (note the speckle), and right: an polychromatic halogen light source. (Hologram courtesy of Pierre Saint Hilaire.)

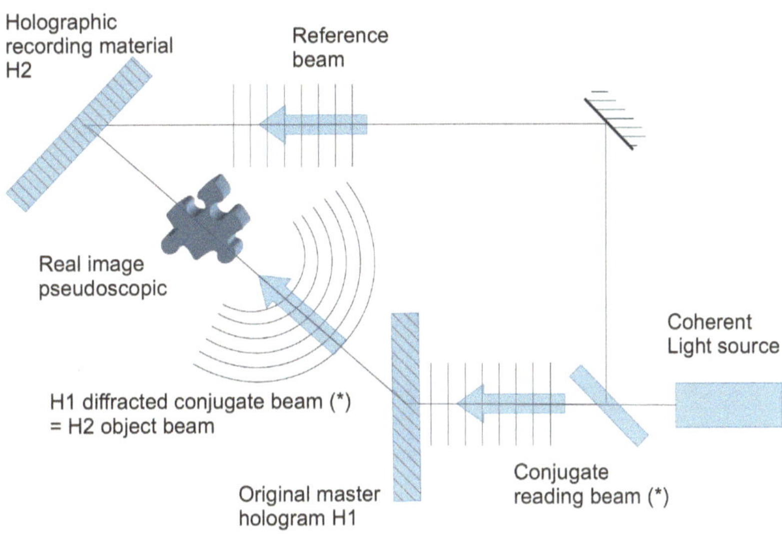

FIG. 1.40 Recording of a transmission transfer hologram such as the image will appear in front of the media.

relief, see Section 5.1). This image is used as the object beam for the transfer hologram, along with another reference beam.

When replaying the transfer hologram, the conjugate of the reference beam is used, which generates a pseudoscopic image of the object, itself pseudoscopic. The double inversion (pseudoscopic of pseudoscopic) restores the original relief.

The parts of the real image that were in front of the material during the recording of the transfer hologram appear in front of the plate, as if there were freely floating in thin air as shown in Fig. 1.41.

Rainbow Hologram (Benton)

To avoid the constraint of using a monochromatic source to read transmission hologram, Stephen Benton invented the rainbow hologram [55]. Rainbow hologram is recorded as a transfer hologram where a horizontal slit is put in front of the master hologram. The slit sacrifices the vertical parallax, to the benefit of being able to read the hologram with white light source. The recording setup for an H2 rainbow hologram is presented in Fig. 1.42.

If the rainbow hologram is read with a monochromatic light source, the image reproduces not only the object but also the slit that was used to record the

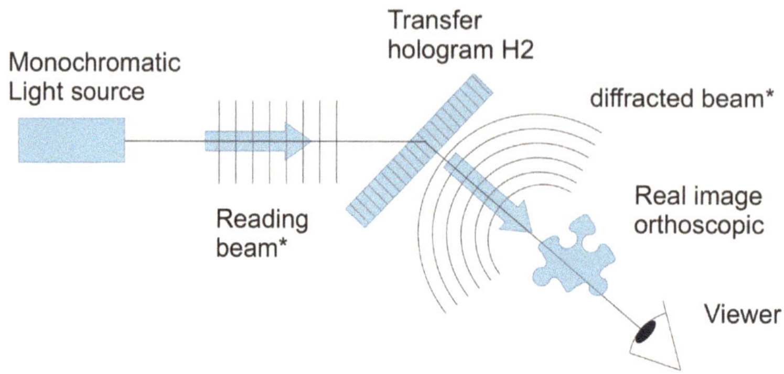

FIG. 1.41 Reading of the transfer hologram H2 that shows how the image is now produced in front of the holographic plate.

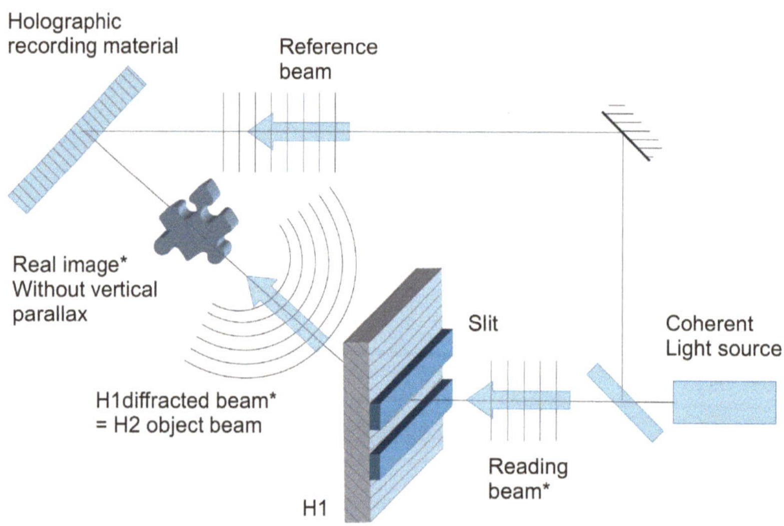

FIG. 1.42 Recording setup for a transfer rainbow hologram: Benton.

hologram. In that case, the eyes of the viewer must align exactly with the slit to observe the object through it, and if the spatial extent is too large, the object appears cropped (see Fig. 1.43).

In the case of a polychromatic reading light source, the different colors are dispersed, and each one reproduces the slit at a different angle (see Fig. 1.44). The observation point can move up and down, catching a different slit, and viewing the object with different colors, hence the name rainbow hologram. Although the vertical extent of the object is restored, there is still no vertical parallax which has been lost during the transfer recording.

Picture of a rainbow hologram is shown in Fig. 1.45 where the effect of the color dispersion is visible.

Holographic Stereogram

Holographic stereograms were invented by DeBitetto to overcome the requirement that other holographic techniques need the actual object to be present in the optical setup to be recorded [56]. The system uses multiple 2D views (pictures) that have been captured at different angles to reconstruct the parallax. Unfortunately, the wavefront (phase) information is lost in the process, so this technique cannot reproduce the accommodation cue of the human eye. For the same reason, objects that are far away from the plane of best focus (which often correspond to the plane of the stereogram) become blurry [57].

In holographic stereogram, the entire image is formed by "pixels" that are themselves holograms

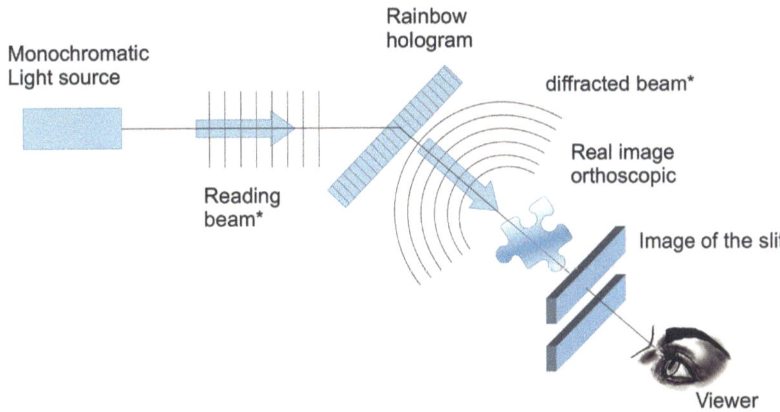

FIG. 1.43 Replaying a rainbow hologram with a monochromatic light source. The viewer see a portion of the object through a slit.

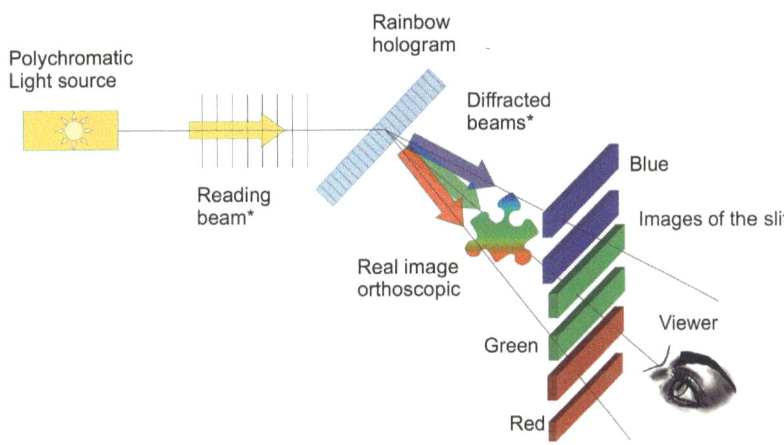

FIG. 1.44 Replaying a rainbow hologram with a polychromatic light source. The entire object is visible with a rainbow color.

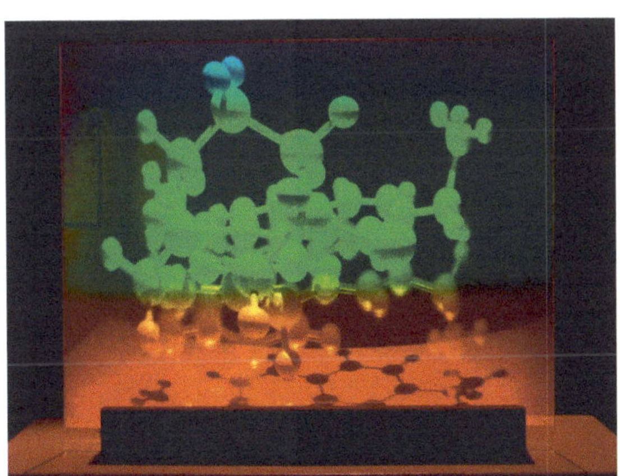

FIG. 1.45 Picture of a rainbow hologram showing the color dispersion. (Coal Molecule hologram from Jody Burns, photographed at the MIT museum by the author.)

(called hogels). The recording of a single hogel is shown in Fig. 1.46 left. A 2D frame of hogel data forms the object beam and is focused by a lens into a small section of the material. The object beam interferes with a reference beam coming at a steeper angle. The reference beam can also come from the other side of the material to obtain a reflection hologram. The operation of recording hogels is repeated, using a different 2D for each one, and to fill the entire surface of the material. The action of recording a holographic stereogram is also called holographic printing.

Once all the holographic pixels have been recorded, the stereogram can be replayed using a reading beam that will recompose the original 2D frames. By comparing the recording and the reading geometry, it can be seen that the spatial extend of image displayed during the recording is perceived by the viewer as angular

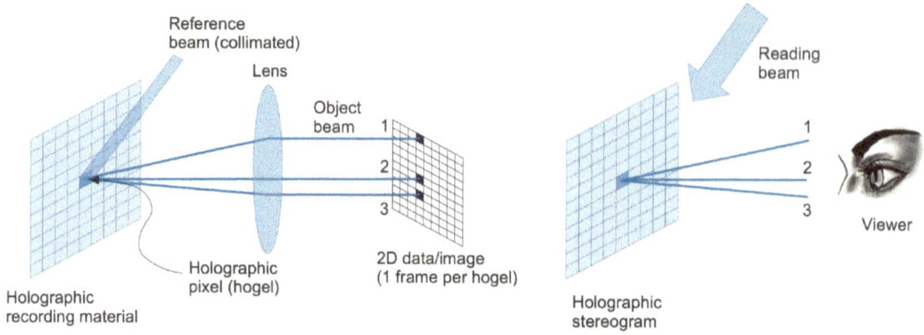

FIG. 1.46 Recording one holographic pixel of on holographic stereogram.

variation. Thus, the viewer experiences parallax when moving in front of the display. This technique is similar to the angular multiplexing used in holographic data storage [58] and is also related to integral imaging [59,60].

For the correct image and parallax to be displayed by the stereogram, some operations need to be performed on the original pictures to obtain the hogel data: spatial and angular coordinates must be inverted. To do so, the first hogel data image is composed of all the first blocks of all the images of the object taken at different angle. The following hodel data images are composed of the successive block from all the original image. This transformation is illustrated for horizontal parallax only in Fig. 1.47.

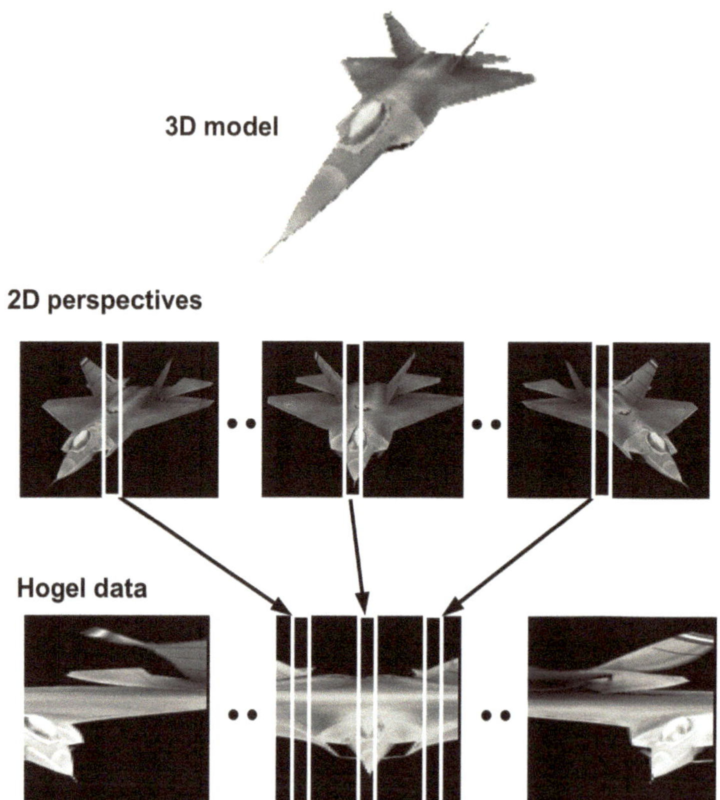

FIG. 1.47 Transformation to obtain the hogel data from the original 2D images in the case of horizontal parallax only.

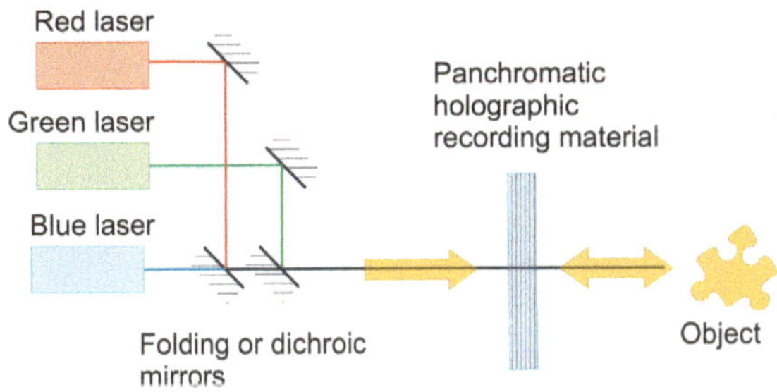

FIG. 1.48 Recording a full color hologram in reflection using three laser sources.

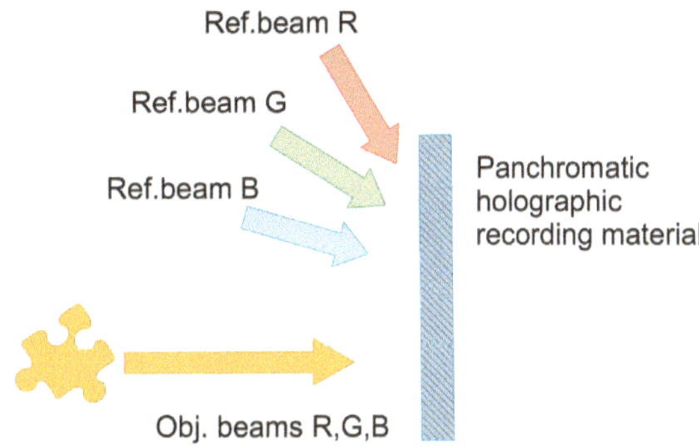

FIG. 1.49 Recording three holograms in transmission using a single laser source but three different angles.

In addition to the parallax, animation can also be recreated using the sterogram method. In this case, images of a moving object are taken and processed similarly as horizontal parallax only hogel data. When the viewer move around the stereogram, (s)he sees the animated sequence played back.

Color Hologram

It is possible to reproduce full color 3D images with hologram by recording three holograms inside the same media using three different lasers: red, green, and blue. In this case, the recording material should be panchromatic, that is, sensitive to all three wavelengths.

Using the reflection geometry, a thick hologram is highly selective in wavelength (see Section 2.3). Illuminated by a polychromatic light source, the hologram only reflects the wavelength it has been recorded with. So, if three holograms have been recorded with sources at the three fundamental colors as shown in Fig. 1.48, the three holograms diffract their own individual colors that mix together to produce a color image. Color can also be controlled by the swelling or shrinkage of the material during the processing, which changes the distance between the Bragg planes.

It also possible to record full color transmission holograms, but it is much more complicated. Indeed, transmission holograms disperse the wavelengths but are angularly selective (see Section 2.3). In addition, transmission holograms have to be read with monochromatic light sources to avoid color blurring (see Fig. 1.39). Considering these properties, three holograms, recorded with three lasers source should be *recorded and read* with sources with different colors.

In addition, if the same reference beam path is used to record the three holograms, each of the three reading beams will be diffracted by the three holograms, and color mismatch will happen. To avoid that the hologram intended for one color diffract the other colors, different angles should be used for the three reference beams during the recording, and the same angle used during the reconstruction. This configuration is shown in Fig. 1.49.

The transmission color hologram should be read with three light sources that have the same wavelength as the recording lasers and are incident at the same angles. In this case, the angular selectivity of the hologram ensures that only the correct beam is diffracted.

Edge-Lit Hologram

Edge-lit holograms are coupling a free propagating beam into a waveguide, or conversely, out coupling a beam that was propagating inside a waveguide into free space. The term waveguide is used here to refer to a medium where the light is kept inside by total internal reflection (TIR).

The name edge-lit hologram comes from the fact that the beam inside the waveguide can be inserted from the edge of the media, revealing the hologram when the material is lit by the edge. This configuration is shown in Fig. 1.50.

The type of geometry shown in Fig. 1.50 has recently gained popularity for solar concentration application [29], augmented reality see-through displays [30,31], and head-up display [32].

To record an edge-lit hologram, it is better to use a prism coupler to insert the beam inside the waveguide, rather than to use the edge of the media. This is because the edge can have a rough surface polish, or be too narrow to allow a comfortable injection. To achieve the TIR angle inside the material from an outside incidence, a prism is index-matched to the waveguide surface as shown in Fig. 1.51.

Holographic Interferometry

In interferometry, the fringe structure produced by the interference of two coherent beams is analyzed to

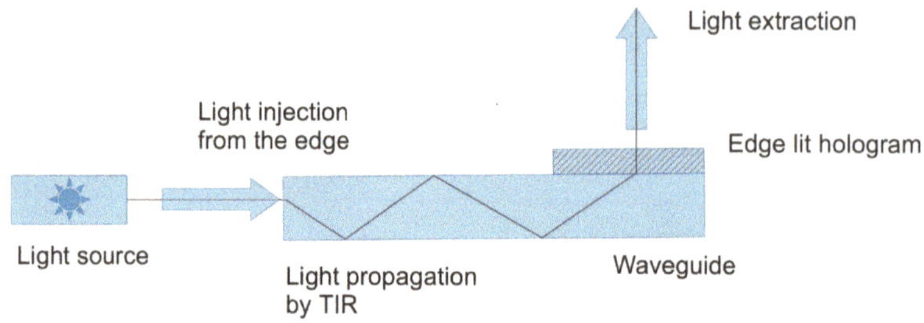

FIG. 1.50 Edge-lit hologram used to extract the light from a waveguide.

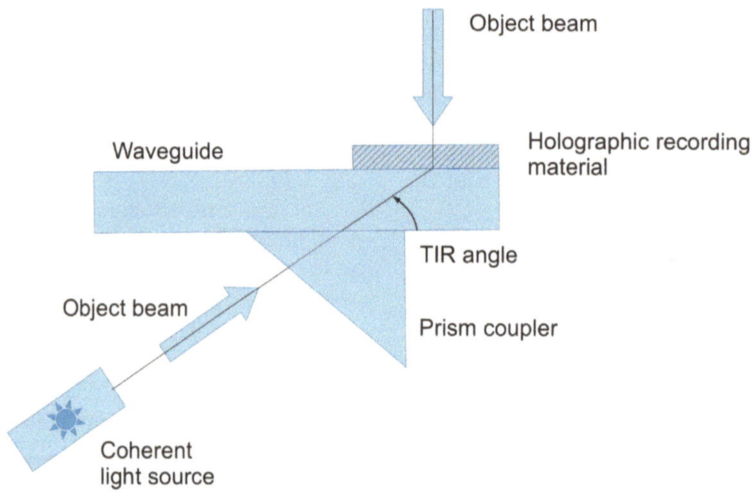

FIG. 1.51 Use of a prism to inject a beam inside the waveguide to record an edge-lit hologram.

retrieve the optical path difference between these two beams. The fringe pattern can be used to determine the shape of an object to a fraction of a wavelength [61]. However, for objects with large difference in optical path, the fringe structure can be so small that it is not distinguishable. To lower the frequency of the fringe structure, the wavefront of the reference beam should have a similar shape as the wavefront of the object beam. Holography can help in that regard by recording and then replaying a specific wavefront. Predetermined wavefront can also be produced using a CGH, so the object is compared with a theoretical shape [62].

More generally, holographic interferometry uses the interference produced between either an object and a hologram, or two holograms. An example of a setup is shown in Fig. 1.52. A hologram with a wavefront similar to the object has been recorded. It is then replayed with a reading beam, and the diffraction is superimposed with the object beam going through the hologram. If the object is deformed between recording and replaying, the deformation is visible through the fringe pattern as shown in Fig. 1.53.

Holography can also be used to record vibration modes of an object since the nodes are fixed and will produce a stable interference pattern that will diffract when the hologram is replayed. Conversely, the antinodes are moving and will not be recorded and subsequently diffract. The observed diffraction pattern is composed of bright zones where the nodes are and dark fringes where the antinodes are located as shown in Fig. 1.54 [64].

The use of dynamic holographic recording materials such as photorefractive crystals opened the door to a large variety of techniques in holographic interferometry for nondestructive testing. These techniques will be further explained in Chapter 6 by Marc Georges.

Phase retrieval

The interferogram recorded by holographic interferometry, or by any interferometric method for that purpose,

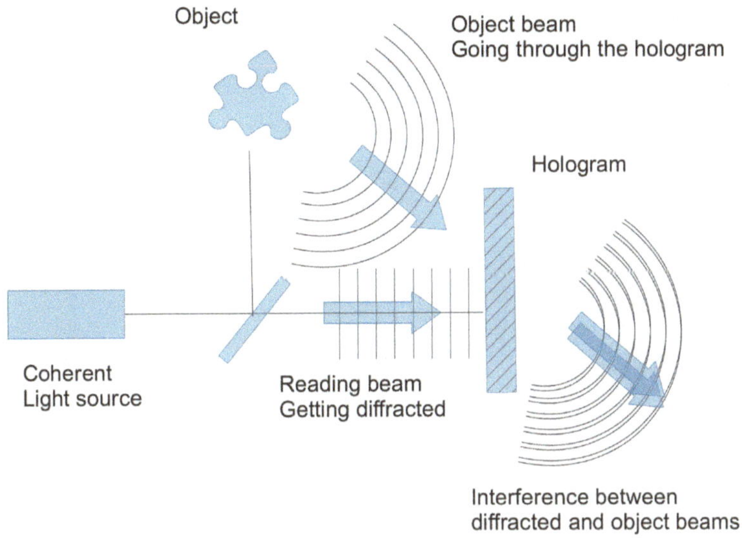

FIG. 1.52 Example of a holographic interferometry setup, where the wavefront directly coming from the object interferes with the wavefront diffracted by a prerecorded hologram. Many variations exist.

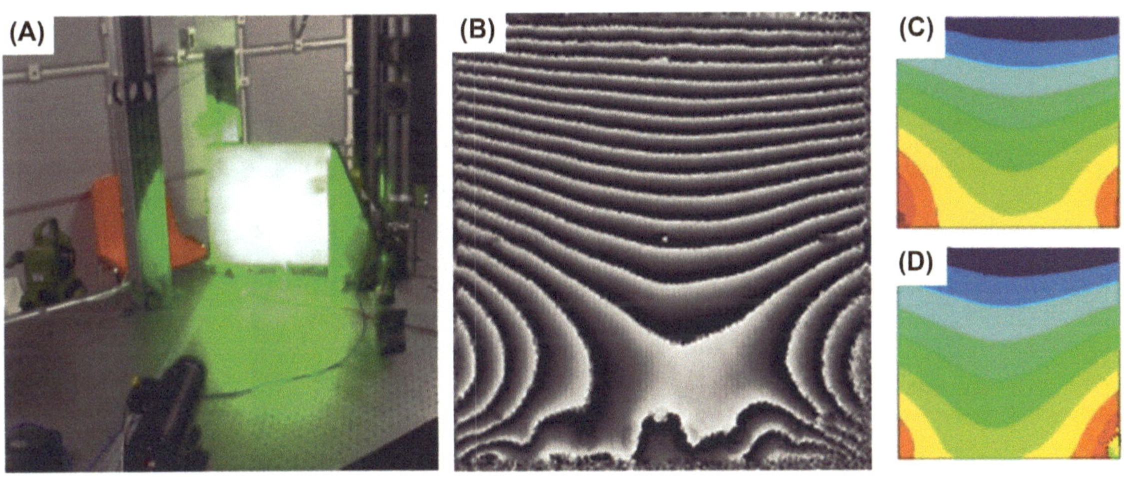

FIG. 1.53 Holographic interferometric measurement of a composite material structure deformed by heat. **(A)** picture of the setup, **(B)** recorded phase map, **(C)** retrieved deformation, and **(D)** computed deformation. (Images courtesy of Marc Georges C. Thizy, P. Lemaire, M. Georges, P. Rochus, J. P. Colette, R. John, K. Seifart, H. Bergander, G. Coe, Comparison between finite element calculations and holographic interferometry measurements, of the thermo-mechanical behaviour of satellite structures in composite materials, in: 10th International Conference on Photorefractive, Effects, Materials, and Devices, Vol. 99, 2005, pp. 700–706.)

is an intensity pattern (see Fig. 1.53B), not the phase map that is needed for numerical wavefront reconstruction (Fig. 1.53C). The most common technique used to retrieve the phase map is phase shifting. In this method, four $\pi/2$ step phase shifts are imposed on the reference beam and an interferogram is recorded for each of these shifts. The intensity map for each interferogram has the form of:

$$I(x,y;\varphi_{ref}) = A_{ref}^2 + A_{obj}^2 + 2A_{ref}A_{obj}\cos(\varphi_{ref} - \varphi_{obj}) \quad (1.91)$$

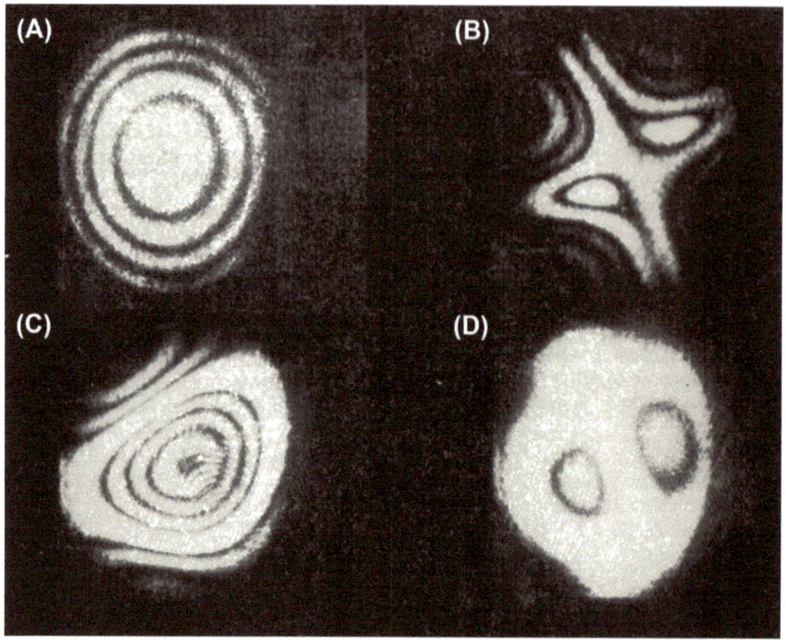

FIG. 1.54 Mode shapes of a membrane vibrating at different frequencies obtained using holographic interferometry recorded with a dynamic photorefractive polymer. (Images from B. L. Volodin, Sandalphon, K. Meerholz, B. Kippelen, N. V. Kukhtarev, N. Peyghambarian, Highly efficient photorefractive polymers for dynamic holography, Optical Engineering 34 (8) (1995) 2213–2223. https://doi.org/10.1117/12.209459.)

The object beam phase is retrieved by using:

$$\varphi(x,y) = \tan^{-1}\left[\frac{I(x,y;3\pi/2) - I(x,y;\pi/2)}{I(x,y;0) - I(xy;\pi)}\right] \quad (1.92)$$

The wavefront at the sensor position has the form:

$$U(x_0, y_0) = A\exp(i\varphi) \quad (1.93)$$

The wavefront at any other position x,y,z can be calculated using the Kirchhoff wave propagation:

$$E(x_z, y_z) = E(x_0, y_0)\frac{\exp(ikr_{z0})}{r_{z0}} \quad (1.94)$$

Similar results can be obtained by phase shifting in increments other than $\pi/2$, frequency shifting, or polarization rotation [66].

Active Phase Stabilization

The recording of a hologram requires the ability to resolve an interference pattern with submicron resolution. Any vibration, air turbulence, thermal expansion, or even sound wave, can shift that pattern and impede the recording of the hologram. For that reason, holographic setups are usually installed on optical table with air cushion to isolate the tabletop from the ground. The setup can also be covered by a box to prevent air movement.

However, depending of the sensitivity of the holographic recording material (see Chapter 2 for definition and details), and the laser power, the exposure can last up to tens of minutes. In this case, passive dampening could not be enough, and active stabilization of the holographic pattern can be necessary.

The phase stabilization of the hologram is done by using an interferometer and monitoring the fringe pattern with photodiode(s). When the pattern moves, a piezoelectric actuator (PZT) moves one of the mirrors composing the setup to bring back the fringes to their original position [67].

Fig. 1.55 shows a setup architecture for active phase stabilization. It is possible to use a portion of the original recording beams to make the interferometer by using a beam splitter element positioned along the bisector of the angle formed by these beams. An aperture is placed in front of the photodiode to limit its field of view to a portion of a fringe. Best sensitivity is obtained when the photodiode observes the midpoint between a dark and a bright fringe, where the intensity

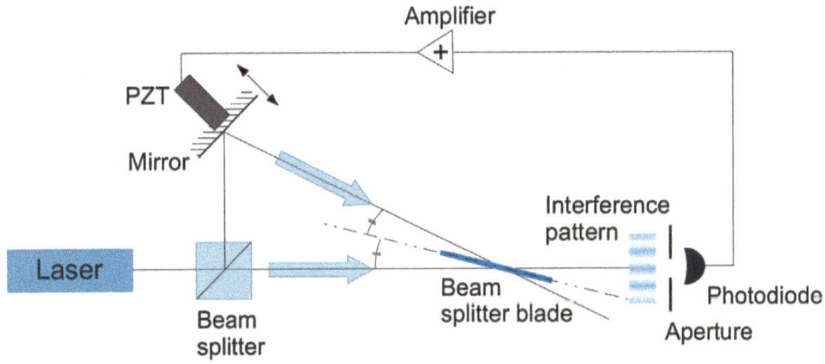

FIG. 1.55 Active fringe stabilization setup. If vibrations alter the interference patter, a signal is send to the piezo transducer that moves the mirror and stabilize the pattern.

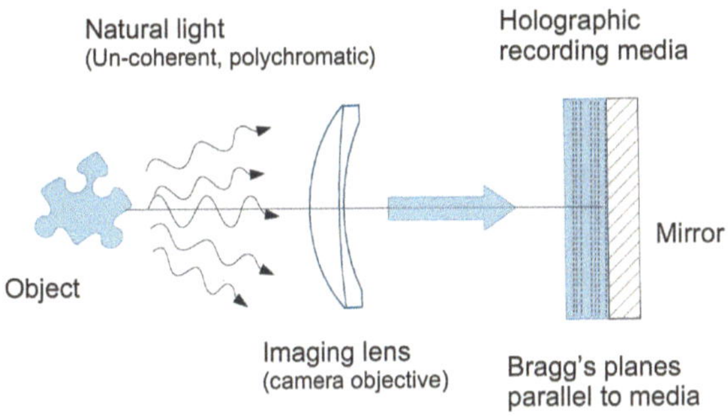

FIG. 1.56 Setup to record a Lippmann photography.

slope is at its maximum. An amplifier is used to increase the signal from the photodiode and drive the piezo transducer that shifts the mirror.

Lippmann photography

Before the invention of the color photographic plates, Gabriel Lippmann came with a solution to reproduce colors that only uses monochromatic media [35,68]. The picture is recorded with the light sensitive material directly in contact with a mirror (see Fig. 1.56). Even though the coherence length of the natural light is extremely short ($\approx 1\ \mu m$), the light going back and forth through the material interfere, creating Bragg's planes, which spacing frequency depends of the light wavelength (they are Fourier transforms of one another). When replayed in front of a white light, the different wavelengths are diffracted back by these Bragg's planes to reproduce the original colors.

Although the Lippmann photography images appear 2D, they are in fact reflection holograms in which the wavelength selectivity is used to reproduce colors. Unlike modern color photography (either analog or digital), that only capture and reproduce the red, green, and blue colors, the Lippmann technique reproduces the exact spectrum of color from the original object, making it a hyperspectral recording method.

The difficulties in Lippmann photography are of several types: the nature of the back mirror, the constraints on the recording material, and the reproduction of the pictures. The back mirror needs to be in intimate contact with the material, there should be no spacing or the coherence length is exceeded. To ensure this close proximity, a layer of mercury is used so that it can easily be removed after exposure to process the material. To record the interference pattern, the material should have a resolution of the order of 200 nm. In the case of silver halide, which is one of the most sensitive recording materials (see Chapter 2 on holographic materials), that resolution requires an extremely small grain emulsion, which decreases the sensitivity. So, minutes of exposures time are required to record a Lippmann photography, which does not suite well life subjects. It is also difficult to duplicate a Lippmann photograph because the only method is to use the exact same process.

BIBLIOGRAPHY

[1] I. Newton, Opticks: or, A Treatise of the Reflexions, Refractions, Inflexions and Colours of Light, 1704.
[2] C. Huygens, Traité de la lumière, 1678.
[3] T. Young, The Bakerian Lecture: experiments and calculations relative to physical optics, Philosophical Transactions of the Royal Society of London 94 (1804) 1–16.
[4] A. Fresnel, The Wave Theory of Light, 1819.
[5] E. Schrödinger, Quantisierung als eigenwertproblem, Annalen der Physik 384 (4) (1926) 361–376.
[6] L.D. Broglie, Waves and quanta, Nature 112 (2815) (1923) 540.
[7] L.D. Broglie, Xxxv. a tentative theory of light quanta, in: Philosophical Magazine Series 6, 47, 278, 1924, pp. 446–458, https://doi.org/10.1080/14786442408634378.

[8] J.C. Maxwell, On physical lines of force part I to IV, The London, Edinburgh, and Dublin Philosophical Magazine and Journal of Science Series, 4, 21 (139) (1861).

[9] D. Gabor, A new microscopic principle, Nature 161 (4098) (1948) 777, https://doi.org/10.1038/161777a0.

[10] D. Gabor, Microscopy by reconstructed wave-fronts, proceedings of the Royal Society A: mathematical, Physical and Engineering Sciences 197 (1051) (1949) 454–487, https://doi.org/10.1098/rspa.1949.0075.

[11] T.H. Maiman, Stimulated optical radiation in Ruby, Nature 187 (4736) (1960) 493–494, https://doi.org/10.1038/187493a0.

[12] N. Leith, J. Upatnieks, Reconstructed wavefronts and communication theory, Journal of the Optical Society of America 52 (10) (1962) 1123–1130.

[13] E.N. Leith, J. Upatnieks, Wavefront reconstruction with continuous-tone objects, Journal of the Optical Society of America 53 (12) (1963) 1377, https://doi.org/10.1364/JOSA.53.001377.

[14] Y. Denisyuk, Photographic reconstruction of the optical properties of an object in its own scattered radiation field, Soviet Physics-Doklady 7 (1962) 543–545.

[15] Y.N. Denisyuk, On the reproduction of the optical properties of an object by the wave field of its scattered radiation, Optics and Spectroscopy 15 (1963) 279–284.

[16] Y.N. Denisyuk, On the reproduction of the optical properties of an object by its scattered radiation II, Optics and Spectroscopy 18 (1965) 152–157.

[17] A.W. Lohmann, D.P. Paris, Binary fraunhofer holograms, generated by computer, Applied Optics 6 (10) (1967) 1739–1748, https://doi.org/10.1364/AO.6.001739.

[18] B.R. Brown, A.W. Lohmann, Computer-generated binary holograms, IBM Journal of Research and Development 13 (2) (1969) 160–168.

[19] E. Leith, A. Kozma, J. Upatnieks, J. Marks, N. Massey, Holographic data storage in three-dimensional media, Applied Optics 5 (8) (1966) 1303–1311.

[20] A. Lohmann, D. Paris, Computer generated spatial filters for coherent optical data processing, Applied Optics 7 (4) (1968) 651–655.

[21] L. Heflinger, R. Wuerker, R.E. Brooks, Holographic interferometry, Journal of Applied Physics 37 (2) (1966) 642–649.

[22] T. Shankoff, Recording holograms in luminescent materials, Applied Optics 8 (11) (1969) 2282–2284.

[23] H. Kogelnik, Coupled wave theory for thick hologram gratings, Bell System Technical Journal 48 (9) (1969) 2909–2947.

[24] D. Brotherton-Ratcliffe, A treatment of the general volume holographic grating as an array of parallel stacked mirrors, Journal of Modern Optics 59 (13) (2012) 1113–1132, https://doi.org/10.1080/09500340.2012.695405.

[25] H. Bjelkhagen, David Brotherton-Ratcliffe, Ultra-realistic Imaging: Advanced Techniques in Analogue and Digital Colour Holography, CRC Press, Taylor & Francis, 2013.

[26] M.G. Moharam, T.K. Gaylord, Three-dimensional vector coupled-wave analysis of planar-grating diffraction, Journal of the Optical Society of America 73 (9) (1983) 1105, https://doi.org/10.1364/JOSA.73.001105.

[27] W. Klein, B. Cook, Unified approach to ultrasonic light diffraction, IEEE Transactions on Sonics and Ultrasonics 14 (3) (1967) 123–134, https://doi.org/10.1109/T-SU.1967.29423.

[28] M. Moharam, L. Young, Criterion for Bragg and Raman-Nath diffraction regimes, Applied Optics 17 (11) (1978) 1757–1759.

[29] J.M. Castro, D. Zhang, B. Myer, R.K. Kostuk, Energy collection efficiency of holographic planar solar concentrators, Applied Optics 49 (5) (2010) 858–870, https://doi.org/10.1364/AO.49.000858.

[30] J. Han, J. Liu, X. Yao, Y. Wang, Portable waveguide display system with a large field of view by integrating freeform elements and volume holograms, Optics Express 23 (3) (2015) 3534, https://doi.org/10.1364/OE.23.003534.

[31] A. Cameron, Optical waveguide technology and its application in head-mounted displays, SPIE Defense, Security 8383 (2012) 1–11, https://doi.org/10.1117/12.923660, and.

[32] C. Bigler, P.-A. Blanche, K. Sarma, Holographic waveguide heads-up display for longitudinal image magnification and pupil expansion, Applied Optics 57 (9) (2018), https://doi.org/10.1364/AO.57.002007.

[33] K. Curtis, L. Dhar, A. Hill, W. Wilson, M. Ayres, Holographic Data Storage: From Theory to Practical Systems, John Wiley & Sons, 2011.

[34] D. Smalley, Q. Smithwick, V. Bove, Anisotropic leaky-mode modulator for holographic video displays, Nature 498 (7454) (2013) 313–317, https://doi.org/10.1038/nature12217.

[35] G. Lippmann, Sur la théorie de la photographie des couleurs simples et composées par la méthode interférentielle, Journal of Theoretical and Applied Physics 3 (3) (1894) 97–107.

[36] R. Magnusson, T. Gaylord, Diffraction efficiencies of thin absorption and transmittance gratings, Optics Communications 28 (1) (1979) 1–3, https://doi.org/10.1016/0030-4018(79)90155-X.

[37] F. Wyrowski, Diffractive optical elements: iterative calculation of quantized, blazed phase structures, Journal of the Optical Society of America A 7 (6) (1990) 961.

[38] G.J. Swanson, Binary Optics technology: theoretical limits on the diffraction efficiency of multilevel diffractive optical elements, Tech. Rep., Lincoln Laboratory MIT, 1991.

[39] P. Picart (Ed.), New Techniques in Digital Holography, Wiley, 2015.

[40] G.T. Nehmetallah, R. Aylo, L. Williams, Analog and Digital Holography with MATLAB, SPIE, 2015.

[41] J.W. Cooley, J.W. Tukey, An algorithm for the machine computation of the complex fourier series, Mathematics of Computation 19 (1965) 297, https://doi.org/10.2307/2003354.

[42] C. Slinger, C. Cameron, M. Stanley, Computer-generated holography as a generic display technology, Computer 38 (8) (2005) 46–53, https://doi.org/10.1109/MC.2005.260.

[43] R.H.-Y. Chen, T.D. Wilkinson, Computer generated hologram with geometric occlusion using GPU-accelerated depth buffer rasterization for three-dimensional display, Applied Optics 48 (21) (2009) 4246, https://doi.org/10.1364/AO.48.004246.
[44] T. Shimobaba, H. Nakayama, N. Masuda, T. Ito, Rapid calculation algorithm of fresnel computer-generated-hologram using look-up table and wavefront-recording plane methods for three-dimensional display, Optics Express 18 (19) (2010) 19504–19509.
[45] A. Greenbaum, W. Luo, T.-W. Su, Z. Göröcs, L. Xue, S.O. Isikman, A.F. Coskun, O. Mudanyali, A. Ozcan, Imaging without lenses: achievements and remaining challenges of wide-field on-chip microscopy, Nature Methods 9 (9) (2012) 889–895, https://doi.org/10.1038/nmeth.2114, arXiv:NIHMS150003.
[46] R.W. Gerchberg, W.O. Saxton, A practical algorithm for the determination of phase from image and diffraction plane pictures, Optik 35 (2) (1972) 237–246, https://doi.org/10.1070/QE2009v039n06ABEH013642.
[47] F.J. Salgado-Remacha, Reducing the variability in random-phase initialized Gerchberg-Saxton Algorithm, Optics and Laser Technology 85 (2016) 30–34, https://doi.org/10.1016/j.optlastec.2016.05.021. URL, https://doi.org/10.1016/j.optlastec.2016.05.021<a name="Line_manuscript_574"/>.
[48] Z. Yaqoob, D. Psaltis, M.S. Feld, C. Yang, Optical phase conjugation for turbidity suppression in biological samples, Nature Photonics 2 (2) (2008) 110.
[49] A. Brignon, J.-P. Huignard, Phase Conjugate Laser Optics, vol. 9, John Wiley & Sons, 2003.
[50] A. Kobyakov, M. Sauer, D. Chowdhury, Stimulated brillouin scattering in optical fibers, Advances in Optics and Photonics 2 (1) (2010) 1–59, https://doi.org/10.1364/AOP.2.000001.
[51] R.A. Fisher, Optical Phase Conjugation, Academic Press, 1983.
[52] B.Y. Zel'Dovich, N.F. Pilipetsky, V.V. Shkunov, Principles of Phase Conjugation, vol. 42, Springer, 1985.
[53] G.S. He, Optical phase conjugation: principles, techniques, and applications, Progress in Quantum Electronics 26 (3) (2002) 131–191. https://doi.org/10.1016/S0079-6727(02) 00004-6.
[54] E.N. Leith, J. Upatnieks, Wavefront reconstruction with diffused illumination and three-dimensional objects, Journal of the Optical Society of America 54 (11) (1964) 1295–1301.
[55] S.A. Benton, Hologram reconstructions with extended incoherent sources in program of the 1969 annual meeting of the Optical Society of America, Jounal of the Optical Society of America 59 (11) (1969), 1545 FE20.
[56] D.J. Debitetto, Holographic panoramic stereograms synthesized from white light recordings, Applied Optics 8 (8) (1969) 1740–1741.
[57] P.S. Hilaire, P. Blanche, Are stereograms holograms? A human perception analysis of sampled perspective holography, Journal of Physics: Conference Series 415 (Isdh 2012) (2013) 012035, https://doi.org/10.1088/1742-6596/415/1/012035.
[58] K. Curtis, L. Dhar, P.-A.P.-A. Blanche, Holographic data storage technology, in: G. Cristobal, P. Schelkens (Eds.), Optical and Digital Image Processing: Fundamentals and Applications, Wiley-VCH, 2011, pp. 227–250, https://doi.org/10.1002/9783527635245.ch11.
[59] G. Lippmann, Épreuves réversibles donnant la sensation du relief, Journal de Physique Théorique et Appliquée 7 (1) (1908) 821–825, https://doi.org/10.1051/jphystap:019080070082100.
[60] X. Xiao, B. Javidi, M. Martinez-Corral, A. Stern, Advances in three-dimensional integral imaging: sensing, display, and applications [Invited], Applied Optics 52 (4) (2013) 546, https://doi.org/10.1364/AO.52.000546.
[61] P. Hariharan, Optical interferometry, Optical Interferometry 339 (2003) 277–288, https://doi.org/10.1016/B978-012311630-7/50019-8.
[62] T. Kreis, Handbook of Holographic Interferometry, vol. 26, 2005, https://doi.org/10.1002/3527604154.
[63] C. Thizy, P. Lemaire, M. Georges, P. Rochus, J.P. Colette, R. John, K. Seifart, H. Bergander, G. Coe, Comparison between finite element calculations and holographic interferometry measurements, of the thermo-mechanical behaviour of satellite structures in composite materials, in: 10th International Conference on Photorefractive, Effects, Materials, and Devices, 99, 2005, pp. 700–706.
[64] N.E. Molin, K.A. Stetson, Measuring combination mode vibration patterns by hologram interferometry, Journal of Physics E: Scientific Instruments 2 (7) (1969) 609–612, https://doi.org/10.1088/0022-3735/2/7/313.
[65] B.L. Volodin, Sandalphon, K. Meerholz, B. Kippelen, N.V. Kukhtarev, N. Peyghambarian, Highly efficient photorefractive polymers for dynamic holography, Optical Engineering 34 (8) (1995) 2213–2223, https://doi.org/10.1117/12.209459.
[66] J.R. Fienup, Phase retrieval algorithms: a comparison, Applied Optics 21 (15) (1982) 2758–2769.
[67] J. Frejlich, L. Cescato, G.F. Mendes, Analysis of an active stabilization system for a holographic setup, Applied Optics 27 (10) (1988) 1967–1976.
[68] H.I. Bjelkhagen, Lippmann photography: reviving an early colour process, History of Photography 23 (3) (1999) 274–280.

CHAPTER 2

Holographic Recording Media and Devices

PIERRE-ALEXANDRE BLANCHE, PHD

HOLOGRAPHY TERMINOLOGY

Holographic recording materials are able to change their optical properties such as their refractive index, or their absorption coefficient, according to the intensity pattern created by two interfering laser beams. Holograms can also be recorded by modifying the surface of the material, transferring the intensity pattern into thickness modulation.

Once the hologram has been recorded and the material is reilluminated by the appropriate light, the modulation previously created diffracts the incoming beam and displays the holographic image.

The distinction between an image-recording material and a holographic recording material is their necessary spatial resolution. For imaging, the material only needs to resolve details down to tens of microns, so the image is detailed and sharp to the human eye. For holography, the material needs to reproduce the modulation obtained by the interference between two laser beams. In this case, the separation between bright and dark regions can be in submicron scale. This is the reason why digital holography had to wait for the development of high-pitch focal-plane-array sensors to become practical.

When the material records the hologram as an absorption modulation, one talks of an amplitude hologram because it is the amplitude term of the light wave that is affected when interacting with the media. When it is the index of refraction of the material that is modulated by the recording, one talks of a phase hologram. In this later case, there is no absorption of the incident light, and the efficiency of the hologram can be larger than for an amplitude hologram.

Holograms can also be recorded by modifying the thickness of the material, such as by scribing or by using lithography. In this case, the hologram is a surface relief structure that modulates the phase of the light wavefront. Surface relief holograms are usually coated with a reflective material, so the difference in light path is double the thickness modulation due to the back and forth travel of the light. The use of a reflective coating also allows to encapsulate the hologram inside a transparent material (glass or polymer) to protect the delicate surface structure. Without such a coating, the index matching between the relief structure and the top material would cancel the light path difference.

Another distinction with holograms is made between the ones that operate in transmission or reflection. With transmission holograms, the incident light goes through the medium and is diffracted on the other side of the material. The Bragg planes in reflection holograms are oriented more or less perpendicular to the surface of the material. For reflection holograms, the light is diffracted back toward the same side of the material as the incident beam. In this case, the Bragg planes are roughly parallel to the surface of the material.

There is an important difference between the spectral and angular properties of transmission and reflection holograms. Transmission holograms are angularly selective but spectrally dispersive. This means that when illuminated with a white light (broad band source) transmission holograms diffract the incident light into a rainbow (the spectrum is dispersed). When transmission holograms are illuminated with a monochromatic source, they will only diffract at a very specific angle of incidence, the Bragg angle (angular selection).

The behavior of reflection holograms is quite the opposite: they are spectrally selective and angularly tolerant. Illuminated with a white light, they only diffract back a specific color, acting like a filter (spectral selection). However, reflection holograms can diffract the incoming light over a large angle of incidence. See Chapter 1, Thick Gratings Characteristics section for a full description of the properties of transmission and reflection holograms.

It also has to be noted that when the back of a transmission hologram is coated with a reflective material, it

Optical Holography-Materials, Theory and Applications. https://doi.org/10.1016/B978-0-12-815467-0.00002-5

first appears to act as a reflection hologram. However, this configuration keeps having the spectral and angular dispersion of a transmission hologram.

There exist two regimes of diffraction: Raman-Nath and Bragg. The Raman-Nath regime is characterized by having a significant amount of energy not diffracted by the hologram (strong zero order), and diffracted into orders higher than 1: ±2,±3, etc. This happens when there is only a short distance of interaction between the light and the hologram. Holograms operating in that mode are also called "thin," even though the actual thickness of the material is only one of several parameters defining the regime. On the other hand, when in the Bragg regime, holograms diffract most of the energy into the first order. For this to happen, the interaction length between the light and the grating should be relatively long, so these holograms are called "thick."

The selection criteria between the Raman-Nath and the Bragg regime is not just about the physical thickness of the material (d); it also involves the wavelength of the light used to read the hologram (λ_0), the material index of refraction (n), the grating spacing (Λ), and the angle of incidence (θ). The Klein and Cook criteria between these two regimes is given by [1]:

$$Q' = \frac{2\pi\lambda_0 d}{n\Lambda^2 cos\theta}, \tag{2.1}$$

where $Q' < 1$ for the Raman-Nath regime and $Q' > 1$ for the Bragg regime.

Another criterion for the energy distribution in higher orders has been devised by the Moharam and Young [2] and involves the index modulation (Δn):

$$\rho = \frac{\lambda_0^2}{n\Delta n\Lambda^2 cos\theta}, \tag{2.2}$$

where $\rho < 1$ for the Raman-Nath regime and $\rho \geqslant 1$ for the Bragg regime.

Given enough energy from a laser, it is definitely possible to write holograms in very peculiar media such as metals [3,4], egg albumin [5], silk [6], or cactus extract [7]. However, these materials hardly qualify as efficient holographic recording materials. This is the reason why it is important to introduce the different figures of merit characterizing a holographic material:

Sensitivity is the most important criteria for holographic material. The sensitivity defines the amount of energy density in $J \cdot cm^{-2}$ units, required to achieve a certain amount of efficiency. The sensitivity figure is generally not divided by the efficiency, but given at a specific efficiency (e.g.,: $J \cdot cm^{-2}$ @ 90%). The sensitivity can also be given as the modulation (either index of refraction (Δn) or absorption ($\Delta\alpha$) according to the energy density. The sensitivity of a material depends of the wavelength used to record the hologram, which brings us to the next figure of merit the spectral sensitivity.

Spectral sensitivity is the range of wavelengths to which the material is responsive and holograms can be recorded. Ideally, it should be a curve of the sensitivity according to the wavelength, but most of the time there is only indication for the main laser lines: krypton red 647 nm, HeNe red 633 nm, argon green 514 nm, doubled YAG green 532 nm, argon blue 488 nm, or cadmium blue 441 nm. When a material has a very limited spectral sensitivity confined in only one color, it is called monochromatic. When the material is sensitive to red, green, and blue, even if only at very specific laser lines, it is called panchromatic. See Chapter 1, Color Hologram section for the recording of color holograms.

Maximum efficiency (η_{max}) is the maximum diffraction efficiency expressed either as the ratio between the first-order diffracted intensity and the incident intensity (external efficiency) or as the ratio between the first-order diffracted intensity and transmitted intensity without hologram (internal efficiency). Note that since the transmitted intensity is already reduced by the absorption, scattering, and Fresnel reflections from the material, the internal efficiency is always larger than the external efficiency. The efficiency has no unit and is often expressed in percent (%).

Maximum modulation is the maximum index (Δn) or absorption modulation ($\Delta\alpha$) that can be achieved with the material. The modulation is calculated from the diffraction efficiency of the hologram using the coupled wave analysis (see Chapter 1, Thick Gratings Characteristics section). The maximum modulation is important to consider when multiplexing several holograms into the same material. In this case, each of the multiplexed holograms takes a fraction of the maximum modulation range. If this modulation is rather small, the maximum efficiency for each hologram is given by $\eta \propto N^{-2}$ (see Chapter 1, Multiplexing section).

Spatial resolution is the range of lateral frequencies the material is able to record and reproduce. When plane waves interfere, they produce intensity fringes spaced according to the grating equation:

$$\Lambda = m\frac{\lambda_0/n}{sin(\theta_d) - sin(-\theta_i)}, \tag{2.3}$$

where Λ is the fringes spacing, m the diffraction order, λ_0 the wavelength of the light, θ_d the diffraction angle, and θ_i the incidence angle, with both angles being defined by the recording geometry. The grating spacing (Λ) ranges from a few hundred line pairs per millimeter ($lp \cdot mm^{-1}$) for low angle transmission hologram, to several thousand for reflection holograms at small wavelengths. More complex hologram such as those projection 3D images can be decomposed as the superposition of a multitude of these plane waves (Fourier decomposition). The material should be able to reproduce these small features in order to reproduce the hologram with maximum angular resolution.

Illumination dynamic range. The interference pattern is composed of an intensity-modulated structure. So the material must be able to react not only to a certain average level of light (sensitivity) but reproduce an entire range of intensity. The dynamic range is given by the maximum and minimum intensity the material can react to. If the minimum intensity is not zero, the intensity ratio between object and reference beams can be adjusted to accommodate the minimum.

Illumination linearity response. Within the dynamic range, it is expected that the material reacts linearly to the intensity. Thus, if the fringe pattern is sinusoidal, the modulation is also sinusoidal. If this is not the case, the phase or intensity modulation inside the material introduces higher spatial frequencies, which translate into more energy directed into higher orders, and a loss of intensity in the first order.

Stability can be understood as the shelf life of the material before exposure, or the lifetime of the hologram once it has been processed, which are different and has to be specified. In any case, the stability of the material depends on environmental conditions such as temperature, humidity, and sunlight (UV) exposure and should be specified. For application in space environment, there is also the effect of thermal cycling, X-ray, and proton radiation.

Absorption spectrum is the absorption coefficient (in cm^{-1} unit) as a function of the wavelength. The absorption coefficient (α) is calculated from the transmittance: $I/I_0 = \exp(-\alpha d)$, where d is the thickness of the material in cm. Unexposed material should have some absorption to interact with the recording light (sensitivity spectrum). However, it is important that, once processed, the material should be as transparent as possible for the waveband of interest. Absorption reduces the external efficiency.

Scattering. In addition to the absorption some loss can be due to scattering. The dispersion is mostly due to inclusions in the material whose size is comparable with the wavelength of light, resulting some Mie scattering. Scattering can be due to microbubble formation such as in gelatin emulsion material, or polymer crystal formation such as in photopolymer. These problems will be detailed in the specific material sections.

Thickness change. When the material is processed after exposure, it can experience some thickness change, either shrinkage or swelling. This volume modification reshapes the Bragg's planes by tilting them and imposing a different grating geometry and frequency. The Bragg's plane tilt changes the diffraction angle, and the modification of the frequency affects the diffracted wavelength. Both changes can be computed according to the grating equation (Chapter 1, Eq. 1.11). For more information and derivation, see the section about "Aberrations in Holograms" in Chapter 1.

Pulse response. The optical energy delivered for recording the hologram can have the form of a long exposure at relatively low intensity (continuous wave) or as a short pulse with high peak power. Laser pulses can be as short as picosecond, femtosecond, or even attosecond. For a given energy, when the duration of the pulse decreases, the peak intensity increases proportionally. Depending on the mechanism involved to record the hologram in the material, there is a peak power, or a pulse duration, for which the recording is less efficient or changes entirely. At that moment, there is a reciprocity failure in the sensitivity of the material, and the efficiency is reduced unless the recording energy is increased [8,9].

PERMANENT MATERIALS

Permanent holographic recording materials are the ones where the diffraction pattern cannot be erased or updated. Most permanent materials need to be processed after the recording of the hologram for the material to diffract the light. The goal of this postprocessing is to enhance the phase or amplitude modulation that has been initiated by the exposure. The postprocessing changes the nature of the material and fixes the hologram permanently. This means that it is not possible to record new hologram with further exposure of the material, and the hologram already present cannot be modified.

Permanent holographic recording materials are the first to have been explored, and their development is closely related to the photographic recording medium. They is still a subject of research today due to the demand for long lasting holographic images, either for the security tag industry (bank notes and anticounterfeit product), art work, holographic optical elements such as notch filters, dispersion gratings and null testers, or, to a lesser extent, holographic data storage.

Silver Halide

The chemistry of silver halide material is similar to the one used in analog black and white photography. Both of them use a colloidal suspension of silver halide crystals such as AgBr, AgCl, and AgI in gelatin. Gelatin is mainly composed of collagen, a natural protein obtained from animal bone or skin. Silver halide emulsion requires a wet chemistry postexposure processing to make the image/hologram appear. This wet process uses a developing agent that enhances the latent image by transforming the entire silver halide crystal that has been exposed into an opaque metallic silver particle. The developer bath is followed by a stop bath to halt the reaction, then a fixer bath that removes the remaining unaltered silver halide crystals. At this point, the picture/hologram is a gray scale (amplitude modulation) reproduction of the intensity distribution that illuminated the material (Fig. 2.1).

There are two differences between the photographic and holographic material: the size of the crystals, which influences both the spatial resolution and the sensitivity; and the bleaching process, which is used in holography to transform the amplitude modulation into phase modulation and enhance the efficiency.

The size of the features to resolve in the case of a hologram (tens of nanometers) is orders of magnitude smaller than in a case of visual photography (tens of microns). Therefore, the grain size of the silver halide crystal used for hologram recording should be much smaller than for visible photography. In this regard, the holographic plates are similar to the ones used for X-ray medical applications. The sensitivity of silver halide is directly proportional to the size of the grain. So, the holographic emulsions are much less sensitive than the photographic one. For example, an emulsion able to resolve 5000 $lp{\cdot}cm^{-1}$ has a grain size of 10 nm and a sensitivity of a few $mJ{\cdot}cm^{-2}$.

When the emulsion is processed to generate an amplitude modulation, its diffraction efficiency is limited to a maximum of 3.7% due to the resulting average absorption (see Chapter 1, Thick Gratings Characteristics section). It is possible to enhance the efficiency up to 100% by transforming this amplitude modulation into a phase modulation. This transformation is done by another wet postprocessing called the bleaching process that is applied in addition to the traditional developer and fixer baths. In this bleaching process, the remaining silver particles are dissolved and leave voids with lower index of refraction than the surrounding gelatin (Fig. 2.2).

The advantage of silver halide emulsion is its very large spectral sensitivity that covers the entire visible region (panchromatic) and even part of the infrared. This spectral sensitivity allows the recording of full color reflection holograms. Silver halide emulsions also have a very good sensitivity in the order or better than $mJ{\cdot}cm^{-2}$. This sensitivity is especially valuable when compared with other holographic recording materials that require much higher energy density. The spatial resolution of silver halide could be made such that it allows the recording of both transmission and reflection holograms. However, there is a trade-off between the spatial resolution and sensitivity. So, a careful selection of the emulsion according to the application is advised.

The disadvantage of silver halide material is its the relative complexity of manufacturing and processing. Indeed, the chemistry, the application of the emulsion, and the careful handling in dark room make silver halide a tricky material to fully master. An example of an automated Dr. Blade coating system to lay down silver halide emulsion is shown in Fig. 2.3. A pump (not visible on the picture) injects the liquid emulsion into a Dr. Blade hollow blade that is moved over the substrate by the gantry system. Translation speed and pump rate can be controlled using the computer on the left. The coating operation should be performed in a clean environment to avoid the contamination of the material with dust particles and in the dark to avoid sensitization of the emulsion before the actual exposure.

Other disadvantages of silver halide are its limited shelf life, the material shrinkage during processing,

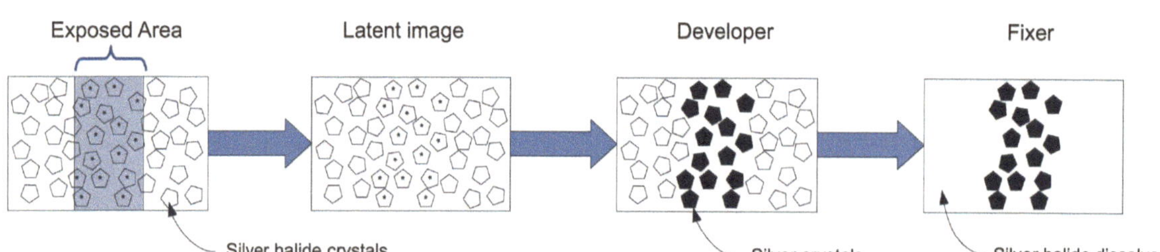

FIG. 2.1 Exposure and postprocessing of silver halide emulsion for amplitude modulation.

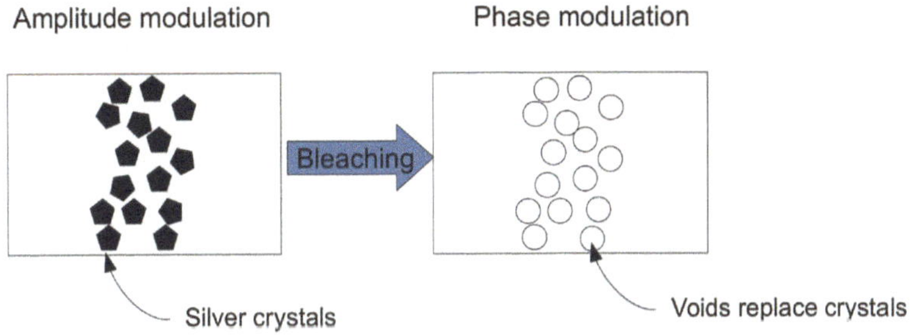

FIG. 2.2 Transformation of the amplitude modulation (absorption) into phase modulation (refractive index) by bleaching.

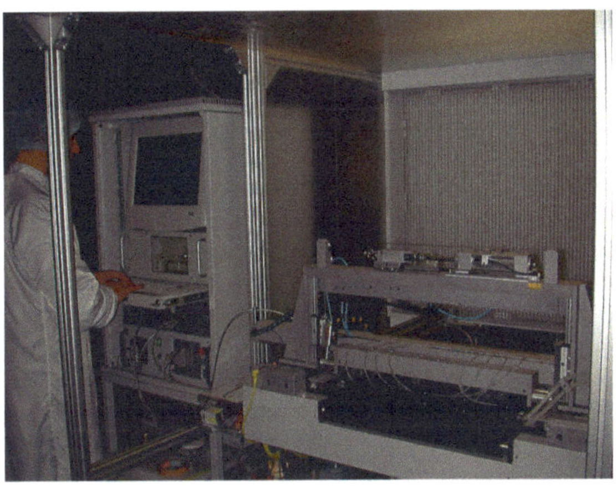

FIG. 2.3 Picture of a fully automated Dr. Blade coating system.

and the possible scattering that occurs due to the bleaching process. This scattering is caused by the relatively large size of the voids left when the silver grains are removed, leaving some haze when the holographic image is rendered.

The shrinkage of the emulsion is particularly inconvenient for the production of reflection holograms. Any material thickness change between recording and display induces chromaticity change, and to a lesser extent, optical aberrations. It is possible to control, and even take advantage of the shrinkage by both pre- and postprocessing, but this requires fine-tuning of the procedure.

For further reading about silver halide emulsion for holography, see Ref. [10].

Dichromated Gelatin

Dichromated gelatin (DCG) is quite similar to silver halide as the medium in both cases is composed of gelatin. Both materials can also produce very efficient phase holograms. However, in the case of DCG, the sensitizer is ammonium dichromate $(NH_4)_2Cr_2O_7$ (potassium dichromate is sometime used $K_2Cr_2O_7$), which is easier to obtain than fine silver halide crystals. The trade-off for this easier manufacturing is that both sensitivity and spectral response of the DCG are more limited than in silver halide emulsion. DCG usually requires exposure in the order of several $mJ \cdot cm^{-2}$ and is only sensitive to the blue and green region of the spectrum, with a response of about 1–5 for these different wavelengths (1@ 488 nm v 5@ 532 nm). Sensitization of gelatin in the red has been demonstrated using other colorants, such as methyl blue, but this requires even higher energy exposure [11].

The small sensitivity means that DCG needs high-power lasers for the recording, or longer exposure time (minutes). Longer exposure time puts high constraints on the stability of the setup, and eventually a phase stabilization system is required to ensure the interference fringes does not shift during the recording (see Chapter 1, Active Phase Stabilization section).

Once the material has been exposed, the dichromate is removed from the emulsion by a fixer bath and rinsed with water. The water also swells the gelatin, which creates larger voids where the gelatin was not exposed. The exposure has cross-linked the collagen molecular chains of the gelatin making it harder. The water is then removed by successive baths of increased concentration of isopropyl alcohol, to end up with a 100% concentration alcohol final bath. Fig. 2.4 shows a picture of a DCG processing facility for high-quality holograms. Air filtration laminar fluxes are set at the one end of the room to remove any dust particles (clean room class 100). Large double-boiler baths containing the chemicals are set on the tables. An oven used for curing the gelatin after processing can be seen on the lower right of the picture. Finally, a set of UV fluorescent tubes on the central table is used to cure the optical glue that encapsulates the final hologram. The room is lightened

FIG. 2.4 Picture of a DCG processing facility for high-quality holograms.

with orange color light to prevent exposure of the material before processing.

After processing, the gelatin film is fully transparent through all the visible spectrum, and it should no longer contain any dichromate. A spectrum from UV to near IR of processed gelatin is shown in Fig. 2.5.

The index modulation of the DCG hologram can be finely tuned by the following:

- The aging of the material before or after exposure, old material is harder and has lower index modulation.
- The exposure energy, very low or high exposure leads to lower index modulation.
- The development process. In this final case, the temperature of the water bath is particularly critical. Higher temperatures produce a higher index modulation.

An index modulation up to 10% of the average index ($\Delta n = 0.14$) can be achieved using DCG. This is one of the highest index modulations for any holographic recording material [12]. Past that point, the size of the voids causes Mie scattering and the gelatin film becomes hazy.

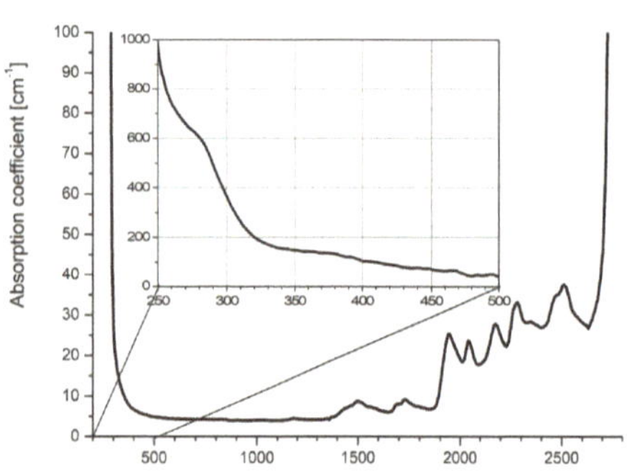

FIG. 2.5 Absorption spectrum of processed gelatin.

The advantage of a high index modulation is that the material can be made extremely thin while keeping a high diffraction efficiency. This combination of thin material and high modulation broadens the dispersion curve and allows a transmission hologram to operate in a large spectral band.

With DCG material, when the developing process did not produce the desired index modulation, the hologram can be reprocessed starting at the water bath, reducing or increasing the modulation by changing the temperature of the water. This is particularly useful when manufacturing dispersing elements where the blaze and superblaze curves (dispersion spectrum) need to be finely tuned.

Yet another advantage of the DCG is its spatial resolution. For silver halide, the resolution is determined by the size of the crystals. For DCG, the resolution is due to the photoreduction of the chromium ions, so resolution up to 10,000 $\text{lp}\cdot\text{mm}^{-1}$ can be achieved.

To avoid water reabsorption by the gelatin once it has been processed, it can be encapsulated between glass plates using optical glue. Once sealed, the gelatin medium is particularly resistant to UV and thermal degradation.

For more information about DCG, see Ref. [13].

Photopolymers

This group of materials are organic synthetic media that undergo a polymerization initiated by the illumination. In addition to this polymerization, the photopolymers also experience a change of refractive index, so the hologram is recorded as a phase modulation. This change of refractive index is not directly due to the polymerization, but rather to the diffusion of the remaining monomers to the polymerized regions. This diffusion operates by osmosis and locally increases the density of the material in the exposed regions, which also increases the index. When the modulation reaches the desired value, the diffusion process can be halted by fully polymerizing the material, which can be achieved by UV illumination or thermal treatment (heating) depending of the material formulation.

The holographic recording process in photopolymer is shown in Fig. 2.6, where the three steps, photopolymerization, diffusion, and final polymerization, are illustrated.

The sensitivity of photopolymers is on the order of 10 $\text{mJ} \cdot \text{cm}^{-2}$ and they can be sensitized to respond to

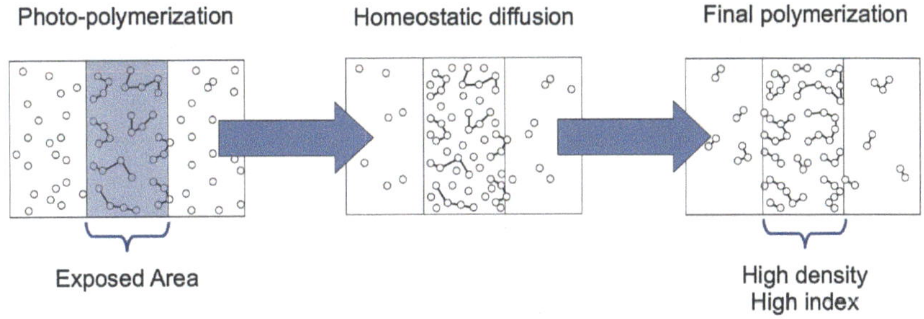

FIG. 2.6 Holographic recording process in photopolymer.

the entire visible spectrum. The spatial resolution can reach up to 4000 $lp \cdot mm^{-1}$, and the amount of shrinkage is minimal between recording and full polymerization, which is extremely useful for accurate color reproduction. Photopolymers also benefit from a prolonged shelf life, and most formulations can be kept in the dark for years without any alteration. Very old samples (5 + years) have shown some crystallization though.

Commercially available photopolymers are the simplest holographic recording material to use. The postprocessing by UV illumination can be accomplished by direct sunlight exposure and does not even require a dedicated source. This process of UV exposure also decomposes the sensitizer molecules and make the material transparent in the visible region. An example of a photopolymer film before and after UV exposure is shown in Fig. 2.7 where the dramatic change in coloration can be seen.

The disadvantage of using commercial photopolymers is the inability to control the thickness of the material, which can be an important factor for controlling the spectral and angular dispersion. In addition, the backing medium that encapsulates the material is not

FIG. 2.7 Photopolymer before (left) and after (right) UV exposure.

under the control of the user, and in some cases it has been reported to induce birefringence.

For more information about photopolymers, see Ref. [14], as well as [15].

Photoresists and Embossed Holograms

The photoresists used for holography recording are the same materials as the one used by the semiconductor industry for lithography such as SU-8. Once exposed to light, the photoresist changes its ability to be dissolved by a solvent. One can distinguish between the positive resist where the parts illuminated become soluble, and the negative resist where the parts illuminated becomes insoluble. For holography, this means that in both cases the hologram recorded is a surface relief type of modulation.

The surface relief modulation can be used as a holographic pattern without any further modification. In this case, the hologram is used in transmission and the optical path length (OPL) difference is given by the thickness modulation (d) times the difference between the index of the material (n) and the air ($OPL = d(n-1)$). As it, the hologram cannot be encapsulated, or one risk to index match the photoresist and cancel the OPL difference.

The surface relief structure can also be coated with a reflective material such as a metal. In this case, the hologram is used in reflection, and the path length difference is twice the thickness modulation ($OPL = 2d$). Once the hologram is coated, it can be encapsulated because there is not problem of index matching anymore.

The most common use for photoresist is to be a master for embossed hologram copy. The process is shown in Fig. 2.8, where the surface relief structure of the photoresist is transferred into a metal by electroplating. During electroplating, a thick layer of metal is grown on the top of the structure. This metal is used as a stamp to reproduce the relief in a heated thermoplastic material by embossing. The thermoplastic material is then

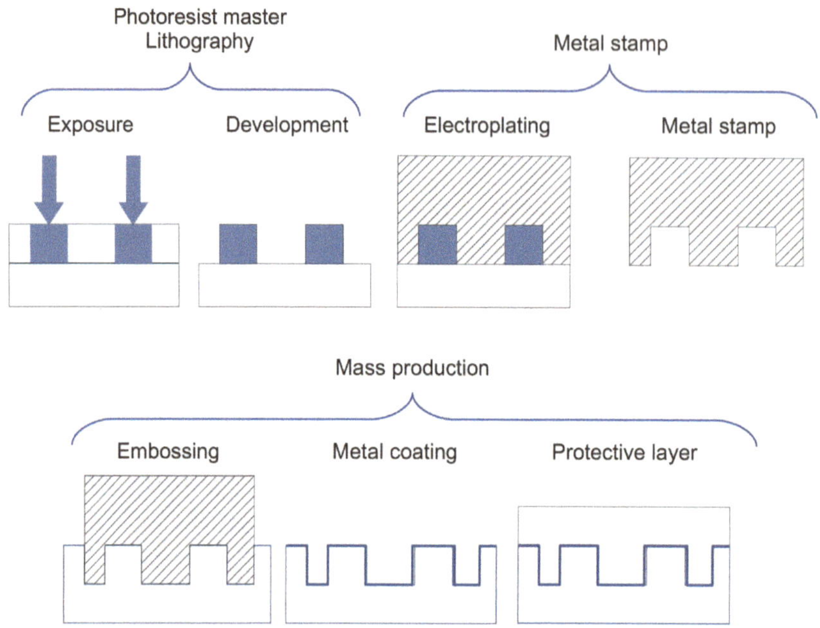

FIG. 2.8 Production of embossed hologram from a photopolymer surface relief structure.

coated with metal and further encapsulated by a protective layer. This process is used for mass producing holograms such as security tags for credit card or as an anticounterfeiting measure for luxury goods.

Because most photoresists have been developed for the photolithographic process, they have been optimized for short wavelengths such as UV light where smaller details can be imaged. Photoresist spectral sensitivity decreases dramatically at longer wavelengths, such as in the visible. The holographic structure can be recorded by the interference of two UV laser beams, but more frequently it will written into the photoresist by a lithographic process, either UV mask exposure, or by direct beam writing.

For more information about photoresist, see Ref. [16].

Photo-Thermo-Refractive Glasses

Photo-thermo-refractive glasses (PTRGs) are inorganic glasses such as $Na_2O-ZnO-Al_2O_3-SiO_2$ doped with silver (Ag), cerium (Ce), and/or fluorine (F) atoms. These glasses are highly transparent in the visible and infrared (from 350 to 2700 nm), but UV exposure below 350 nm followed by thermal precipitation of crystalline phase produces a decrease of the refractive index. This phenomenon can be used to record phase volume holograms.

The maximum value of the refractive index modulation for PTRG is quite small (10^{-3}) compared with other holographic materials, but this is compensated by the very large material thickness and high transparency. Modulation time thickness ($\Delta n \cdot d$) in PTRG is enough to create very efficient volume hologram with diffraction efficiency exceeding 99%.

The photosensitivity of PTRG for 325 nm irradiation followed by 3 h of development at 520°C is about 1.5×10^{-3} $cm^2 \cdot J^{-1}$ which means that a standard exposure for high-efficiency hologram recording is in the order of hundreds of $mJ \cdot cm^{-2}$. The spatial resolution of PTRG materials is particularly large ranging from 0 (continuous) up to 10,000 $lp \cdot mm^{-1}$.

PTRG materials are particularly important because they are useful for manufacturing the fiber Bragg gratings that can be found in telecommunication fibers and fiber lasers. The fiber Bragg grating is used as a notch filter to select a particular bandwidth and reject all the others. Based on this principle, optical add-drop multiplexers can be built.

Another major advantage of the PTRG materials is their very high damage threshold. Being particularly transparent, they can withstand nanosecond pulses with an energy exceeding tens of $mJ \cdot cm^{-1}$ (measured at 1064 nm, and 10 Hz repetition rate), and an average power exceeding 100 kW cm^{-2} of CW irradiation (measured at 1085 nm). This allows for creation of holographic beam combiners for high-energy lasers.

For more information about PTRGs, see Ref. [17].

Holographic Sensors

The spectral characteristic of the light diffracted by a hologram is extremely sensitive to the index modulation (Δn) and the spacing between the Bragg planes (Λ). This can be seen from the Bragg's Eq. (2.3) that can be further differentiated to find the spectral dispersion:

$$\frac{\Delta\lambda}{\lambda} = \frac{\Delta n}{n} + \frac{\Delta\Lambda}{\Lambda} + \cot\theta\Delta\theta \tag{2.4}$$

Therefore, if a recorded hologram is altered by its environment by swelling, or shrinking, the characteristics of the diffracted spectrum will change. This spectral, or color, change can be used as a sensor to detect the component responsible for the thickness change.

Gelatin emulsion, for example, is very well known to absorb the air humidity and swell. Swelling separates the Bragg planes already recorded and makes the hologram color shifts toward the red part of the spectrum and eventually disappear in the IR. It is possible to then restore the original color of the hologram by dehydrating the gelatin by heating it up.

Based on this example of the gelatin, several other polymer matrices have been used to make the hologram sensitive to a wide variety of components. One can cite poly(2-hydroxyethyl methacrylate) (pHEMA), poly(-acrylamide) (pAAm), or poly(vinyl alcohol) (PVA), and poly(dimethylsiloxane) (PDMS). These matrices are porous to a selection a solvents, and their absorption will make them swell.

It is also possible to dope the porous polymer matrix with other molecules such as crown ethers that will bind with specific metal ions present in the solution and make the hologram sensitive to that specific ion.

Another example is the addition of 3-(acrylamido) phenylboronic acid (3-APB) into pAAm matrix. The 3-APB molecule can bound with glucose, which makes the hologram a potential sensor for monitoring bodily fluid sugar.

This technique can be generalized to any chemical reagent sensitive to specific molecules. Once incorporated into the porous matrix that has been cross-linked to form a hologram (see the case for the gelatin in Dichromated Gelatin section), the spectral dispersion is affected by the presence of those particular molecules, and the hologram can be used as a sensor.

For more information about holographic sensors, see Ref. [18].

REFRESHABLE MATERIALS

In this section, we will introduce the holographic recording materials where the hologram can be erased, and the material reused to record a new diffraction pattern. This class of material is not used to permanently display the hologram, but rather, to present it momentarily, analyze the diffraction, change the recording setup, and write a new interference figure.

These refreshable materials can further be distinguished between those which are dynamic and do not need postprocessing for the hologram to appear, and those which need postprocessing. The advantage of the dynamic materials is that they do not need to be moved away from their original position in the recording setup before being read. This allows for a greater stability and reproducibility in the experiment, especially in holographic interferometry. The advantage of the nondynamic material is that the hologram is more stable and will last longer even during the exposure by the reading beam.

Photochromic Materials

The word photochromism describes a material that changes its absorption coefficient due to illumination. This change of absorption can be used to record amplitude holograms. However, the diffraction efficiency of phase holograms is much larger, and the Kramer-Kronig relationship specifies that a change of absorption always goes along with a refractive index change, eventually at a shifted wavelength. So, it is better to use the photochromic material for its index modulation than its absorption modulation. To do so, the writing wavelength is tuned near the absorption peak, but the reading of the hologram is performed in a more transparent region of the spectrum, where the index modulation is enhanced.

The photochromic effect is a macroscopic, observable modification of the material property, and can be generated by many different microscopic processes. It can be permanent: such as photobleaching; or reversible: like in photoisomerization. Photochromism can happen in inorganic material, such as doped glass, or inorganic molecules, such as azoic dyes.

In photochromic glasses, the SiO_2 matrix is doped with a metallic compound such as silver halide crystals. Under light excitation, the transparent silver halide molecules undergo a decomposition into metallic silver particles, which are opaque. Because the halogen molecules are trapped inside the glass and cannot escape, they can recombine with the silver after the illumination is gone, and the glass retrieves its initial absorption profile. For more information about photochromic glasses, see Refs. [17,19].

Photochromic glasses are not often used for dynamic holography due to their slow response time and low sensitivity. See Photo-Thermo-Refractive Glasses section for the PTRG material.

On the other hand, there exist several types of organic molecules that react to illumination by reversibly changing their absorption spectrum and refractive index. The most commonly used in optics and more specifically for holography recording are spiropyrans, diarylethenes, and azobenzenes [20]. When a photon is absorbed by one of these molecules, it undergoes a conformation change such as isomerization.

Some molecules such as bacteriorhodopsin have several metastable excitation states that can be addressed with different wavelengths. By taking advantage of this particularity, the hologram can be recorded at one wavelength, read nondestructively with another, and erased by a third [21]. The selection of the different wavelengths is such that the recording is done in a strong absorption region of the molecule, the reading away from it in a transparent part of the spectrum, and the erasing excites the molecule into an unstable conformation that decays back into the stable form. This process is illustrated in Fig. 2.9.

Persistent Spectral Hole Burning

When cooled down at cryogenic temperature (liquid helium $\approx 4K$), photochromic materials can experience an inhomogeneous broadening of their absorption spectra. This means that each of the absorption centers (molecules or ions) acquire a narrow spectrum that is shifted in frequency due to the interaction with the host matrix (either glass or polymer). The material spectrum still spans a large bandwidth, but individual centers possess a narrow line as shown in Fig. 2.10.

The observation of the inhomogeneous broadening requires very low temperature. This is because at room temperature the spectrum broadening of individual centers is due to the interaction with phonons and other excitation. At cryogenic temperature, these interactions are minimized, revealing the inhomogeneous broadening of the material.

If the host matrix was perfect at cryogenic temperature, each absorption center would be in the exact same state, and the spectrum would be a single narrow line. In these conditions, no inhomogeneous broadening happens and no spectral hole burning would be possible. There will be only one spectral line.

The hole burning itself happens when the inhomogeneous broadened photochromic material is illuminated with a narrow bandwidth source such as a laser.

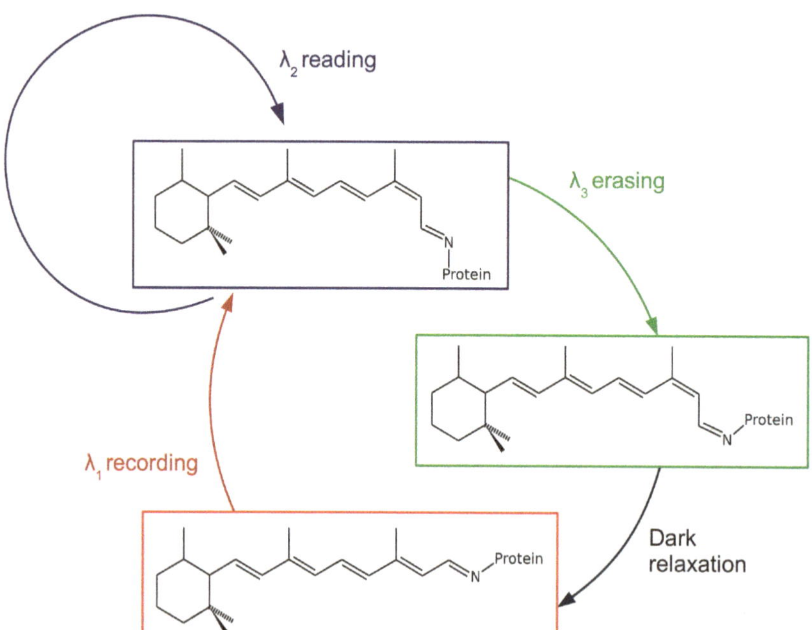

FIG. 2.9 Molecular structures of the various isomeric forms of bacteriorhodopsin and the different wavelengths used for writing, reading, and erasing a hologram.

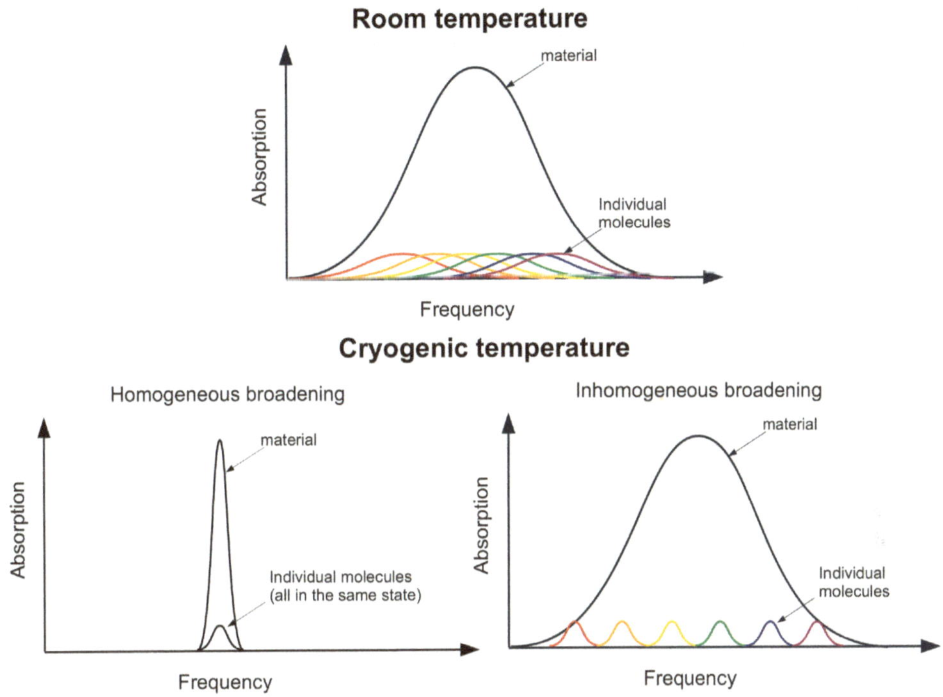

FIG. 2.10 Spectral broadenings. Top: at room temperature, the broadening is due to phonon interaction and the spectrum of individual center is large. Bottom left: at cryogenic temperature, in a perfect lattice, all the centers are in the same state and the spectrum of the material would be narrow. Bottom right: the spectrum of individual center is narrow, but due to the interaction with the lattice, each spectrum is shifted in frequency.

The source will only excite the centers with a spectrum overlapping the light frequency. Once photoexcited, the molecule or ion composing the center will have a different absorption spectrum and leave a notch in the original absorption band as shown in Fig. 2.11.

An important parameter for spectral hole burning material is the ratio between the width of the notch that can be formed, Γ_{ZPL} (or zero phonon line), and the width of the material spectrum due to the inhomogeneous broadening, Γ_{inh}. This ratio indicates the number of individual frequencies that can addressed in the material to record information and can be as large as 10^6.

Spectral hole burning materials can either be permanent or not. It is reversible if the photoexcited state of the absorption centers can decay back to the original

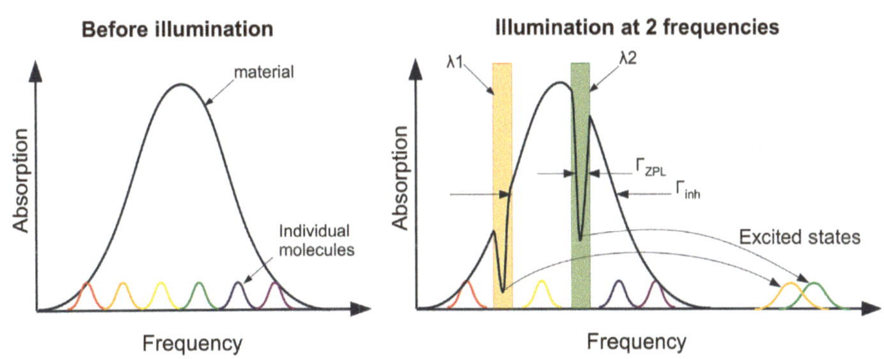

FIG. 2.11 Spectral hole burning principle: a narrow band source only excites the centers that have a spectrum overlapping the source frequency. The center excited states have a different spectrum, which leave a notch in the material initial spectrum at the same frequency as the source.

state. The temporal behavior is strongly dependent on the material, the temperature, as well as the mechanisms of relaxation that bring back the excited state to its original state. Hours of dark decay time have been demonstrated.

The interest of spectral hole burning for holography comes from the possibility to use massive wavelength multiplexing. Each individual spectral notch that can be created is an opportunity to write a different and independent hologram. The use of the frequency domain to record the hologram is a new dimension for encoding the information in addition of time and space, which increases dramatically the quantity of information that can be stored in the material. Spectral hole burning give the opportunity to increase the capacity of holographic data storage and can also be used to design very high-density correlation filters.

For more information about persistent spectral hole burning and its application for holography, see Ref. [22].

Polarization-Sensitive Materials

Some of the photochromic materials are sensitive to the light polarization and can be used to record not only amplitude-modulation holograms but also polarization-modulation holograms. This mechanism is also refer as orientational hole burning in comparison with spectral hole burning.

Polarization holograms are formed when two coherent beams with orthogonal polarization interfere. In that case, the intensity is constant, but the polarization vector changes along the period of the grating. For left and right circular polarized recording beams, the modulation is a linear polarized vector whose direction is rotating around the bisector of the beams propagation. For linear s- and p-polarized recording beams, the modulation is elliptical polarization changing from left to right directions [23]. These cases are shown in Fig. 2.12.

Sensitivity of photochromic molecules to polarization can be explained by the preferential absorption when the electric field is aligned to the axis of delocalized electronic orbital. This is the case of the azobenzene dyes that become oriented perpendicular to the light electric vector after multiple trans-cis photoisomerization and cis-trans relaxation cycles. During each cycle, the molecule can rotate by a finite and random amount. Over a large number of excitation-relaxation cycles, the molecule distribution becomes predominantly organized with the main axes orthogonal to the polarization. The orientation is a statistical process and is not driven by any torque, which means that it is rather inefficient, explaining the slow dynamic time of the material in comparison with the very fast isomerization process. For more information on polarization holography and polarization-sensitive materials, see Ref. [24].

In some materials, the multiple transitions also induce a molecular migration that can leave a surface relief modulation in addition to the volume phase hologram [25].

Even though the trans- and cis-forms of the azobenzene have different absorption spectra, the photoorientation process leaves the molecules in their relaxed form (trans). The hologram formed in this case is a phase modulation due to the birefringence induced by the anisotropic molecular distribution.

Sensitivity of polarization-sensitive azobenzene is on the order of mJ$\cdot$cm^{-2}, and the index modulation achievable is rather large, on the order of 10^{-2}.

Photorefractive Materials

The literal meaning of a photorefractive material is the one that changes its refractive index under illumination. However, the scientific community has come to define the photorefractive process very specifically as the reversible and dynamic change of the index due to an electronic process. So, photorefractive materials should not be confused with photochromic even though the observable macroscopic effect is similar.

The photorefractive effect is a multiple-step process that starts with the absorption of photons and the

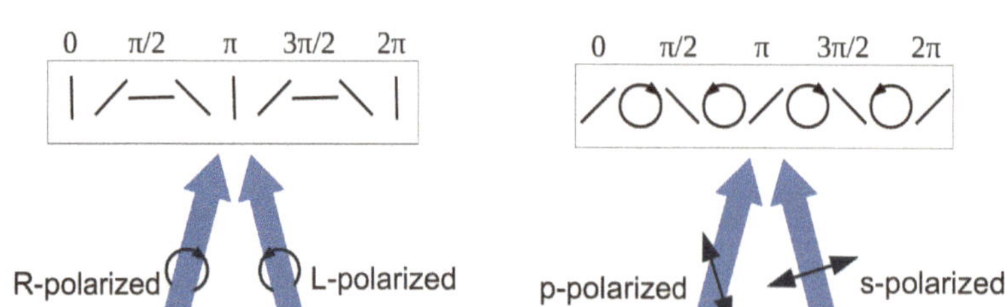

FIG. 2.12 Polarization gratings formed by the interference of, left: left and right circular polarized coherent beam; right: s- and p-polarized coherent beams.

generation of electric charges inside the material. This aspect is similar to the photovoltaic process. Both electron and hole charges are created because there is conservation of the material electrical neutrality. However, in photorefractive materials, the mobility of the charge carrier depends on its sign. One type of charge carriers migrate inside the material, whereas the other type stays in the localization where it was created. There are several charge transport mechanisms such as diffusion, drift (under external electric field), and photovoltaic that explain this migration. The mobile charge carriers are eventually trapped in the dark regions of the material, and the local charge distribution creates a space-charge electric field between the illuminated and dark regions of the material. This space-charge field modulates the index of refraction by either nonlinear electro-optic effect (in inorganic crystals) or molecular orientation (in organic materials). The different steps leading to the photorefractive process is shown in Fig. 2.13.

The photorefractive effect is characterized by a phase shift φ between the intensity pattern and the index modulation. This phase shift is responsible for the self-diffraction of the recording beams in a two-beam coupling experiment.

Photorefractive materials can be organized into two different categories: inorganic crystals and organic compounds. Inorganic crystals are grown at high temperature, and their composition is imposed by their stoichiometry and crystalline structure. Some inorganic dopants, such as metallic atoms, can be incorporated to modify the electro-optical properties such as absorption band or carrier mobility. Examples of inorganic crystals include the silenites ($Bi_{12}SiO_{20}$, $Bi_{12}GeO_{20}$), strontium-barium niobate ($Sr_xBa_{(1-x)}Nb_2O_6$), barium titanate ($BaTiO_3$), lithium niobate ($LiNbO_3$), as well as semiconductors such as GaAs, InP, and CdTe. For more information on photorefractive inorganic crystals, see Ref. [26].

Photorefractive organic compounds are mixtures of several organic molecules exhibiting specific function to achieve the photorefractive effect. They are mainly composed of a photoconductive polymer matrix that allows the charge transport. Such matrices are PVK: poly (N-vinylcarbazole), or PATPD: poly (acrylic tetraphenyl-diaminobiphenol). To enhance the sensitivity in the visible the matrix is doped with a sensitizer that is responsible for the charge photogeneration. The most used sensitizer that covers a large band in the visible region is a derivative of the C_{60} fullerene molecule, PCBM: phenyl-C61-butyric acid methyl ester. The index modulation is provided by chromophores, rod-like molecules with large dipolar moment and polarizability, such as DMNPAA: 2,5-dimethyl-4-nitrophenylazoanisole, or 7-DCST: 4-azacycloheptylbenzylidene-malonitrile. To increase the mobility of the chromophores that need to

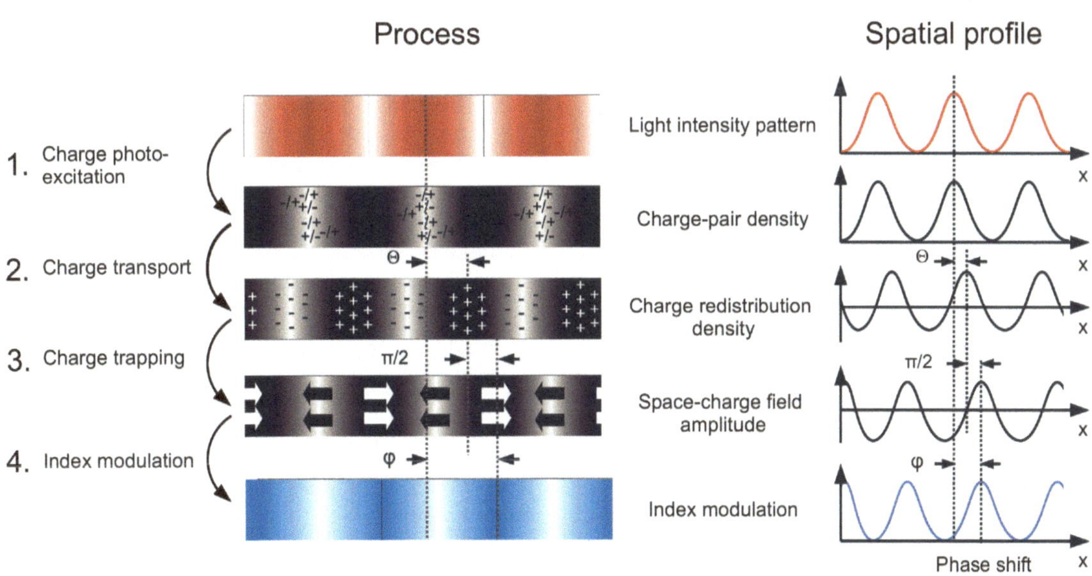

FIG. 2.13 Photorefractive process starting with the intensity distribution of the interference pattern, charge photogeneration, charge transport and distribution after migration and trapping, space-charge field, and final index modulation.

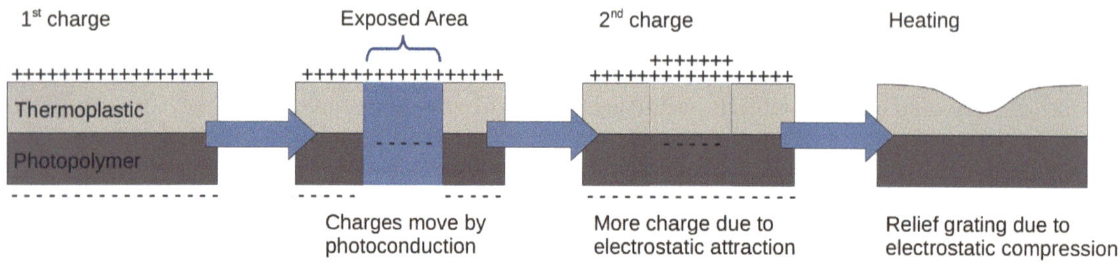

FIG. 2.14 The photo-thermoplastic process: charging the material, exposure drives the charges inside the photopolymer, the second charging increases the electrostatic attraction where the material has been exposed, heating the thermoplastic modulates the surface relief.

orient in the space-charge field, the glass transition of the polymer matrix is lowered with plasticizers, low-molecular-weight molecules such as ECZ: N-ethylcarbazole, or BBP: n-butyl benzyl phtalate. For more information on photorefractive organic materials, see Ref. [27].

The optical properties of the photorefractive crystals and organics are strongly dependent on their nature and cannot be generalized. Their sensitivity ranges from $\mu J \cdot cm^{-2}$ to $mJ \cdot cm^{-2}$, and their spectral response varies through all the visible up to the infrared.

Because of their very specific properties, photorefractive materials are used in numerous applications, such as phase conjugation [28], coherent beam amplification [29], imaging through turbid or scattering media [30], dynamic holographic imaging [31], and holographic interferometry [32].

Photo-Thermoplastic Process

The term photo-thermoplastic does not describe a material, but a material process. Thermoplastic polymers have their structural rigidity altered by temperature. By heating a sheet of material, it becomes supple and can be deformed. This property can be exploited to record a surface relief hologram, which is encoded in the deformation of the surface of the sheet. To do so, the thermoplastic material is coated on the top of a photoconductor. The surface of the photoconductor is charged by corona discharge, which attracts opposite charge on the top of the thermoplastic. When illuminated, the charges that were on the top of the photoconductor migrate through the material and came in contact with the thermoplastic. After illumination, the photoconductor is charged a second time. Because of this second charging, there is now more charge in the region that was illuminated than where the material was not exposed to light. At that moment, the thermoplastic material is heated up, and the electrostatic attraction between the charges squeezes the film, modulating its thickness. To erase the surface relief, the material is evenly charged and heated. These processes are illustrated in Fig. 2.14.

Nowadays, the photo-thermoplastic effect is not widely used due to the delay there exists between the recording and the use of the hologram. The material also has a limited number of uses before losing its ability to record a hologram. For more information about the photo thermoplastic process, see Refs. [33,34].

ELECTRONIC DEVICES

Twenty years ago, a technologic transition happened from analog to digital photography. This transition was possible thanks to the development of the focal plan array detector and the personal printer. Today, we are assisting to the same kind of transformation for holography. More and more often, the analog recording medium is substituted for an electronic device. This transformation is made possible by the high resolution of both the detectors and the microdisplay devices. As it was stated in the introduction, when photography requires micrometer resolution, holography, which records the wavefront features, requires nanometer precision. Needless to say, this is much more difficult to achieve.

The advantages of the electronic devices are numerous. They are dynamic and do not require post-processing to capture the hologram, they can be used over a very large wavelength bandwidth, they allow computer manipulation of the holographic pattern for direct interpretation or eventual transformation, they are easy to use, and they are reusable with no preprocessing.

The figure of merit for the recording and reproduction of holograms by electronic devices is the space-bandwidth product (SBP), which is a measure of the rendering capacity of an optical system. It is defined as the product of the spatial frequency bandwidth and the spatial extent of the image. In other words, at constant SBP you can either capture/reproduce a small image at high resolution or a large image at low resolution. The transformation from one to another can simply be performed by a magnifying lens.

Although very useful and convenient, electronic devices are still suffering from a coarse resolution and pixel count compared with the recording materials (small SBP), but this might be solved in the very near future. A more fundamental problem is their inability to record and reproduce volume holograms. Detectors only record the interference along a plane, and the current microdisplays only produce surface relief modulation. For a volume hologram to be recorded, the phase needs to be sensed over a nonnegligible thickness, and reproduced likewise. This excludes both the sensing and reproduction of reflection hologram by electronic devices.

Even with these limitations, electronic devices are used in a large variety of holographic setups, from 3D display to microscopy and adaptive optics. They are promising the same bright future for holography as the one they created for photography.

Focal Plane Array Detector

The function of detecting the intensity modulation generated by the interference of the coherent beams is provided by the focal plane array detector (FPAD). Two technologies can be used: either charge-coupled devices (CCDs), or complementary metal-oxide-semiconductors (CMOSs).

These devices have been popularized by digital photography and are very well known, even outside the scientific community. It is only recently, however, that the pixel pitch and count made them interesting for holography. A commercial grade FPAD can now have 20 million pixels with a 4-micron square size. This size of pixel corresponds to a resolution of 125 lp mm^{-1}, which is far from the several thousand that analog materials offer but is enough for detecting interference fringes for in-line holography or low angular separation beams at long wavelength.

Scientific grade FPAD can have a pixel size down to 2.2 μm for a total of approximately 120 million pixels on a single sensor. The problem with decreasing the pixel size is the same as with the silver halide emulsion: when the size shrink, there are fewer and fewer photons interacting with the pixel (crystal), and the sensitivity decreases dramatically. This is particularly true for the CMOS technology where the pixel surface is shared between the sensing area and the electronic amplification and logic. This shared real estate reduces the fill factor of the CMOS devices.

Once the interference pattern has been detected by the FPGA, it can be used to reconstruct the 3D light field (phase and amplitude) using a computer or displayed by another electronic device to reproduce the hologram.

For more information about the use of FPAD for holography, see Ref. [35].

Acousto-Optic Modulator

A sound wave is a compression wave propagating inside a material. This compression (and dilatation) modulates the index of refraction and if the frequency is carefully selected, can be used to diffract the light. Acousto-optic modulators (AOMs), use this principle to generate diffraction gratings. They are composed of a piezoelectric transducer (PZT) that generates the wave, which is coupled to a transparent medium, generally glass, quartz, or other crystal. Two configurations are shown in Fig. 2.15. The first one is the commonly used Bragg cell, where the sound wave generated by

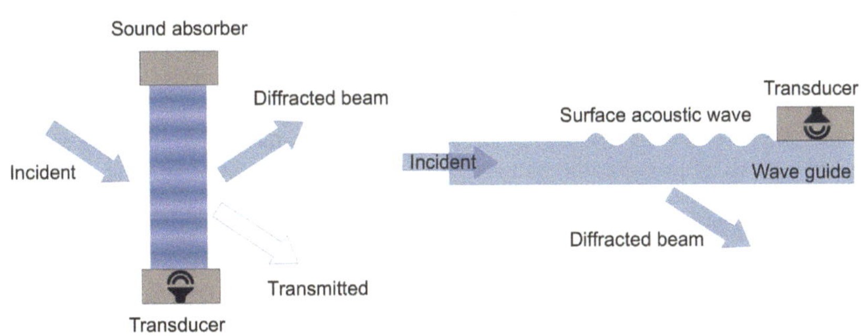

FIG. 2.15 Acousto-optic modulator configurations. Left: Bragg cell where the sound wave generated by the transducer modulates the index inside the material. Right: leaky-mode mode coupling where the transducer produces a surface acoustic wave on top of a waveguide.

the transducer travels inside the material. The second is the one where the transducer is put on top of a waveguide and generates a surface acoustic wave. The wave diffracts the light out of the waveguide, so this configuration is called leaky-mode coupling.

There are two modes of operation for the Bragg cell: static and traveling wave. When the acoustic wave is traveling, it shifts the diffracted beam frequency due to the Doppler effect. This shift is about 1 GHz due to the speed of sound in the material. To avoid this shift, a sound reflector can be used at the other end of the material. This reflection generates a standing wave like in musical wind instruments.

By changing the frequency and the amplitude of the vibration at the PZT, it is possible to change the diffraction angle and amplitude. AOM can achieve a diffraction efficiency up to 99%.

The sound wave can only generate diffraction gratings that redirect the incident beam but do not change its wavefront. Used simply as is, AOMs would not be able to form a 3D image. However, if instead of using a fixed frequency, the AOM is driven with a sum of many frequencies, it becomes similar to a one-dimensional arbitrary holographic pattern: any waveform can be decomposed as a sum of sinusoids.

The advantage of AOMs is their very large space-bandwidth product compared with the spatial light modulator. An AOM can produce larger image size and larger view angle than other technologies [36].

Owing to the very fast reaction time, AOMs are driven by a signal ranging from tens to hundreds of MHz. AOMs are used for Q-switching pulsed lasers, for pulse shaping, in telecommunication to modulate the optical signal, so it carries the information, and in spectroscopy for frequency control.

For more information about AOMs, see Ref. [37].

Spatial Light Modulator

Spatial light modulators (SLMs) are dynamic pixelated electronic devices where each pixel can be individually controlled to change the amplitude or the phase of an incoming light beam. Initially developed for the display industry, SLMs are also called microdisplays because of the small size of the pixels. Thanks to that small size, they can also be used to display holographic patterns, which diffract the incident visible light over an appreciable angle. Larger pixels reduce the angle of diffraction according to the Bragg equation and need to be used at larger wavelengths.

Microdisplays, where the light is directly emitted from the pixels such as in light-emitting diodes (LEDs) and thin film transistors (TFTs), cannot be used for holography because the phase of the different sources cannot be controlled individually. A microdisplay system where the phase of the light emitter could be controlled would be the equivalent of a phase array radar for the visible and is currently the object of active research.

Because SLMs are dynamic and refreshable at will, they are extremely attractive for dynamic holography applications such as optical tweezers, optical switching, nonmechanical scanner, wave front correction, and holographic interferometry.

Of course, 3D display is also a potential application, but the space-bandwidth product of commercially available SLMs (number of pixels over the pitch) is still too small to generate a comfortable 3D image, even with a 4K UHD resolution: 4096×2160 pixels. Systems tiling several SLMs together to increase the SBP have been demonstrated though [38].

Two types of SLMs can be distinguished according to their mode of operation: liquid crystal on silicon (LCoS) and micro-opto-electro-mechanical systems (MOEMS).

Liquid crystal on silicon

Liquid crystals (LCs) are liquid form materials composed of birefringent molecules that self-align. Owing to their large dipole moment, the LC molecules are also sensitive to externally applied electric field. When an electric field is applied the molecules rotate, which changes the optical properties of the material: birefringence amplitude and axis, as well as the refractive index.

In LCoS SLMs, a layer of LC is deposited on top of a CMOS structure forming individual cells where the voltage can be applied independently. This is the same technology used in LCD and LC television. When voltage is applied, the LCs induce phase retardation in the polarized incident light as large as a few wavelengths, which is enough to produce a fully modulated (2π) holographic phase pattern. This phase modulation can take multiple values (usually 8 bits = 256 levels), which allow the LCoS to reproduce continuous phase holograms.

It has to be noted that at maximum spatial frequency, the pattern is only represented by 2 pixels per period, and only binary phase can be displayed. To

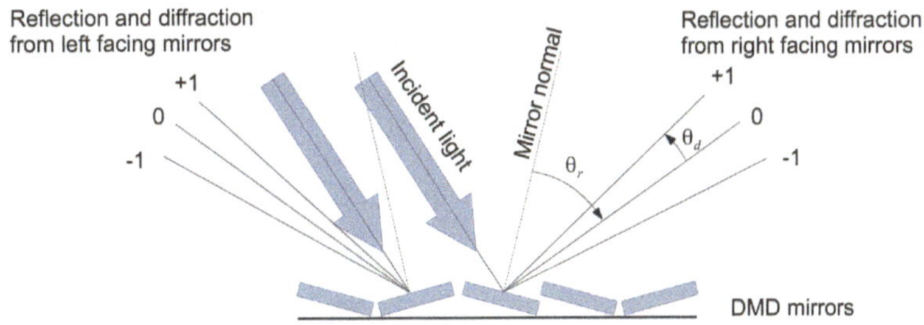

FIG. 2.16 Geometry of diffraction from the digital micromirror device (DMD).

reproduce multilevel functions, several pixels need to be used and the frequency is reduced.

To increase the reflectivity of the device, a layer of aluminum or dielectric mirror is coated between the LC and the CMOS electronic. The mirror can be ordered according to the wavelength at which the device is supposed to be used (IR, visible, or specific laser line) which give a reflectivity larger than 90%.

Amplitude modulation LCoS displays use a polarizer in front of the SLM so the phase retardation is converted into amplitude modulation following Malus's law: $I = I_0\cos^2\rho$, where ρ is the angle between the polarization vector and polarizer direction. This mode of operation is not interesting for holography as amplitude holograms only reach 10% in the first order.

A phase only LCoS is more interesting for holography due to the higher diffraction efficiency (up to 100%). In this case, the device modulates the index of refraction directly, according to the main axes of the birefringence.

Typical LCoS pixel pitch is a few microns, which offers a maximum diffraction angle of a few degrees in the visible (4 μm pitch $\approx 4°$ in the visible). The pitch is limited by the field bleeding from one cell to another and it would be hard to reduce it further in the future. An LCoS benefits from a fill factor (active pixel area over pixel size) as large as 90%. The refresh rate is limited by the viscoelastic relaxation of the molecules but can increase up to a few hundred Hz. Unfortunately, this refresh rate does not allow for the time multiplexing or very fast reconfiguration sought in some holographic applications.

For more information about LCoS SLM and holographic application, see Ref. [39].

Micro-opto-electro-mechanical systems

Micro-electro-mechanical systems (MEMSs) are devices where a micron-scale mechanical feature can be activated by an applied voltage. When this device is also used to interact with light, it is named MOEMS with the introduction of the term "opto."

The best known of these MOEMSs, due to its commercial success, is the Texas Instruments DMD (digital micromirror device), which is also known as the DLP (digital light processor). The DMD is used in display applications such as televisions and projectors. It is composed of an array of micron size mirrors: 13 μm for the 0.7″ XGA, down to 5.4 μm for the 1080p "pico." The size of the DMD follows the resolution standard for display: XGA = 1024 × 768, 1080P = 1920 × 1080, WQXGA = 2560 × 1600, and 4K UHD = 3840 × 2160.

The DMD is a binary device, and the mirrors can only take two orientations: tilted left or right by 12° for the large DMD, or 17° for the pico, according to the surface normal. Accordingly, the DMD can be used to display binary amplitude holograms. The incident light is reflected by the mirrors, and because the pattern is composed of left and right tilted mirrors, there are two reflection directions θ_r (or zero orders). In each of these directions, there are multiple diffraction orders (±1,±2, ...) due to the diffraction by the structure created by the mirror orientation (see Fig. 2.16). The maximum angle of diffraction (θ_d) around the 0 orders is given by the pixel pitch and is about 2° for 13 μm.

The maximum diffraction efficiency for the binary amplitude hologram displayed by the DMD is 10.1% as predicted by the Fourier decomposition theory (see Chapter 1, Thin Gratings Characteristics section, Fig. 2.13). However, the big advantage of the DMD is its refreshing rate, which can be up to 20 kHz (100 times

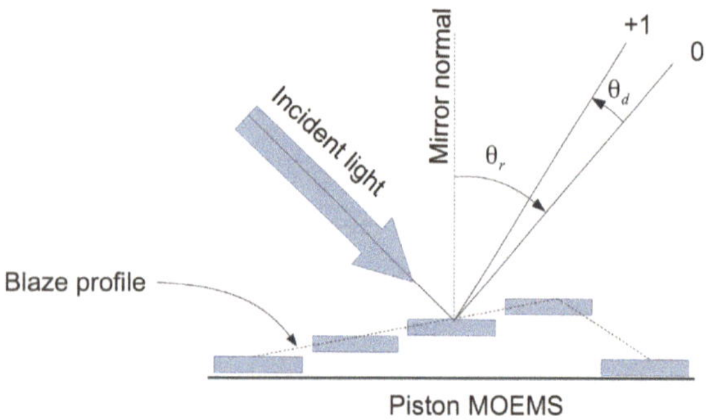

FIG. 2.17 Blazed grating formed by a piston micro-electro-mechanical system.

faster than an LCoS). This allows the DMD to be used in applications that are not accessible with LCoS SLMs, such as circuit switch for data center [40].

In display application, the fast refresh rate allows the DMD to display numerous intensity levels (10 bit gray scale), although it is a binary device that can only display black (light reflected away from viewer) or white (light reflected toward viewer). By oscillating the mirror at different frequencies, the intensity directed toward the viewer is modulated. This strategy only works because the eye integration time is rather slow ($\approx$24 Hz) compared with the 20 kHz of the device, and it cannot be used for applications requiring smaller integration time.

Another diffractive MOEMS worth mentioning is the grating light valve (GLV), which is composed of parallel thin ribbons (few micron pitch) that can take two positions: up or down. The phase difference generated by these two positions can be used to diffract the incident light with 40% efficiency (binary phase hologram). The GLV has been shown to have a refresh rate of up to 50 GHz (2000 times faster than the DMD) but also has some severe limitations for holography: it is a one-dimensional array and only one or two ribbons can be activated at a time [41].

Other types of MOEMS are under development specifically for holography. To overcome the limited diffraction efficiency of the DMD, new MOEMS modulate the phase. Such MOEMS are combining both the efficiency of the LCoS and the speed of the DMD.

One such device is an analog piston MOEMS where the micromirrors are moved up or down according to the voltage applied (see Fig. 2.17) [42]. Like in the GLV, this movement creates a phase difference between the mirrors, which diffracts the light. The difference from the GLV is that the mirrors can be positioned at multiple levels instead of just two. This allows display a gray scale phase hologram, which can have a diffraction efficiency as high as 100% (see Chapter 1, Fig. 2.14).

Another approach to obtain a phase shift is to structure the top of the pixels with a diffraction grating, and to apply a shift in between the pixels (see Fig. 2.18). This technique offers the advantage that the phase shift does not depend of the reading wavelength. Instead, the phase shift is relative to the grating spacing, so a 2π phase shift can be obtained with small displacement, even for long wavelengths such as IR used in telecommunication [43].

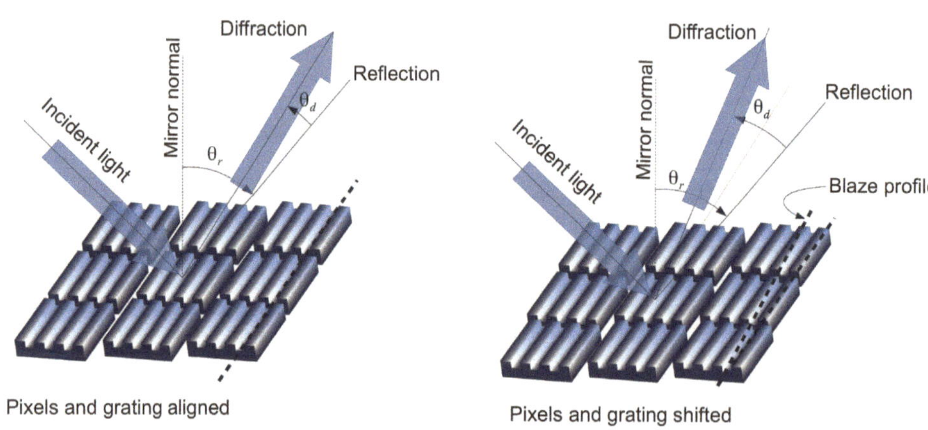

FIG. 2.18 Blazed grating formed from a piston micro-electro-mechanical system. (For more information about MOEMS and holographic application, see Ref. [44].)

BIBLIOGRAPHY

[1] W. Klein, B. Cook, Unified approach to ultrasonic light diffraction, IEEE Transactions on Sonics and Ultrasonics 14 (3) (1967) 123–134, https://doi.org/10.1109/T-SU.1967.29423.

[2] M. Moharam, L. Young, Criterion for Bragg and Raman-Nath diffraction regimes, Applied Optics 17 (11) (1978) 1757–1759.

[3] Q.-Z. Zhao, J.-R. Qiu, X.-W. Jiang, E.-W. Dai, C.-H. Zhou, C.-S. Zhu, Direct writing computer-generated holograms on metal film by an infrared femtosecond laser, Optics Express 13 (6) (2005) 2089–2092, https://doi.org/10.1364/OPEX.13.002089.

[4] R. Berlich, D. Richter, M. Richardson, S. Nolte, Fabrication of computer-generated holograms using femtosecond laser direct writing, Optics Letters 41 (8) (2016) 1752–1755, https://doi.org/10.1364/OL.41.001752.

[5] P. Pérez-Salinas, N.Y. Mejias-Brizuela, A. Olivares-Perez, A. Grande-Grande, G. Páez-Trujillo, M.P. Hernández-Garay, I. Fuentes-Tapia, Holograms With Egg Albumin 6912 (2008), 6912 – 6912 – 8, https://doi.org/10.1117/12.762307.

[6] M. Cronin-Golomb, A.R. Murphy, J.P. Mondia, D.L. Kaplan, F.G. Omenetto, Optically induced birefringence and holography in silk, Journal of Polymer Science Part B: Polymer Physics 50 (4) (2012) 257–262.

[7] A. Olivares-Pérez, S. Toxqui-López, A. Padilla-Velasco, Nopal cactus (*Opuntia ficus-indica*) as a holographic material, Materials 5 (11) (2012) 2383–2402, https://doi.org/10.3390/ma5112383.

[8] K.M. Johnson, L. Hesselink, J.W. Goodman, Holographic reciprocity law failure, Applied Optics 23 (2) (1984) 218–227.

[9] P.-A. Blanche, B. Lynn, D. Churin, K. Kieu, R.A. Norwood, N. Peyghambarian, Diffraction response of photorefractive polymers over nine orders of magnitude of pulse duration, Scientific Reports 6 (2016) 29027.

[10] H.I. Bjelkhagen, Silver-Halide Recording Materials: For Holography and Their Processing, second ed., Springer, 1993.

[11] C. Solano, R.A. Lessard, P.C. Roberge, Methylene blue sensitized gelatin as a photosensitive medium for conventional and polarizing holography, Applied Optics 26 (10) (1987) 1989–1997.

[12] P.-A. Blanche, P. Gailly, S.L. Habraken, P.C. Lemaire, C.A.J. Jamar, Volume phase holographic gratings: large size and high diffraction efficiency, Optical Engineering 43 (2004), 43 – 43 – 10, https://doi.org/10.1117/1.1803557.

[13] C.G. Stojanoff, A review of selected technological applications of dcg holograms, Proceedings of SPIE 7957 (2011), 79570L–79570L–15, https://doi.org/10.1117/12.874192.

[14] J. Guo, M.R. Gleeson, J.T. Sheridan, A review of the optimisation of photopolymer materials for holographic data storage, Physics Research International 2012 (2012) 803439.

[15] H. Berneth, F.-K. Bruder, T. Fäcke, R. Hagen, D. Hönel, D. Jurbergs, T. Rölle, M.-S. Weiser, Holographic recording aspects of high-resolution bayfol® hx photopolymer, Proceedings of SPIE 7957 (2011) 79570H.

[16] A. del Campo, C. Greiner, Su-8: a photoresist for high-aspect-ratio and 3d submicron lithography, Journal of Micromechanics and Microengineering 17 (6) (2007) R81.

[17] L. Glebov, Photochromic and Photo-Thermo-Refractive Glasses, John Wiley & Sons, Inc., 2002, https://doi.org/10.1002/0471216275.esm060.

[18] A.K. Yetisen, I. Naydenova, F. Da Cruz Vasconcellos, J. Blyth, C.R. Lowe, Holographic sensors: three-dimensional analyte-sensitive nanostructures and their applications, Chemical Reviews 114 (20) (2014) 10654–10696, https://doi.org/10.1021/cr500116a.

[19] R. Wortmann, P. Lundquist, R. Twieg, C. Geletneky, C. Moylan, Y. Jia, R. DeVoe, D. Burland, M.-P. Bernal, H. Coufal, et al., A novel sensitized photochromic organic glass for holographic optical storage, Applied Physics Letters 69 (12) (1996) 1657–1659.

[20] H. Dürr, H. Bouas-Laurent, Photochromism: Molecules and Systems, Gulf Professional Publishing, 2003.

[21] A. Seitz, N. Hampp, Kinetic optimization of bacteriorhodopsin films for holographic interferometry, The Journal of Physical Chemistry B 104 (30) (2000) 7183–7192.

[22] Moerner, W. E. (Ed.), Springer Berlin Heidelberg, Berlin, Heidelberg, 1988.

[23] K. Kuroda, Y. Matsuhashi, R. Fujimura, T. Shimura, Theory of polarization holography, Optical Review 18 (5) (2011) 374.

[24] L. Nikolova, P.S. Ramanujam, Polarization Holography, Cambridge University Press, 2009.

[25] K. Harada, M. Itoh, T. Yatagai, S.-i. Kamemaru, Application of surface relief hologram using azobenzene containing polymer film, Optical Review 12 (2) (2005) 130–134.

[26] P. Günter, J.-P. Huignard, Photorefractive Materials and Their Applications, Springer, 2007.

[27] P.-A. Blanche (Ed.), Photorefractive Organic Materials and Applications, Springer, 2016.

[28] P. Günter, Holography, coherent light amplification and optical phase conjugation with photorefractive materials, Physics Reports 93 (4) (1982) 199–299.

[29] A. Goonesekera, D. Wright, W. Moerner, Image amplification and novelty filtering with a photorefractive polymer, Applied Physics Letters 76 (23) (2000) 3358–3360.

[30] Z. Yaqoob, D. Psaltis, M.S. Feld, C. Yang, Optical phase conjugation for turbidity suppression in biological samples, Nature Photonics 2 (2) (2008) 110.

[31] P.-A. Blanche, A. Bablumian, R. Voorakaranam, C. Christenson, W. Lin, T. Gu, D. Flores, P. Wang, W.-Y. Hsieh, M. Kathaperumal, et al., Holographic three-dimensional telepresence using large-area photorefractive polymer, Nature 468 (7320) (2010) 80.

[32] M.P. Georges, V.S. Scauflaire, P.C. Lemaire, Compact and portable holographic camera using photorefractive crystals. application in various metrological problems, Applied Physics B 72 (6) (2001) 761–765.

[33] L.H. Lin, H. Beauchamp, Write—read—erase in situ optical memory using thermoplastic holograms, Applied Optics 9 (9) (1970) 2088—2092.
[34] A. Chirita, N. Kukhtarev, T. Kukhtareva, O. Korshak, V. Prilepov, I. Jidcov, Holographic imaging and interferometry with non-bragg diffraction orders on volume and surface-relief gratings in lithium niobate and photo-thermoplastic materials, Journal of Modern Optics 59 (16) (2012) 1428—1433.
[35] U. Schnars, W.P.O. Jüptner, Digital recording and numerical reconstruction of holograms, Measurement Science and Technology 13 (9) (2002) R85.
[36] D.E. Smalley, Q. Smithwick, V. Bove, J. Barabas, S. Jolly, Anisotropic leaky-mode modulator for holographic video displays, Nature 498 (7454) (2013) 313.
[37] R.G. Hunsperger, Acousto-optic modulators, in: Integrated Optics, Springer, 2009, pp. 201—220.
[38] C. Slinger, C. Cameron, M. Stanley, Computer-generated holography as a generic display technology, Computer 38 (8) (2005) 46—53.
[39] W. Osten, N. Reingand, Optical Imaging and Metrology: Advanced Technologies, John Wiley & Sons, 2012.
[40] M. Ghobadi, R. Mahajan, A. Phanishayee, N. Devanur, J. Kulkarni, G. Ranade, P.-A. Blanche, H. Rastegarfar, M. Glick, D. Kilper, Projector: agile reconfigurable data center interconnect, in: Proceedings of the 2016 ACM SIGCOMM Conference, ACM, 2016, pp. 216—229.
[41] J.I. Trisnadi, C.B. Carlisle, R. Monteverde, Overview and applications of grating-light-valve-based optical write engines for high-speed digital imaging, in: MOEMS Display and Imaging Systems II, vol. 5348, International Society for Optics and Photonics, 2004, pp. 52—65.
[42] T.A. Rhoadarmer, S.C. Gustafson, G.R. Little, T.-H. Li, Flexure-beam micromirror spatial light modulator devices for acquisition, tracking, and pointing, in: Acquisition, Tracking, and Pointing VIII, vol. 2221, International Society for Optics and Photonics, 1994, pp. 418—431.
[43] B.-W. Yoo, M. Megens, T. Sun, W. Yang, C.J. Chang-Hasnain, D.A. Horsley, M.C. Wu, A 32× 32 optical phased array using polysilicon sub-wavelength high-contrast-grating mirrors, Optics Express 22 (16) (2014) 19029—19039.
[44] P.-A. Blanche, P. Banerjee, C. Moser, M.K. Kim, Special section guest editorial:special section on the interface of holography and mems, Journal of Micro/Nanolithography, MEMS, and MOEMS 14 (4) (2015) 041301.

CHAPTER 3

The Gerchberg-Saxton Phase Retrieval Algorithm and Related Variations

TOM D. MILSTER, BSEE, PHD

INTRODUCTION

The Gerchberg-Saxton (GS) algorithm is a famous technique that has been used in many applications for nearly 50 years. It is basically a way to solve for an unknown parameter where a second parameter resulting from a unitary transformation is known. Applications include image recovery, adaptive optics, calculation of computer-generated holograms (CGHs), and many others. In this chapter, the basic geometry of the GS algorithm is reviewed, the original algorithm is described, and an argument for why it converges is presented. Early work on improving the algorithm is listed, and several variations of the algorithm are given. Algorithms for producing CGHs with speckle-free images, CGHs that form arbitrarily polarized images, and broad-wavelength CGHs are discussed. Lastly, recent algorithms using multiple diversity planes are described.

As a starting point, the original GS algorithm geometry is described with respect to how an image formed by an optical system and the light incident onto the system are related. In the special case of narrow-band laser illumination, there is a Fourier transform relationship between the light in the exit pupil of the optical system and the image.[1] With the geometry shown in Fig. 3.1, a mathematical description of this relationship is

$$\begin{aligned} U_0(x_0,y_0) &= -\frac{je^{jkz_0}}{\lambda z_0}\exp\left[j\frac{2\pi\rho_0^2}{\lambda z_0}\right]\iint_{\infty} U_s(x_s,y_s) \\ &\quad\times \exp[-j2\pi(\xi x_s+\eta y_s)]dx_s dy_s \\ &= -\frac{je^{jkz_0}}{\lambda z_0}\exp\left[j\frac{2\pi\rho_0^2}{\lambda z_0}\right]F_\xi[U_s(x_s,y_s)], \end{aligned} \tag{3.1}$$

where $\xi = \frac{x_0}{\lambda z_0}$, $\eta = \frac{y_0}{\lambda z_0}$, k = 2 pi / lambda, z_0 is the distance from the exit pupil to the image, $U_s(x_s,y_s)$ is the complex electric field in the exit pupil, $U_0(x_0,y_0)$ is the complex electric field in the image, $\rho_0^2 = x_0^2 + y_0^2$, and $F_\xi[U_s(x_s,y_s)]$ is a shorthand notation for a Fourier transform evaluated at $\xi = \frac{x_0}{\lambda z_0}$ and $\eta = \frac{y_0}{\lambda z_0}$. The complex electric fields have both amplitude and phase, where

$$\begin{aligned} &U_0(x_0,y_0) = A_0(x_0,y_0)\exp[j\phi_0(x_0,y_0)] \\ &\text{and} \\ &U_s(x_s,y_s) = A_s(x_s,y_s)\exp[j\phi_s(x_s,y_s)]. \end{aligned} \tag{3.2}$$

The classic problem for phase retrieval is to estimate $\phi_s(x_s,y_s)$ from only knowledge of the image irradiance,

$$I_0(x_0,y_0) = C_I|U_0(x_0,y_0)|^2 = C_I A_0^2(x_0,y_0), \tag{3.3}$$

where $C_I = \frac{1}{2}cn\varepsilon_0$, c is the speed of light in vacuum, n is the real refractive index, and $\varepsilon_0 = 8.85 \times 10^{-12}\ \mathrm{Fm}^{-1}$ is the permittivity of free space. As the phase of the image $\phi_0(x_0,y_0)$ is not known in the measurement, the Fourier transform relationship of Eq. (3.1) cannot be used directly to calculate $\phi_s(x_s,y_s)$. If a random or arbitrary phase is simply assigned to $\phi_0(x_0,y_0)$, the error between the calculated and actual value of $\varphi_s(x_s,y_s)$ is large when Eq. (3.1) is used to find $\phi_s(x_s,y_s)$. The goal of phase retrieval algorithms is to minimize this error, which is usually accomplished through several iterative cycles.

A popular method for estimation of $\phi_s(x_s,y_s)$ is the GS algorithm [1]. Since the algorithm was published in 1972, there have been many variations and related algorithms that have been discussed. Because of the limited space in this chapter, only a few applications and results are highlighted. Instead, the space is used to highlight differences between representative algorithms.

THE ORIGINAL GERCHBERG-SAXTON ALGORITHM

The GS algorithm takes advantage of the Fourier transform relationship of Eq. (3.1) and nonlinear constraints through several iterative cycles. To simplify the

[1]See the scalar theory of diffraction in Chapter 1 for more details.

Optical Holography-Materials, Theory and Applications. https://doi.org/10.1016/B978-0-12-815467-0.00003-7

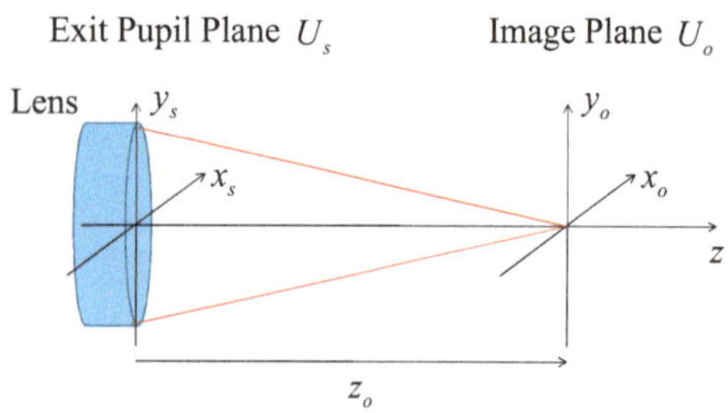

FIG. 3.1 Basic layout of the relation between the electric field U_s at the exit pupil of an optical system and the electric field U_0 at an image plane. The distance between the exit pupil and the image plane is z_0. A simple Fourier transform relationship relates the two field distributions.

calculations, phases $\phi_0(x_0,y_0)$ and $\phi_s(x_s,y_s)$ are part of the calculation, but phase terms and amplitudes in front of the Fourier transform of Eq. (3.1) are ignored. Amplitudes $A_s(x_s,y_s)$ and $A_0(x_0,y_0)$ are used to impose nonlinear constraints, and $A_0(x_0,y_0) = \sqrt{I_0(x_0,y_0)}$ is used to eliminate C_I. With these simplifications, the Fourier transform relationships become

$$A_0(x_0,y_0)\exp[j\phi_0(x_0,y_0)] = F_\xi[A_s(x_s,y_s)\exp[j\phi_s(x_s,y_s)]]$$

and

$$A_s(x_s,y_s)\exp[j\phi_s(x_s,y_s)] = F_x^{-1}[A_x(x_0,y_0)\exp[j\phi_0(x_0,y_0)]], \quad (3.4)$$

where $F_x^{-1}[\ ]$ is the inverse Fourier transform.

The heart of the GS algorithm is the iterative cycle, as shown in Fig. 3.2. In the figure, spatial dependencies of the U, A, and ϕ variables are not shown explicitly, but they are understood. The goal of the calculation is to find a good estimate for ϕ_s in the pupil plane by knowing only image irradiance I_x and pupil irradiance I_s distributions.

The original GS algorithm starts by working with image-plane data, where the number of iteration cycles k starts at unity. The phase ϕ_0 is unknown, and importantly, it is a free variable in the calculation. An initial pixel-by-pixel random assignment of ϕ_0 across the full range of phase values assures that the inverse Fourier transform of $U_0(k)$ contains modulation information across its entire domain. This type of phase assignment is called a *random diffuser*. The square root of the measured image irradiance $\sqrt{I_0}$ is assigned to the amplitude of $U_0(k)$, which provides the first nonlinear constraint. Typically, I_0 has significant value only inside regions that define the *signal window*. Regions outside of the signal window, but still within the range of calculation are set to zero in this step and define the *region of compact support*. The inverse Fourier transform provides mapping into the pupil plane, where the mathematical operation is usually carried out with an inverse fast Fourier transform [2].

Once in the pupil plane, the amplitude of the Fourier transform is replaced with the square root of the measured irradiance $\sqrt{I_s}$, if it is known, for the second nonlinear constraint. If $\sqrt{I_s}$ is unknown, the typical case is to set all amplitude pixel values to a constant or unity. The phase $\phi_s(k)$ of the inverse Fourier transform result $g_s(k)$ is assigned to $U_s(k)$. $U_s(k)$ is then defined, and the Fourier transform, which is usually calculated with a fast Fourier transform (FFT), maps the data back onto the image plane as $U_0^{test}(k)$. At this point, macroscopic mean error ε is typically calculated by

$$\varepsilon = \frac{\sum_{1}^{N_{pix}} \left[\left|U_0^{test}\right| - \sqrt{I_0}\right]^2}{N_{pix}}, \quad (3.5)$$

where N_{pix} is the number of image pixels. If ε is below a predetermined threshold ε_{th}, the iterations are stopped and the current value of $\phi_s(k)$ is the estimate of the true phase ϕ_s. The mean error ε is often used as a measure of convergence of the algorithm.

If $\varepsilon > \varepsilon_{th}$, the next iteration starts by replacing the amplitude of $U_0^{test}(k)$ with $\sqrt{I_0}$ and assigning the phase of $U_0^{test}(k=1)$ to $\phi_0(k)$. The loop continues as before, and the iterations stop only when $\varepsilon \leq \varepsilon_{th}$ or when the maximum number of iteration cycles is reached.

CONVERGENCE OF THE ORIGINAL GERCHBERG-SAXTON ALGORITHM

The original 1972 paper gives a good conceptual argument for why the GS algorithm converges. In this section, the main points of this argument are discussed. The starting point is iteration k at the point where the field U_s is calculated in the upper left of Fig. 3.3. A phasor diagram is displayed for a complex electric field value at a single pixel in the pupil plane, for which the magnitude is $\sqrt{I_0}$ and the phase angle is ϕ_s. The black circle is the range of possible U_s phasor tip locations across all iterations.

The macroscopic Fourier transform $F_\xi[\ ]$ relates all of the pixels in the pupil plane to pixels in the image plane, and the resulting complex electric field phasor U_0^{test} at one of the image-plane pixels is shown in the upper right of Fig. 3.3. The GS algorithm nonlinear amplitude constraint is applied to produce $U_0 = U_0^{test} + \delta_0$, where $|U_0| = \sqrt{I_0}$.

The inverse Fourier transform $F_x^{-1}[\ ]$ is applied to U_0, and the resulting field of the image plane phasor g_s is shown in the lower left portion of Fig. 3.3 for the same image pixel as the starting point in the upper

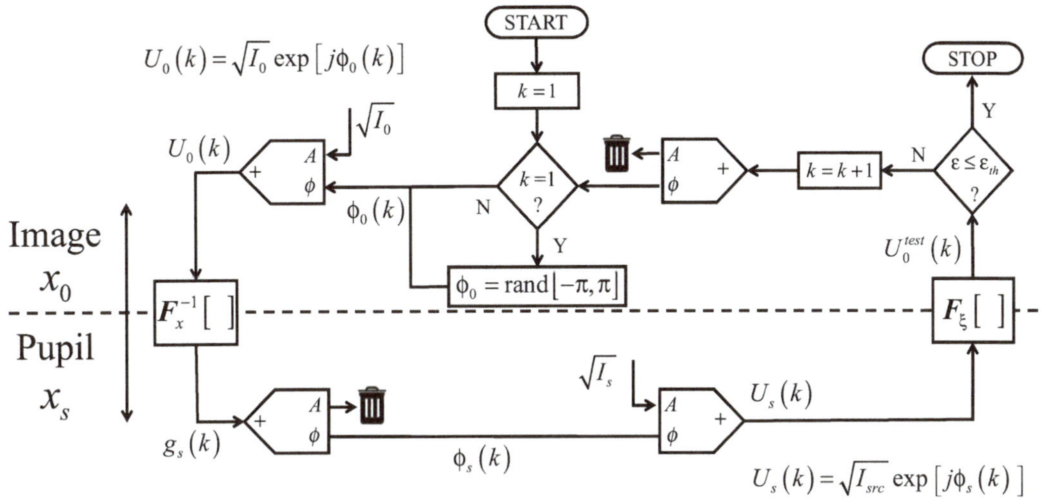

FIG. 3.2 The original Gerchberg-Saxton (GS) algorithm is displayed in a flowchart-like layout. Image-plane and pupil-plane calculations are separated by Fourier and inverse Fourier transforms. In one iterative cycle, nonlinear amplitude constraints in both image space and pupil space are applied to force convergence for estimating the phase ϕ_s. The image phase ϕ_0 is a free parameter and is a necessary part of converging to a solution.

left of the figure. The new estimate for the pupil field at that pixel is found by applying the nonlinear amplitude constraint for I_s so that $U_s' = g_s + \delta_s$. Note that the phasor difference Δ_s between the original U_s and the new estimate U_s' exhibits a magnitude that, by inspection, is greater than or equal to $|\delta_s|$. That is,

$$|\Delta_s| \geq |\delta_s|. \tag{3.6}$$

Continuing to the next step of the iteration loop, $F_\xi[\ ]$ is applied to U_s' for calculation of $U_0^{test\prime}$, which is shown in the lower right portion of Fig. 3.3 as a phasor at the pixel corresponding to the same image plane pixel as is used in the upper right diagram. The new estimate for the image field at that pixel is found by the same procedure as described for the upper-right diagram, where $U_0' = U_0^{test\prime} + \delta_0'$. In an analogous way

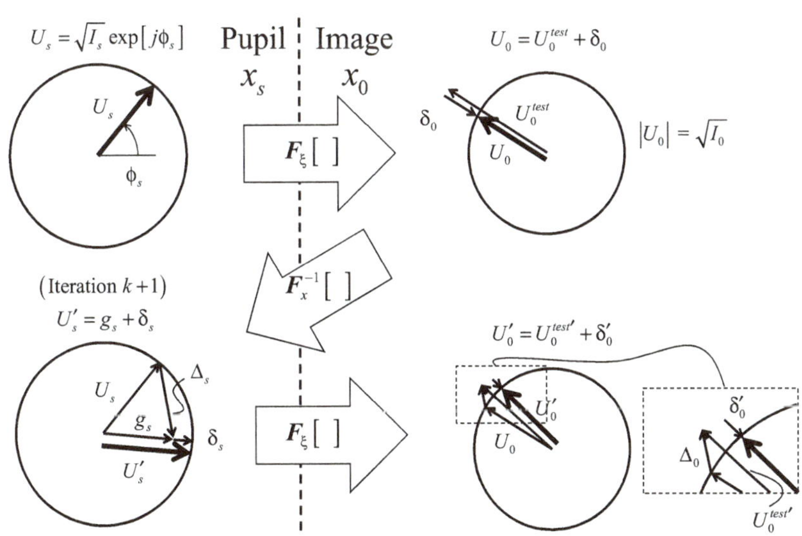

FIG. 3.3 Convergence of the GS algorithm is explained by examining one pixel in the pupil space and one pixel in the image space. Radii of the black circles represent nonlinear amplitude constraints at those points. In the text it is shown that, because $|\Delta_s| \geq |\delta_s|$ and $|\Delta_0| \geq |\delta_0'|$, mean error ε must decrease or stay the same in each iteration.

to the pupil-plane discussion, note that the phasor difference Δ_0 between U_0 and the new estimate U_0' exhibits a magnitude that, by inspection, is greater than or equal to $|\delta_0'|$. That is,

$$|\Delta_0| \geq |\delta_0'|. \tag{3.7}$$

The mean error ε of Eq. (3.5) at this point in the calculation is simply the summation

$$\varepsilon' = \frac{\sum_1^{N_{pix}} |\delta_0'|^2}{N_{pix}}. \tag{3.8}$$

To show convergence of the algorithm, the condition $\varepsilon' \leq \varepsilon$ must be true. In other words, it must be shown that

$$\sum_1^{N_{pix}} |\delta_0'|^2 \leq \sum_1^{N_{pix}} |\delta_0|^2. \tag{3.9}$$

Now the relationships between Δ's and δ's in the pupil plane and the image plane are examined in detail. Since at the pupil plane

$$\begin{aligned} g_s &= F_x^{-1}[U_0] \\ &= F_x^{-1}\left[U_0^{test}\right] + F_x^{-1}[\delta_0] \\ &= U_s + F_x^{-1}[\delta_0], \end{aligned} \tag{3.10}$$

and also

$$g_s = U_s + \Delta_s, \tag{3.11}$$

it must be that

$$\Delta_s = F_x^{-1}[\delta_0]. \tag{3.12}$$

Similarly, it can be shown that

$$\Delta_0 = F_\xi[\delta_s]. \tag{3.13}$$

Application of Parseval's theorem to Eqs. (3.12) and (3.13) states that

$$\sum_1^{N_{pix}} |\Delta_s|^2 = \sum_1^{N_{pix}} |\delta_0|^2, \tag{3.14}$$

and

$$\sum_1^{N_{pix}} |\Delta_0|^2 = \sum_1^{N_{pix}} |\delta_s|^2. \tag{3.15}$$

Eqs. (3.6) and (3.7), along with Eqs. (3.14) and (3.15), respectively, are combined to yield

$$\sum_1^{N_{pix}} |\Delta_s|^2 = \sum_1^{N_{pix}} |\delta_0|^2 \geq \sum_1^{N_{pix}} |\delta_s|^2, \tag{3.16}$$

and

$$\sum_1^{N_{pix}} |\Delta_0|^2 = \sum_1^{N_{pix}} |\delta_s|^2 \geq \sum_1^{N_{pix}} |\delta_0'|^2. \tag{3.17}$$

Note that Eqs. (3.16) and (3.17) are easily combined to yield the condition of Eq. (3.9), so it is true that *the mean error ε must decrease or remain constant through each iteration in the original GS algorithm.*

INITIAL IMPROVEMENTS TO THE ORIGINAL GERCHBERG-SAXTON ALGORITHM

It was found that, under certain conditions, the original GS algorithm stagnates [3]. To improve convergence of the algorithm, Fienup et al. introduced the *error-reduction algorithm* shown in Fig. 3.4 that is designed for estimation of an object irradiance distribution from a single irradiance measurement I_s in the pupil plane [4]. Here, the x_0 variable describes locations in an object plane instead of an image plane, but both object-pupil and pupil-image relationships are based on the same Fourier transform pair. The goal is to estimate the object f_0 with only a single measurement of I_s, where f_0 is real and nonnegative. The motivation for astronomical studies is that imaging through turbulence and with high photon noise can severely degrade an image, but it does not seriously degrade the pupil distribution. In this case, measurement of the pupil distribution and application of a phase retrieval algorithm are used to reconstruct the object. The difference between Fig. 3.4 and the original GS algorithm is the object-space amplitude constraint given as

$$U_0(k) = \begin{cases} \left|U_0^{test}\right| \exp[j\phi_0(k)] & x_0 \notin \gamma \\ 0 & x_0 \in \gamma \end{cases}, \tag{3.18}$$

where γ is the set of pixels that violate additional constraints, like nonnegative A and a loose restriction on the compact support in the image plane. As iterations continue, $g_0(k)$ approaches f_0, but generally at a slow rate [4].

A variation on the error-reduction algorithm is the *input-output algorithm*, which replaces Eq. (3.18) with

$$U_0(k) = \begin{cases} \left|U_0^{test}\right| \exp[j\phi_0(k)] & x_0 \notin \gamma \\ g_0(k-1) - \beta\left|U_0^{test}\right| & x_0 \in \gamma \end{cases}, \tag{3.19}$$

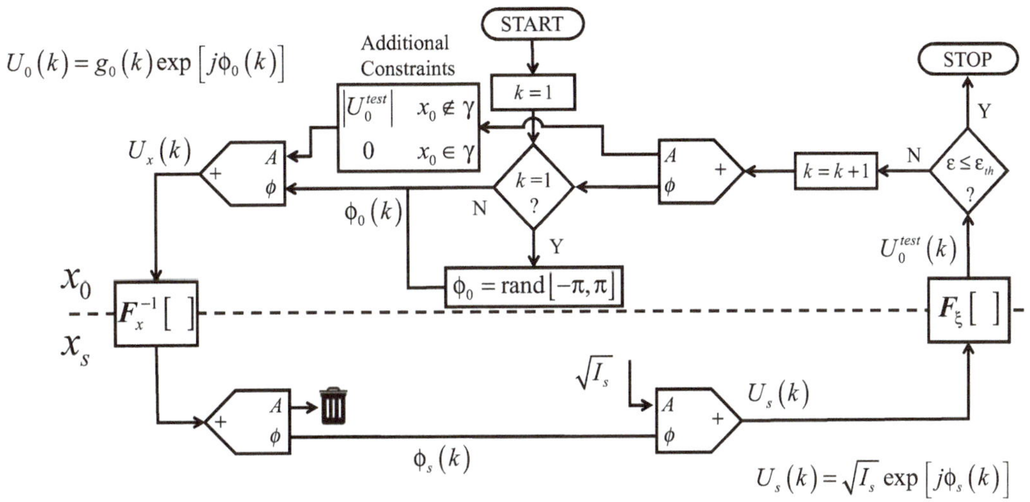

FIG. 3.4 Initial improvements to the GS algorithm included imposing additional constraints on the image amplitude, including setting the image amplitude equal to the input in the region outside the region of compact support γ.

where β is a constant feedback parameter. Values of β that work well are between 0.5 and 1. Generally, the input-output algorithm is more successful in terms of rapid convergence than the error-reduction algorithm for single-irradiance-plane measurement problems. Hybrid approaches, where 20 to 50 iterations of the input-output algorithm followed by 5–10 iterations of the error-reduction algorithm, are suggested in Ref. [3]. A comparison of the convergence for several different algorithms circa 1982 is found in Ref. [4].

COMPUTER-GENERATED HOLOGRAMS

Additional GS-like algorithms are discussed in this section as applied to the problem of CGHs, including the Wang algorithm for rapid convergence and reduced-speckle images, using regions of no interest (RONIs) to reduce speckle noise, the Dallas algorithm to produce arbitrary polarization and irradiance in the image plane, and the Vorndran algorithm for use with broad-wavelength sources.

A useful application of the classic GS algorithm is for calculation of CGHs. The goal is to use a CGH illuminated by a laser to produce a desired image distribution I_0. The conceptual experiment is shown in Fig. 3.5, where a pixelated-phase spatial light modulator (SLM) is illuminated by a laser beam with source irradiance distribution I_{src} at the CGH. The SLM surface is reimaged onto the optical system exit pupil. By setting $\sqrt{I_s} = \sqrt{I_{src}}$ in Fig. 3.2, phase ϕ_s resulting from the original GS algorithm can be used directly to load into the SLM as its phase map to produce the desired image. This technique is also commonly applied to produce laser light pointers with arrow, company logo, spaceship, or other image distributions using mass-produced transmissive CGHs in glass or plastic illuminated with inexpensive laser diodes.

However, the original GS algorithm is often slow to converge, and images from the CGHs generated with the original GS algorithm contain speckle noise resulting from the random diffuser. An example algorithm that addresses both of these issues is described by Wang et al. [5]. The design image I_0 for an annular ring to be used as an illuminator with a lithographic projector lens is shown in Fig. 3.6(A). The ideal ring has uniform amplitude inside the ring with inner radius σ_{inner} and outer radius σ_{outer} and zero in the center and outside. The areas of ideal zero amplitude are the *region of compact support*, which is designated by γ in the algorithm. The resulting image from a CGH design using the original GS algorithm is shown in Fig. 3.6(C), and an irradiance trace along the diameter of that image is shown in Fig. 3.6(F). Clearly, the original GS algorithm does a good job of defining the boundaries of the ring, but speckle-like noise on the ring itself is severe. The Wang algorithm described by Fig. 3.7 is used to produce the much smoother image inside the ring in Fig. 3.6(B) at the

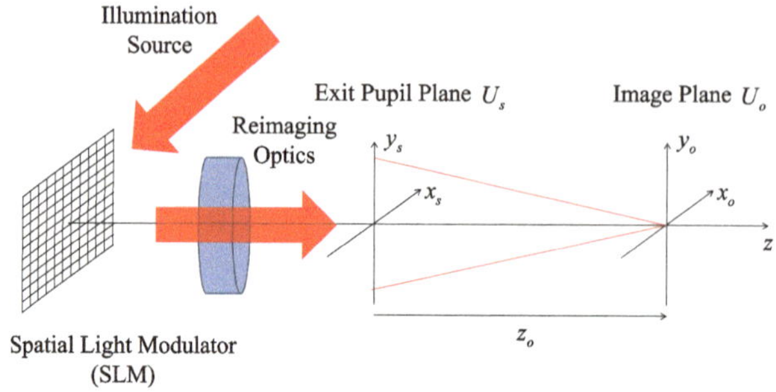

FIG. 3.5 The basic layout for computer-generated holograms (CGHs) is shown, where a pixelated spatial light modulator (SLM) is illuminated by a laser beam and reimaged into the exit pupil of the optical system to form the pupil field U_S. In practice, the SLM could be an electronically addressable display, like a liquid-crystal-on silicon (LCoS) chip, or it could be a mass-produced transmissive glass or plastic relief structure.

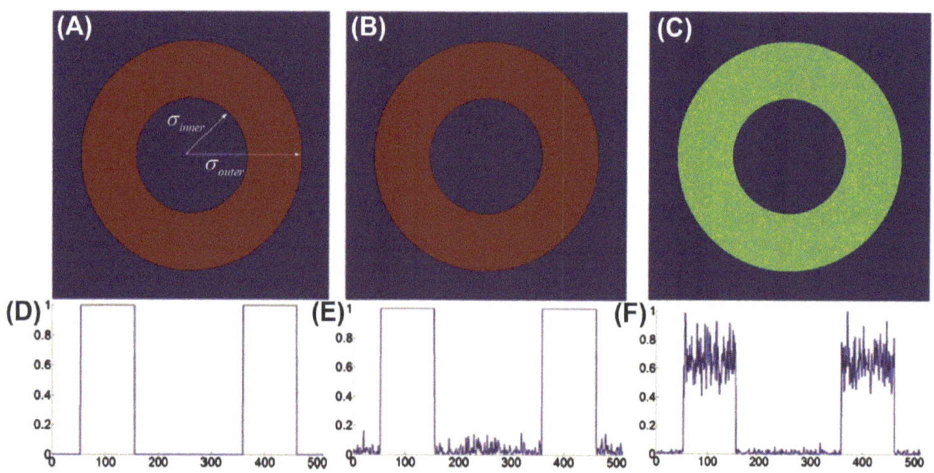

FIG. 3.6 An illustrative problem from Ref. [5], where the Wang algorithm of Fig. 3.7 is used to design a CGH that displays a ring image. **(A)** Design image of the ring; **(B)** Simulation result using the Wang algorithm; **(C)** Simulation result using the GS algorithm and **(D)**, **(E)**, and **(F)** are line profiles of **(A)**, **(B)** and **(C)**, respectively [5].

penalty of increased noise in the region of compact support that does not significantly affect performance of the instrument.

The Wang algorithm shown in Fig. 3.7 contains two primary differences from the original GS algorithm. First, gradient decent phase estimation is included in ξ space, where the kth estimate of CGH phase $\phi_s(k)$ is given by

$$\phi_s(k) = \angle U_s(k) - [\angle U_s(k-1) + \alpha(k)h(k)], \tag{3.20}$$

where ($\angle$) means "the phase of". In Eq. (3.20),

$$h(k) = \angle U_s(k) - \angle U_s(k-1) \tag{3.21}$$

is the directional gradient. The acceleration coefficient $\alpha(k)$ is found from

$$t(k) = \angle U_s(k) - \phi_s(k), \tag{3.22}$$

and

$$\alpha(k) = \frac{\sum t(k)t(k-1)}{\sum t(k-1)t(k-1)}. \tag{3.23}$$

As Figure 3.8(A) [5] describes the improved phase convergence by using larger phase jumps for each iteration. Image space x_0 includes a one-time change to allow nonzero pixel values in the region of compact support. Initially, Flag = 0, and the algorithm continues in image space as the original GS algorithm. If the value

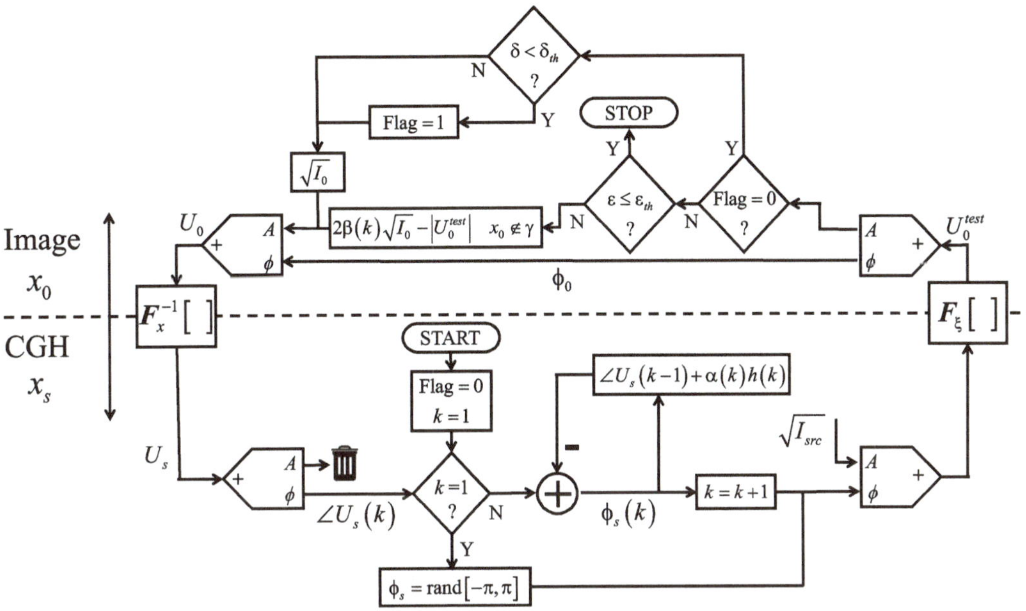

FIG. 3.7 A flowchart of the Wang algorithm [5]. A two-step amplitude constraint is applied in image space, so that when the initial GS-like convergence slows, the algorithm switches over to a difference amplitude constraint based on convergence parameter β. In CGH (pupil) space, an estimation technique is applied to increase speed of the convergence and reduce the number of iterations required.

of δ, which is the average change of amplitude in the image and is defined by

$$\delta = 1/N_{pix}\sum_{1}^{N_{pix}}\left|\left|U_0^{test}(k)\right| - \left|U_0^{test}(k-1)\right|\right|, \tag{3.24}$$

falls below a threshold δ_{th}, then for the next iteration the amplitude is allowed to change in the region of compact support, and the amplitude in the ring is

$$2\beta(k)\sqrt{I_0} - \left|U_0^{test}\right| \quad x_0 \notin \gamma, \tag{3.25}$$

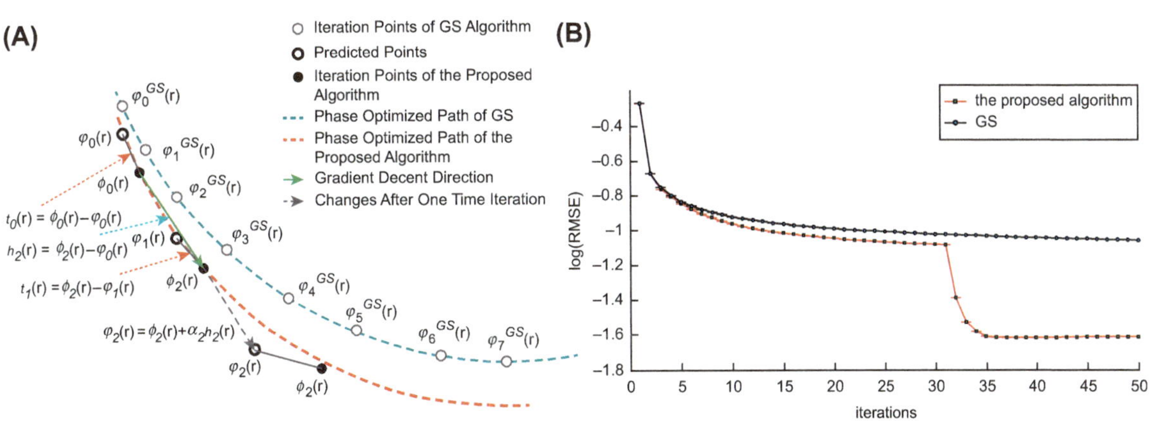

FIG. 3.8 CGH phase and image amplitude convergence of the Wang algorithm [5]. **(A)** The phase estimation technique of Fig. 3.7 is displayed conceptually and compared to convergence of the GS algorithm; and **(B)** The two-step amplitude constraint shows how the switch to a difference amplitude constraint improves the convergence and mean error ε.

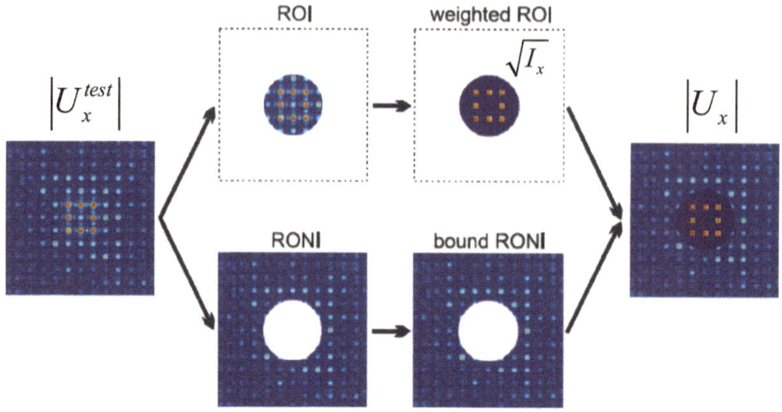

FIG. 3.9 Engström et al. describe the use of a region of no interest (RONI) that is used as a free-amplitude and phase variable space for the iterative GS calculation, so that amplitude constraints in the region of interest (ROI) result in greatly reduced speckle noise [6].

if the convergence threshold ε_{th} has not been reached. Eq. (3.25) is a version of the input-output algorithm shown in Eq. (3.19), and it includes a feedback parameter that varies with iteration number that is calculated in the signal window and is defined by

$$\beta(k) = \frac{\sum |U_0^{test}|}{\sum I_0}, \quad x_0 \notin \gamma. \tag{3.26}$$

The convergence behavior of this algorithm is shown in Fig. 3.8(B), where ε versus iteration number for the original GS algorithm and the Wang algorithm is shown. Notice that at iteration 31, the value of Flag changes to 1, and at iteration 32, ε drops sharply and levels at a much lower value than the original GS algorithm in succeeding iterations.

A related amplitude constraint is used for calculating CGHs that produce multiple focus spots in three-dimensional laser trapping [6]. Although calculation of the desired image-plane field involves multiple defocused spots and some spacing restrictions, the basic iteration algorithm is very similar to the original GS with the exception of a peripheral *region of no interest* (RONI), as shown in Fig. 3.9. The RONI is a region of the image plane outside the *region of interest* (ROI), where the ROI includes the signal window and the region of compact support. Irradiance in the RONI does not influence ε. In essence, RONI is a "waste" region for the calculation. To add a RONI to the calculation, the number of pixels is usually increased by zero padding to maintain sampling and bandwidth inside the ROI. Usually, the FFT pixel number on a side of the array doubles or triples for RONI calculations. In Fig. 3.9, a weighting factor is applied in the ROI that has the effect of improving uniformity of the focused spots. As shown in the second frame of the upper row, the desired and weighted irradiance forms the ROI. Adding RONIs to CGH calculations for other applications could significantly improve image quality in the ROIs.

To this point in the discussion of CGH calculation using GS-like algorithms, the goal is to form a desired irradiance distribution in an image plane or in a three-dimensional neighborhood around an image plane. The irradiance can be either binary or gray scale, because gray scale images do not present any significant additional burden to the algorithm. However, the GS algorithm is capable of much more than a simple irradiance image. For example, the Dallas algorithm shown in Fig. 3.10 is a two-channel algorithm that is designed to produce a spatially varying polarization state, as well as a spatially varying irradiance distribution [7]. As shown in the upper left-hand corner of Fig. 3.10, the target polarized image is a ring that is tangentially polarized. By denoting horizontal linear polarization with ($\perp$) and vertical linear polarization with ($\|$), the necessary projection of the two linear states is

$$\begin{aligned} \sqrt{I_0^{\|}} \exp\left(j\phi_0^{\|}\right) &= -\sin\theta_0 g(r_0, \theta_0) \\ \sqrt{I_0^{\perp}} \exp\left(j\phi_0^{\perp}\right) &= \cos\theta_0 g(r_0, \theta_0), \end{aligned} \tag{3.27}$$

where θ_0 is the angular coordinate in the image plane and $g(r_0, \theta_0)$ is the common irradiance envelope, which is a ring in this case.

The Dallas algorithm is a two-channel algorithm, where two parallel channels, one for each polarization state, are calculated throughout most of an iteration cycle. The goal of the algorithm is to calculate $\phi_s^{\|}$ and $\phi_s^{\perp}$ for implementation in a projection device, which is

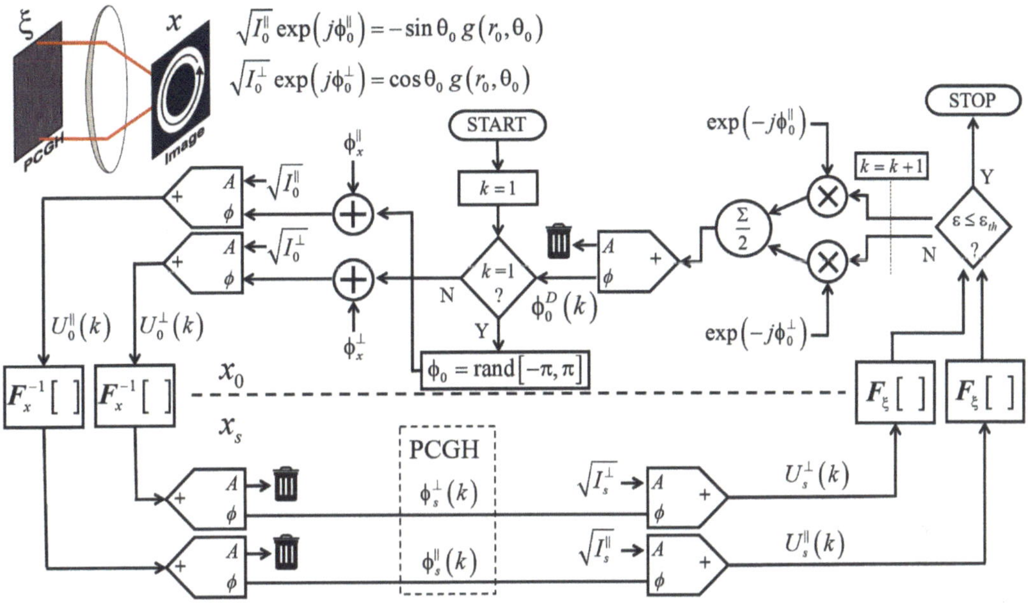

FIG. 3.10 The Dallas algorithm uses two parallel channels and a common diffuser in image space in order to design a polarization computer-generated hologram (PCGH) that projects an arbitrary amplitude and polarization state distribution to the image [7].

called a polarization computer-generated hologram (PCGH). Several clever examples of devices for implementing PCGHs are described in Ref. [7]. A direct application of PCGHs is for advanced lithographic projection cameras, where control of the polarization state as a function of position in the illumination pupil, as well as the irradiance distribution, is important for optimizing printing of circuit patterns on semiconductor wafers. A critical point in the algorithm is calculation of the common diffuser ϕ_0^D, which is necessary to maintain the phase relationship of the orthogonal projections. The mean error calculation ε is now based on an average of the overall polarization state and irradiance distribution in each channel.

Another algorithm extends application of CGHs to nonlaser, broad wavelength range operation. In the Vorndran algorithm, as shown in Fig. 3.11, the target image irradiance distribution is wavelength specific with separate regions of compact support for each of the sample wavelengths in the calculation [8]. The algorithm consists of an inner loop of calculations for each wavelength, which is followed by calculation of an average CGH surface profile. This profile is the starting point for the next GS iteration cycle. As shown in Fig. 3.10, the algorithm starts with defining a random diffuser, like the original GS algorithm, for each wavelength out of the set of N_p wavelengths. The signal windows are initially defined with unit amplitude for each wavelength inside $x_0 \notin \gamma_p$. For each wavelength, the inverse Fresnel transform is used to propagate the field back to the CGH. The CGH in this application is not far enough away from the image plane to warrant a Fourier transform, but the Fresnel transform adequately describes the propagation. The Fresnel transform equations are related to the Fourier transform pairs by

$$Fr_x^{-1}[U_0(x_0,y_0)] = F_x^{-1}\left\{U_0(x_0,y_0)\exp\left[-j\frac{\pi}{\lambda z_0}(x_0^2+y_0^2)\right]\right\}, \tag{3.28}$$

and

$$Fr_\xi[U_s(x_s,y_s)] = F_\xi\left\{U_s(x_s,y_s)\exp\left[j\frac{\pi}{\lambda z_0}(x_s^2+y_s^2)\right]\right\}, \tag{3.29}$$

where z_0 is the propagation distance.[2]

In the CGH plane, the first iteration cycle is used to set initial ideal heights of the CGH surface profile given by

$$h(k,\lambda_p) = \frac{\lambda_p \phi_s(k,\lambda_p)}{2\pi[n(\lambda_p)-1]}, \tag{3.30}$$

[2]See Chapter 1 for more details on the Fresnel diffraction integral.

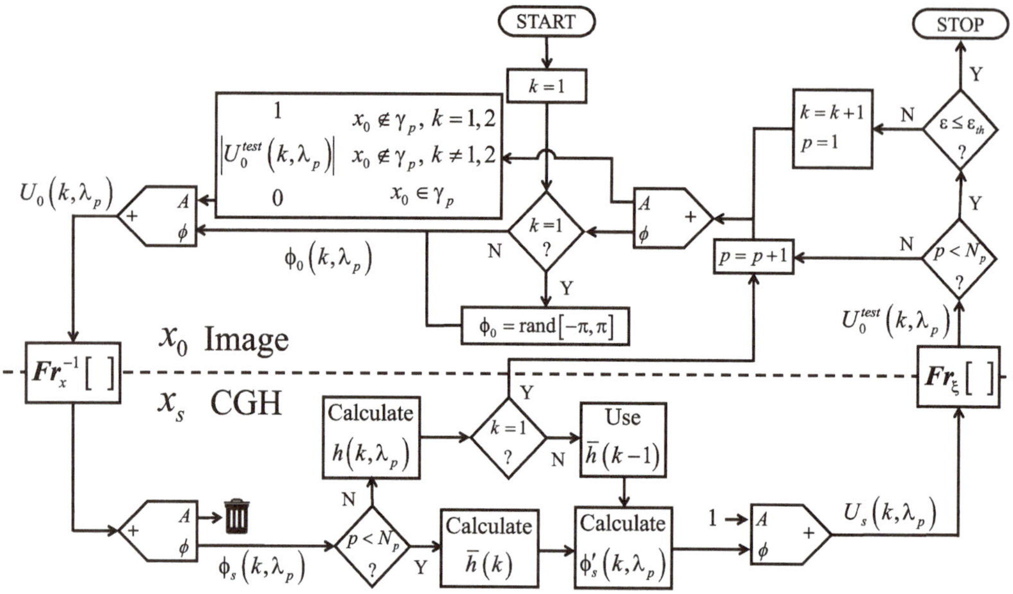

FIG. 3.11 The Vorndran algorithm is used to define separate regions of compact support γ_p for a set of design wavelengths in a broad-bandwidth photovoltaic CGH application [8]. Calculation of the CGH surface profile involves averaging possible ideal surface heights from each wavelength, which include multiple harmonics of optical path difference.

where $n(\lambda_p)$ is the refractive index of the CGH substrate. This height distribution is different for each λ_p. Owing to the harmonic nature of light, an equivalent optical path difference for each wavelength is obtained if the surface height contains multiple wavelengths of additional optical path difference, where

$$h(k, \lambda_p, m) = \frac{\lambda_p \phi_s(k, \lambda_p)}{2\pi[n(\lambda_p) - 1]} + \frac{m\lambda_p}{[n(\lambda_p) - 1]}, \tag{3.31}$$

and m is an integer. These values are stored over a range of m for each wavelength. After all N_p wavelengths are used in the $k = 1$ iteration loop, the stored results of Eq. (3.31) are used to select the smallest range over all wavelengths for each pixel. The final pixel height $\overline{h}(k)$ is a weighted average over this range, where the weights bias the result to wavelengths of highest importance.

For the second and subsequent iteration cycles, the previous value of $\overline{h}(k-1)$ is used to calculate intermediate values of

$$\phi'_s(k, \lambda_p) = \frac{2\pi}{\lambda_p}\overline{h}(k-1)[n(\lambda_p) - 1], \tag{3.32}$$

which form the input phase for the Fresnel transform back to the image space. The iteration loop continues with an amplitude constraint based only on the region of compact support, where the amplitude values within the signal widow are due to $\left|U_0^{test}(k, \lambda_p)\right|$.[3] Because of the initialization for $\overline{h}$ in the $k = 1$ loop, the GS-like iterations actually start with $k = 2$. An application of the algorithm was used to design, construct and test a spectrum-splitting CGH for use with an optimized photovoltaic array, and the results showed reasonably good performance [9] (Fig. 3.11).

MULTIPLE-PLANE DIVERSITY ALGORITHMS

There are several new algorithms that take advantage of multiple images, rather than the two-plane algorithms discussed in previous sections. This section describes two examples. The first example uses defocus as a *phase diversity* to collect images over multiple defocus values, and then it uses serial calculations to reconstruct the desired phase distribution. The second example is a general algorithm that uses multiple images in parallel, which can use any form of phase diversity.

The Ilovitsh et al. algorithm in shown in Fig. 3.12 [10]. Although [10] discusses how to use the algorithm for synthetic aperture radar, this section discusses only

[3]Although not specifically mentioned in Ref. [8], it is assumed that the values of $\phi_0(2, \lambda_p)$ are equal to the same random distributions used for $k = 1$.

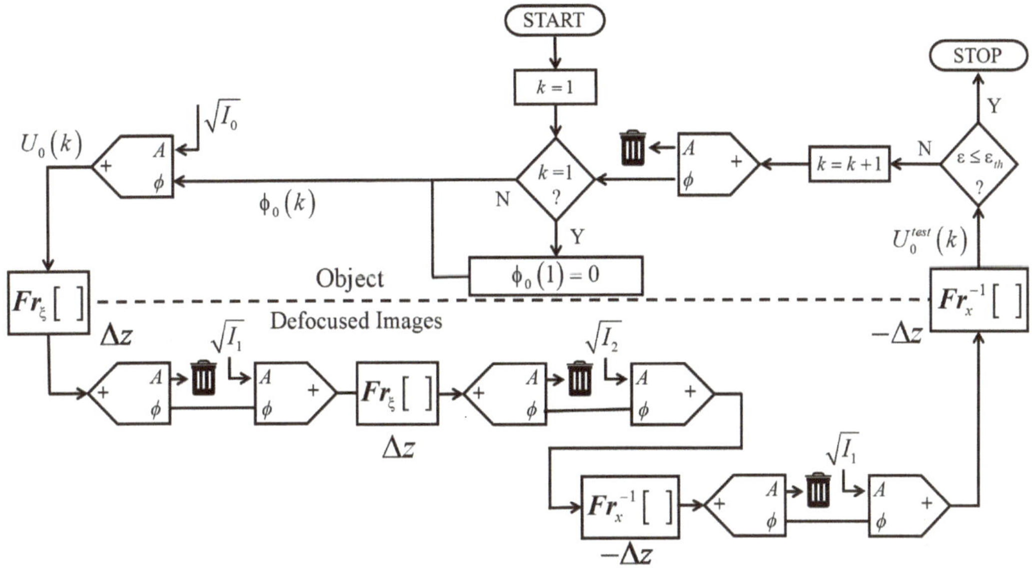

FIG. 3.12 The Ilovitsh algorithm uses multiple defocus image planes in a defocus diversity application that are applied serially in each iteration loop to estimate the object phase ϕ_0 [10].

the basic algorithm, because of the way that multiple image data are used. The technique purportedly avoids stagnation problems. The reference dataset includes three images at different defocus distances. I_0 is the in-focus image, and I_1 and I_2 are defocused images. I_1 is separated axially from I_0 by Δz, and I_2 is separated axially from I_1 by and additional Δz. The goal of the algorithm is to estimate the object phase ϕ_0. Initially, $\phi_0 = 0$, and the amplitude of the measured in-focus image $\sqrt{I_0}$ is applied to form $U_0(k)$. A Fresnel transform is used to propagate the field to the first defocused plane through distance Δz, where the amplitude is replaced by the measured amplitude at that plane $\sqrt{I_1}$. A second Fresnel transform is then used to propagate the to the next defocus plane at an additional distance Δz, where the amplitude is replaced by the measured amplitude $\sqrt{I_2}$. The field is then back propagated a distance $-\Delta z$ with an inverse Fresnel transform, where the field amplitude is replaced by $\sqrt{I_1}$. The final inverse Fresnel transform is used to propagate the light back to the object plane. At this point, the error in the image amplitude is evaluated, and if $\varepsilon \leq \varepsilon_{th}$, the calculation is stopped. If $\varepsilon > \varepsilon_{th}$, iteration loops continue until the threshold condition is reached.

A new algorithm introduced by Gerchberg uses multiple image-plane data in parallel, as shown in Fig. 3.13 [11]. The goal of the algorithm is to estimate the pupil phase ϕ_s that is incident onto the optical system. The phase diversity for each image, which is indicated by ϕ_1, ϕ_2, etc. in Fig. 3.13, are known and produce *phasorgrams* $\sqrt{I_1}\sqrt{I_2}$, etc. A primary difference between this algorithm and others is the averaging of individual back-propagated pupil fields with Σ/N. In addition, the phase diversity for each image is subtracted before the back propagation. Calculation of image error ε involves comparison of the phasorgrams with amplitudes of the calculated values from the estimated $\phi_s(k)$ and application of the phase diversity and a Fourier transform. As with [5], the algorithm is purportedly very robust and avoids stagnation issues. Any type of phase diversity can be used, including defocus and random phase.

CONCLUSION

There are many interesting and useful variations of the GS algorithm that exist today, which are outcomes of the need for improved convergence, smoother image structure, and solutions of complicated problems. There was not room in this chapter to list a wider scope of applications for the GS algorithm, but the interested scientist or student can easily search for that information. Although the utility of the GS algorithm is widespread throughout science and engineering, it is likely that future developments of inverse problems for phase retrieval, CGH design, image reconstruction, microscopy, and adaptive optics will involve new GS variations. In fact, components of algorithms presented in this chapter will likely form some of those new variations.

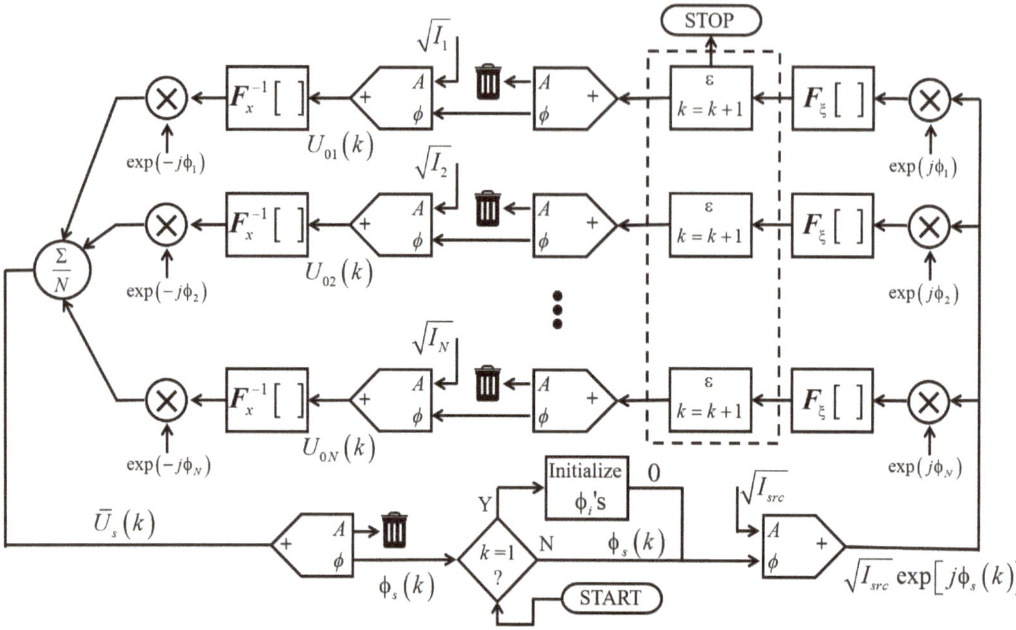

FIG. 3.13 The Gerchberg algorithm uses multiple diversity phase filters ϕ_1, ϕ_2, etc. in parallel with field averaging in each iteration cycle to converge on an estimate of the pupil phase ϕ_s [11]. Several types of phase diversities are suggested, including focus diversity and random phase diversity. In addition, the algorithm also works with amplitude sampling, like what can be used with X-ray systems.

REFERENCES

[1] R.W. Gerchberg, A practical algorithm for the determination of phase from image and diffraction plane pictures, Optik 35 (1972) 237–246.

[2] J.W. Cooley, J.W. Tukey, An algorithm for the machine calculation of complex Fourier series, Mathematics of Computation 19 (90) (1965) 297–301.

[3] J.R. Fienup, C.C. Wackerman, Phase-retrieval stagnation problems and solutions, JOSA A 3 (11) (1986) 1897–1907.

[4] J.R. Fienup, Phase retrieval algorithms: a comparison, Applied Optics 21 (15) (1982) 2758–2769.

[5] H. Wang, W. Yue, Q. Song, J. Liu, G. Situ, A hybrid Gerchberg–Saxton-like algorithm for DOE and CGH calculation, Optics and Lasers in Engineering 89 (2017) 109–115.

[6] D. Engström, A. Frank, J. Backsten, M. Goksör, J. Bengtsson, Grid-free 3D multiple spot generation with an efficient single-plane FFT-based algorithm, Optics Express 17 (12) (2009) 9989–10000.

[7] I. Matsubara, Y. Unno, W. Dallas, T.D. Milster, U.S. Patent No. 9,116,303, U.S. Patent and Trademark Office, Washington, DC, 2015.

[8] S. Vorndran, J.M. Russo, Y. Wu, S.A. Pelaez, R.K. Kostuk, Broadband Gerchberg-Saxton algorithm for freeform diffractive spectral filter design, Optics Express 23 (24) (2015) A1512–A1527.

[9] S.D. Vorndran, L. Johnson, T. Milster, R.K. Kostuk, Measurement and analysis of algorithmically-designed diffractive optic for photovoltaic spectrum splitting, in: 2016 IEEE 43rd Photovoltaic Specialists Conference (PVSC), IEEE, June 2016, pp. 3513–3517.

[10] A. Ilovitsh, S. Zach, Z. Zalevsky, Optical synthetic aperture radar, Journal of Modern Optics 60 (10) (2013) 803–807.

[11] Gerchberg, R. W. (2005). U.S. Patent No. 6,906,839. Washington, DC: U.S. Patent and Trademark Office.

CHAPTER 4

Holographic Television

V. MICHAEL BOVE, JR., SB, SM, PHD

INTRODUCTION

Display holography was originally developed in the era when television was becoming widespread around the world, thus early holographers were inevitably asked (as current holographers continue to be), "But can you make the images move?" Considering holographic television, we will confine our discussion to true holograms, with a diffraction pattern somehow updated at video rates. This is an important distinction nowadays, when the term "hologram" is commonly applied to the entire range of autostereoscopic displays and even to some 2D displays such as a digital reemergence of the old magician's illusion called the Pepper's Ghost.

The idea of holographic television is not new, and pioneering holographers set out to try to answer the above question. In 1965, only about a year after Leith and Upatnieks published the technique for the off-axis transmission hologram, they gave a presentation at a conference of the Society of Motion Picture and Television Engineers on the practical requirements for holographic television capture and transmission [1]. In addition to an analysis of the bandwidth needed, they also proposed two candidate display technologies: an electrostatically deformed thin oil film and laser-written photochromic glass. A year later, a team at Bell Labs was able to transmit a still holographic interference fringe pattern using a television system, by capturing it with a vidicon camera and using a cathode-ray tube to expose a photographic transparency that was then illuminated by a laser to reconstruct an image [2]. CBS Laboratories in 1972 took a similar approach in capturing an interference pattern with a vidicon, but their receiver was a device called a Lumatron, in which an electron beam in real time wrote a phase hologram onto a reusable plastic material. They further modified the system such that multiple frames could be written onto a set of substrate sheets, allowing the transmission of a short video sequence [3]. These early projects were the beginning of a long search for the ideal light modulation technology for holographic television, an exploration that continues to this day.

The 1970s also saw a remarkable attempt at holographic cinema, led by Komar at the Cine and Photo Research Institute (NIKFI) in Moscow. A 70-mm film was shot as holographic frames using pulsed red, green, and blue lasers; alternatively incoherent-light multiview images captured using a lens array were converted to hologram frames during a film printing process. The resulting holographic film was projected onto a special holographic screen that created multiple view zones so that more than one viewer could see the film simultaneously [4]. This system did not involve electronic processing or transmission of holograms and seems not to have significantly influenced the design of later holographic video systems.

A somewhat more modern approach was taken by investigators in the early 1990s, when liquid-crystal light modulators designed for video projectors became available. Illuminated by coherent light, these provided the output means for several analog holographic television transmission experiments [5,6]. All of the above-described systems captured an interference pattern between reflected scene light and a reference beam on a standard image sensor; Poon et al. took a different approach in their optical scanning holography system, in which the scene was x-y scanned with a time-varying Fresnel zone plate pattern generated by the interference of a plane wave and a spherical wave with a time-varying frequency shift. Reflected light picked up by a single photodetector was then processed in the analog domain using the frequency shift signal to produce an output signal representing a hologram of the scene [7].

The assumption was made in the early years of holographic television research (and is sometimes still incorrectly made by those who are not familiar with the field) that a holographic display device requires that the scene capture must also be done holographically,

Optical Holography-Materials, Theory and Applications. https://doi.org/10.1016/B978-0-12-815467-0.00004-9

illuminating the scene with a laser and interfering the reflected object light with a reference beam on an electronic image sensor. Such an approach entails obvious impracticalities, including the difficulty of illuminating large—or outdoor—scenes with coherent light and the fact that the image sensor must generally be the same size as the output device, although holographic capture is currently practiced in some specialized applications such as holographic microscopy where these constraints do not prove problematic.

Computational advances many orders of magnitude beyond what was imagined in the 1960 and 1970s have meant that the diffraction pattern for the display need not be captured, but instead can be computed, even in real time. Doing so begins with enough information about the scene such that it is possible to calculate diffraction fringe patterns that will reconstruct a light field of the scene (or possibly a stereogram that will be perceived as a light field of the scene). A variety of approaches are currently feasible and have the added advantage that a high-data-rate interference pattern need not be transmitted but instead can be computed at the display from a more compact scene representation.

If the scene to be displayed already exists in the form of a standard computer graphics data format such as texture-mapped polygon mesh or point cloud (as might be the case with a video game or a virtual reality scene) then generation of the fringe patterns for the frames of a holographic video are a straightforward application of computer-generated holography techniques as discussed elsewhere in this volume. A fortunate side effect of the availability of consumer-grade structured-light or time-of-flight depth sensing cameras (which were generally targeted at human-computer interaction rather than general scene capture) is that it has also become inexpensive to create point-cloud models for real scenes, at least relatively small indoor ones. The author's research group in 2011 demonstrated that dynamic point-cloud models of people could be captured in real time and transmitted over the Internet for holographic display on an acousto-optic spatial light modulator (SLM) (the Mark II system described below) as well as other diffractive display technologies, to create video-rate holographic telepresence [8].

As image sensors have increased in resolution, so has the practicality of using them to capture integral images of scenes, in which an array of lenses samples a scene from a range of ray angles, in effect trading off the camera's spatial resolution for angular sampling. While integral images are closely tied to light field displays, which effectively reverse the camera's operation by projecting observations of the scene at the angles observed by the camera, a hologram located at the back focal plane of the lens array can be computed simply as the Fourier transform of the elemental images [9], meaning that an integral camera provides an interesting option for scene capture for a holographic display as well. Yamamoto et al. demonstrated a real-time holographic television system in which RGB holograms resulting from this process were displayed on liquid-crystal SLMs illuminated by red, green, and blue lasers [10]. Readers might be wondering, since the Fourier transform could be computed in the opposite direction as well, whether a holographic interference pattern might be converted to imagery for a light field display. Several researchers have indeed reported doing so, but none appears to have built a real-time video system based on this principle [11–13].

A related approach entails capturing conventional stereograms with an array of cameras and converting them to a holographic stereogram using standard methods for holo-stereogram computation. Blanche's group at the University of Arizona in 2010 captured stereograms with an array of 16 cameras and converted them to holograms for a display based on refreshable photorefractive polymer at a rate of 0.5 frames/second (which was limited by the writing of the material with a pulsed laser, not by the capture or computation speed) [14]. (Fig. 4.1) A related possibility is to use fewer or irregularly spaced cameras, apply machine-vision algorithms to generate a 3D scene model from the images, and then compute a hologram from the model. One example of this has been reported in Ref. [15] although not in real-time video operation; computation on GPUs might well permit such an approach to be performed at video rates.

SPATIOTEMPORAL BANDWIDTH

As discussed in the introductory chapter, a static light field of a given spatial extent, angular view zone, and resolution will require a hologram to have a minimum space-bandwidth product (SBP) to reconstruct it properly. In the spatial domain, this SBP can be understood as the product of the maximum spatial frequency (or equivalently the pixels per unit distance in a discretized hologram) and the spatial extent of the diffraction pattern, and consequently its units can be expressed as cycles or as a pixel count. Optics can be used to reallocate the available SBP (e.g., making a smaller light field with a wider view angle) but cannot escape the SBP requirement of the subject matter.

FIG. 4.1 Camera array for capture of images used in University of Arizona holographic telepresence experiment, and views of hologram displayed on rewritable photorefractive polymer. (University of Arizona College of Optical Sciences).

To display a satisfactory moving hologram to a human observer, it is necessary to update the diffraction pattern at a rapid enough rate to provide the perception of smooth motion and also to avoid visible flicker of the image. Under most viewing circumstances and for most imagery the first of these goals requires a significantly lower frame rate than the second; it has been known since the earliest days of cinema that the illusion of smooth motion can be created with about 16 frames/second, whereas elimination of flicker requires at least twice that rate [16]. This observation led to the development of approaches such as the multibladed cinema projector shutter, in which a frame of film is flashed more than once to reduce flicker perception; in the case of a dynamically refreshed hologram something similar might be done in which the illumination can flash more than once per hologram update.

For a dynamic holographic light field viewed by a human, we can derive a new requirement for spatiotemporal bandwidth product (STBP), in which we multiply the SBP by the frame rate for smooth motion. Notable here is that scanning optics could allow trading between spatial and temporal cycles, in that for example a SLM with a higher-than-needed update rate could be used to image across a larger hologram plane than its active area.

To give a sense of the sorts of numbers involved here, consider a hypothetical holographic mobile phone with a screen 100 mm by 60 mm, and a full-parallax viewing angle of 30 degrees. Because we are interested in an order-of-magnitude answer, we will simplify a bit by neglecting the incidence angle of the illumination and by assuming a single illumination wavelength of 550 nm. Then from the grating equation (11) of chapter 1, we have

$$\Lambda = \frac{\lambda}{sin(30°)}$$

which gives us a grating spacing of 1.1 μ and thus a maximum pixel size of 0.55 μ (the same as λ) as each grating line requires a minimum of two pixels to represent. Thus, the screen will have approximately 182,000 by 110,000 pixels for a total pixel count of 2×10^{10} pixels. Updating this screen at even 15 frames/second gives a data rate of 3×10^{11} pixels s^{-1}. For comparison, an HDMI 2.0 cable can carry 18 Gbit/second, so if these are 8-bit pixels, more than 133 HDMI connections would be needed to carry the data (although inside a phone it would make little sense to convert pixel data into HDMI signals and back just to travel a few centimeters—this comparison is made just to give a more tangible sense of the rates involved).

For another comparison, let us consider a "4K" (3840 by 2160) light modulator chip as might be found in a current video projector. If its maximum update rate is 60 frames/second, it can deliver 5×10^8 pixels s^{-1}, so our hypothetical phone above requires 600 times the spatiotemporal bandwidth of these devices.

Digital micromirror light modulators are one-bit devices and produce the appearance of grayscale by changing the duty cycle of the mirrors, but if one is willing to accept the noise and other artifacts of a binary hologram, the frame update rate can be impressively high. A current Texas Instruments device as of this writing, the DLP9500, has a resolution of 1920 by 1080 and a mirror update rate of 1700 Hz, thus providing 3.5×10^9 (binary) pixels s^{-1}.

A detailed review of spatiotemporal bandwidth issues in holographic and other light field displays is given by Yamaguchi in Ref. [17].

What is crucial to remember here is that unlike with nondiffractive displays, we cannot make a bigger screen by keeping the pixel count constant and enlarging the pixels; the result would be a much smaller diffraction angle (and thus view angle). Thus, scaling up the size of dynamic holographic light fields requires finding practical ways to scale up the STBP; this has been a primary focus of innovation in the field of holographic video.

Light Modulation and Displays

Obtained by whatever means, a diffraction fringe pattern must be physically instantiated on some device with dynamically controllable phase and/or transmittance/reflectance, which will be illuminated with coherent light to create a holographic light field. This device is called SLM. Ideally, an SLM should be able to control both phase and amplitude of the light (or equivalently the real and imaginary parts), but in the absence of that capability, a phase-only modulator is more than five times as diffraction-efficient than an amplitude-only modulator and thus would be a better choice for a holographic display [18].

Although in years past it was difficult to compute in real time a diffraction pattern for display on the best available SLMs, computation (particularly using GPUs) is no longer the limiting factor on the quality of holographic video. The challenge, rather, is that the best available SLMs do not have nearly enough STBP or have too much bandwidth available in one dimension and not enough in another, and the engineering challenge in making a holographic video display is developing a practical strategy to overcome this situation. Ref. [19] provides further analysis of the spatial and temporal bandwidth requirements for SLMs to reproduce a given dynamic hologram.

After the pioneering 1989 "Mark I" system built by Benton's group at the MIT Media Lab [20], a variety of approaches were tried to create computer-driven holographic displays; an overview of many of these systems can be found in Refs. [21,22]. Mark I was based on a single-channel acousto-optic modulator (AOM) with an analog bandwidth of 50 MHz, which can be understood as an STBP of 1×10^8 pixels s^{-1}. In such a device, electrical signals are converted to acoustic waves in the bulk of a transparent medium (tellurium dioxide, in this case) and create a phase diffraction pattern. Because the fringes travel with the speed of sound, a counter-rotating mirror must be used to create a stationary image; this mirror also provided scanning along a horizontal line while another scanner ran in the vertical direction to create a horizontal-parallax-only image that filled a 25-mm cube and had a 15-degree view angle. This system was architecturally quite similar to the first use of an AOM for electronic display, the Scophony television set of the 1930s [23]. A three-channel AOM later replaced the single-channel one, and was illuminated with three lasers, to turn Mark I into a full-color system [24]. (Fig. 4.2)

The Mark II display, designed by St. Hilaire [25], parallelized the AOM such that 18 scan lines could be displayed simultaneously, giving a 30 degrees view angle and a light field volume of 150 mm × 75 mm × 150 mm (width × height × depth). Still, the images had only 144 vertical scan lines and thus were not quite standard-definition television quality (Fig. 4.3).

FIG. 4.2 Stephen A. Benton with the full-color version of the Mark I holographic video display, circa 1992. (MIT Media Lab).

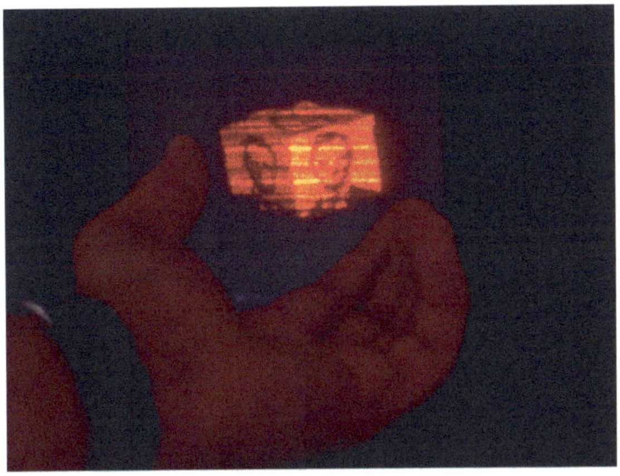

FIG. 4.3 Image from the MIT Mark II holographic video display. (MIT Media Lab).

Poon et al. in 1993 used an electron-beam addressed SLM, in which deflection coils cause an electron beam to scan across a crystal of lithium niobate, changing the refractive index of the material through the Pockels effect and creating a holographic image in reflected coherent light [26].

While LCD and LCOS SLMs are commonly used in holographic displays, usually as phase modulators, at least one group has combined a pair of modulators, one controlling phase and one amplitude, to display complex-valued Fresnel holograms [27]. An approach to generating complex holograms with a single modulator by combining multiple position-shifted amplitude holograms has also been reported [28]. Several teams have used digital micromirror devices to display holograms [29,30].

In general, standard light modulator technologies originally developed for 2D display (e.g., LCD, LCOS, DLP), even if optimized for holographic SLM applications, do not have nearly enough STBP to produce a large light field with a wide view angle, or may have the STBP distributed in unusable ways (e.g., a much higher refresh rate than needed but not nearly enough spatial pixels). Most attempts to make large wide-angle holographic video displays have used one or more of four strategies to get around this problem: spatial tiling, scanning, image storage, or eye tracking.

Spatial tiling is straightforward, provided that the multiple SLMs used are designed such that they can be arrayed side-by-side with minimal gaps (unfortunately this is rarely the case). Tiling modulators on a plane [31] increases the spatial extent of the light field, while the view angle can be increased by tiling them on a curved surface [32]. Reference [33] shows how two sets of curved SLM arrays can be arranged with a beamsplitter so as to cover the view angle gaps that occur when the active areas of the devices do not extend all the way to the edges of the device packages.

Scanning is similar to tiling but uses a single high-refresh-rate SLM imaged by scanning optics so as to cover a larger spatial and/or angular extent. An interesting variation of this straightforward approach to reallocating temporal bandwidth to the spatial dimension is to use a cylindrical lens to squeeze the SLM's pixels in the horizontal direction to create a wider diffraction angle (but a narrower light field), then to scan horizontally to restore the width of the light field, resulting in a display with only horizontal parallax [34]. Recently, several groups have investigated scanning in a circular direction to create displays in tabletop orientation with a full 360-degree view zone [35,36]. A novel mechanism for scanning was developed by a group at Seoul National University, who used a holographic optical element to replicate the SLM's output in multiple locations through multiple diffracted orders, and selected one at a time by the use of electrically controlled shutters [37].

A collaboration between Disney Research and the University of Cambridge developed a technique called "angular tiling," a combination of tiling and scanning—and indeed also a combination of holography and integral imaging—that uses a fast SLM to display a series of small-view-angle subholograms, and a scanning mirror to direct these to a lenslet array behind an output transform lens. The result is that the STPB of a fast but low-SBP SLM can be used to enable a light field of increased size and increased viewing angle [38].

Scanning can also be combined with a rewritable storage medium to retain the diffraction pattern, a process sometimes referred to as optical tiling. Such an approach not only offers the ability to reallocate the SLM's STBP but also permits the SLM's pixels to be demagnified in size to increase the view angle and enables the SLM to be illuminated with a lower power (and possibly incoherent) source than the full diffraction pattern, permitting SLM technologies to be used that would not support a very bright display if viewed directly. The QinetiQ display system imaged a high-refresh-rate 1024-by-1024 SLM 25 times through a lens and shutter array onto a liquid-crystal optically addressed SLM to create a 5120-by-5120 diffraction pattern (Figs. 4.4, 4.5) The module that generated this hologram could then be tiled to create larger light fields [21]. The University of Arizona photorefractive polymer system discussed earlier also used tiling but in a manner closer to holographic printing; a "hogel" of sliced scene

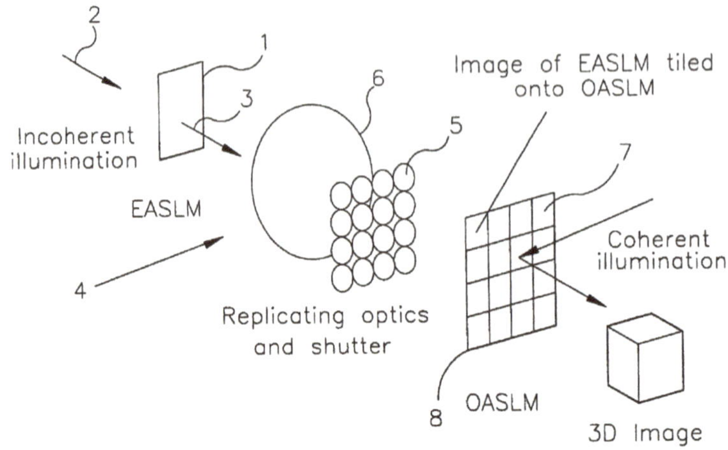

FIG. 4.4 Optically tiled modulator architecture for the QinetiQ display. (Douglas Payne and Christopher W. Slinger, Holographic Displays, US patent 6753990B1, Jun. 22, 2004).

FIG. 4.5 Image from the QinetiQ display. (From Maurice Stanley, Mark A.G. Smith, Allan P. Smith, Philip J. Watson, Stuart D. Coomber, Colin D. Cameron, Christopher W. Slinger, Andrew Wood, 3D electronic holography display system using a 100 mega-pixel spatial light modulator, Proceedings of SPIE 5249, Optical Design and Engineering, (18 February 2004); https://doi.org/10.1117/12.516540).)

view data was presented on an SLM illuminated with coherent light and imaged onto the storage material where it interfered with a reference beam to create the diffraction pattern, after which the SLM image was moved to a new position and the process repeated [39].

Holographic video researchers dream of something like a flat-panel monitor with small enough pixels to create wide view zones, but even the finest-pitch large display panels currently available have pixels too large to enable such a scenario. Current fine-pitch display panels as of this writing are in the range of 150–200 pixels in.$^{-1}$ (as usually listed in spec sheets), which converts to a pixel size of 0.17–0.13 mm. This is more than two orders of magnitude coarser than the hypothetical 30 degrees screen discussed above and would create a diffraction angle of only a fraction of a degree. However, such a panel might be combined with eye tracking and a steerable backlight, such that large, narrow-view-zone holographic light fields can be directed to the positions at which the tracking sees the viewer's eyes. This principle was applied in the SeeReal display [40].

Acoustic modulation, as used in the early MIT Media Lab systems, has a number of advantages over other types of SLMs, particularly in that the construction of the modulators is comparatively extremely simple, and there is no sharp-edged pixel structure to cause higher-order images. It also has the significant disadvantages of requiring compensation for the moving acoustic patterns, primarily being useful for horizontal-parallax-only displays, and (in the case of bulk AOMs) the need to make the modulator physically large if it has multiple channels to avoid interchannel cross talk. A different architecture, the acousto-optic guided-wave device, avoids at least the last of these problems. In a guided-wave device, parallel waveguides are created under the surface of a transparent piezoelectric material, and coherent light traveling as a guided mode in the waveguides is outcoupled when it encounters the hologram in the form of surface acoustic waves generated by electrical transducers. These devices exhibit some potentially valuable properties including the fact that the outcoupled light has its polarization rotated from the light in the waveguide, consequently it is possible to use a polarizer at the output to block zero-order and other unwanted light from the holographic light field. Even more intriguing is that devices can be made such that the acoustic frequency bands that cause outcoupling of red, green, and blue light

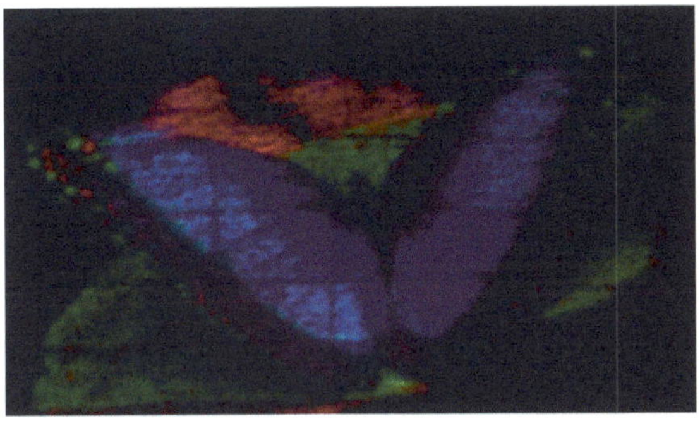

FIG. 4.6 Holographic stereogram generated by the author's group using an acousto-optic guided-wave device.

are essentially nonoverlapping, enabling simultaneous illumination by all three colors and driving the display with a full color hologram signal created by single-sideband modulating the holograms for the three components to the appropriate frequencies and summing them together (typically full color images—irrespective of the modulator technology—would have to be created either by time-sequential illumination or by having parallel modulators or channels for each color). Each color channel has a 50-MHz useable analog bandwidth per hogel, although many hogels can be placed on one horizontal line and many waveguides can be arrayed vertically. It is also important to note that in these devices, unlike other AOMs, the light encounters the hologram at a glancing angle rather than perpendicularly and as a result of the foreshortening of the diffraction fringes the resulting diffraction angle is approximately three times that suggested by the STBP figure for the device. The author's group has demonstrated full-color holographic stereograms at video rates on a Scophony-architecture display using such a modulator (Fig. 4.6) [41]. Ongoing work on such modulators (a collaboration between the MIT Media Lab and Smalley's group at BYU) focuses on eliminating the need for scanning optics, making the device operate as a see-through holographic display panel. In this version, the modulator architecture is modified such that light out-couples at multiple locations along the waveguide (rather than just at the end, as in the original version), and pulsed laser illumination is used to mitigate the motion of the acoustic diffraction pattern [42].

HOLOGRAPHIC AUGMENTED AND VIRTUAL REALITY

There is an obvious experiential connection between display holography and augmented or virtual reality; indeed some companies and journalists have mislabeled stereoscopic VR images as holograms. But a technical connection has also emerged in recent years, as the demands of producing a compelling and comfortable VR or AR experience have caused system designers to use holographic technology in various ways.

A challenge for wearable AR display designers comes from the fact that most display devices are not see-through, and the ones that are typically do not create a virtual image at a distance far behind them so that eyes can focus on it. Thus, it is necessary to design a means to relay an image from a microdisplay and imaging lens to a see-through optical element in front of the eye. Volume diffraction gratings can be used to couple a pair of stereoscopic images into and out of see-through image waveguides (Fig. 4.7); an example of a system using this approach is the Sony SmartEyeglass [43]. The Microsoft HoloLens [44] uses a similar arrangement but with separate stacked waveguides and gratings for red, green, and blue. Multiple transparent holographic elements can also be stacked not for combining colors but for creating a compressive light field display where rays from different layers are added together incoherently [45].

True holographic wearable displays are still rare, but several approaches have been taken by researchers. A group at Samsung has developed a display following the above gratings-and-light guide architecture but with a coherently illuminated SLM creating a holographic light field rather than an incoherent 2D image [46]. A group at Microsoft Research has built several prototype holographic near-eye displays based on reflective LCOS SLMs sequentially illuminated with red, green, and blue lasers; one version demonstrated the ability to correct for astigmatism by introducing vision correction into the hologram [47].

The guided-wave device described in the previous section can be applied to augmented-reality displays, whether wearable or at a distance from the viewer in a "head up" application [48]. Because the light outcoupling from the waveguide is continuing to travel in a forward direction, a tilted volume phase grating under the waveguide redirects the angular fan of the light symmetrically to the surface normal of the display (Fig. 4.8).

Other types of SLMs have also been explored in vehicular head-up display applications. Analogously to wearable AR displays, in a head-up display for a windshield it is typical for a 2D light modulator to be imaged through a lens that enables the viewer to focus on an image some distance past the windshield's surface, and reflected off the windshield from below. Such an arrangement has a single available depth plane and requires optical

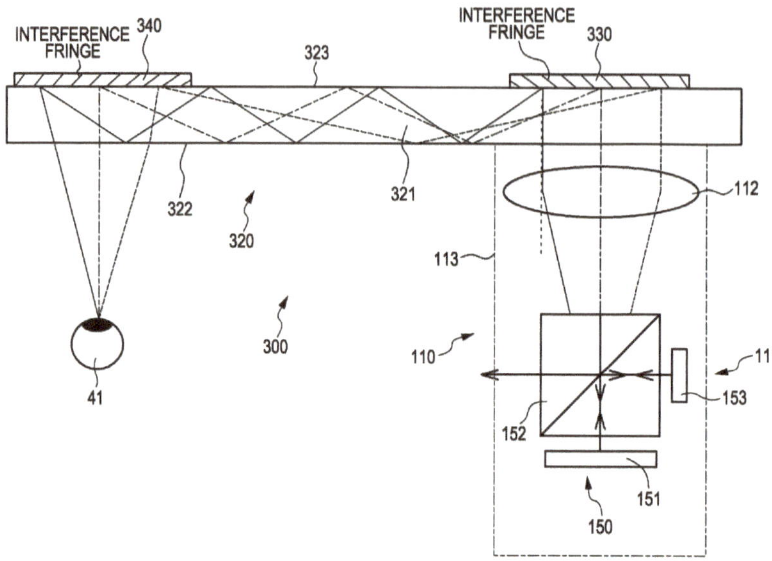

FIG. 4.7 Use of holographic optical elements in a Sony wearable augmented reality display. 2D image from microdisplay 150 is coupled by volume grating 330 into see-through light guide 321 and outcoupled toward the eye by volume grating 340 [42].

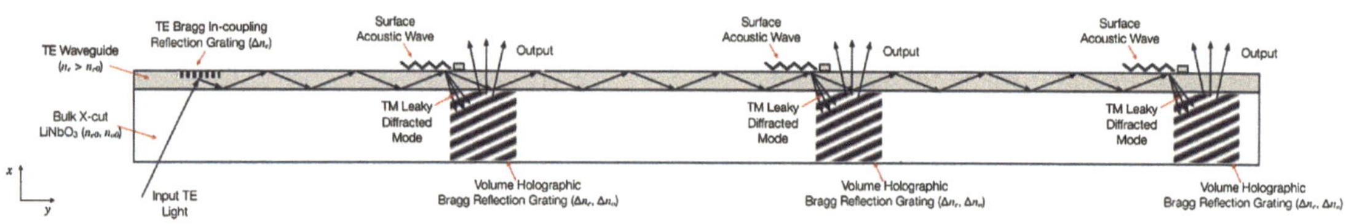

FIG. 4.8 Acousto-optic guided-wave device for see-through horizontal-parallax-only holographic display (portion of one line shown).

correction by a freeform lens for image distortion created by the curved windshield and the shallow projection angle. A team at DAQRI instead used an LCoS phase modulator displaying a holographic fringe pattern, creating 3D imagery and allowing the distortion correction to be computed as part of the generation of the hologram in place of the freeform lens [49].

REFERENCES

[1] E.N. Leith, J. Upatnieks, B.P. Hildebrand, K. Haines, Requirements for a wavefront reconstruction facsimile system, Journal of the Society of Motion Picture and Television Engineers 74 (10) (October 1965) 893–896.

[2] L.H. Enloe, J.A. Murphy, C.B. Rubinstein, Hologram transmission via television, Bell System Technical Journal 45 (2) (Feb. 1966) 335–339.

[3] R.J. Doyle, W.E. Glenn, Remote real-time reconstruction of holograms using the Lumatron, Applied Optics 11 (5) (1972) 1261–1264.

[4] V.G. Komar, Progress on the holographic movie process in the USSR, Proc. SPIE Three-Dimensional Imaging 0120 (1977).

[5] K. Sato, K. Higuchi, H. Katsuma, Holographic television by liquid-crystal device, Proceedings of SPIE Practical Holography VI 1667 (1992) 19–31.

[6] N. Hashimoto, K. Hoshino, S. Morokawa, Improved real-time holography system with LCDs, Proceedings of SPIE Practical Holography VI 1667 (1992) 2–7.

[7] T.-C. Poon, M.H. Wu, K. Shinoda, Y. Suzuki, Optical scanning holography, Proceedings of the IEEE 84 (5) (May 1996) 753–764.

[8] J. Barabas, S. Jolly, D.E. Smalley, V.M. Bove Jr., Diffraction specific coherent panoramagrams of real scenes, Proceedings of SPIE Practical Holography XXV 7957 (2011) 795702.

[9] R. Oi, T. Mishina, K. Yamamoto, M. Okai, Real-time IP-hologram conversion hardware based on floating point DSPs, Proceedings of SPIE Practical Holography XXIII 7233 (2009) 723305.

[10] K. Yamamoto, T. Mishina, R. Oi, T. Senoh, T. Kurita, Real-time color holography system for live scene using 4K2K video system, Proceedings of SPIE Practical Holography XXIV 7619 (2010) 761906.

[11] B. Javidi, S.-H. Hong, Three-dimensional holographic image sensing and integral imaging display, Journal of Display Techology 1 (2) (December 2005) 341–346.

[12] A.Ö. Yöntem, L. Onural, Integral imaging based 3D display of holographic data, Optics Express 20 (22) (2012) 24175–24195.

[13] X. B. Dong, L.-Y. Ai, E.-S. Kim, Integral imaging-based large-scale full-color 3-D display of holographic data by using a commercial LCD panel, Optics Express 24 (4) (2016) 3638–3651.

[14] P.-A. Blanche, A. Bablumian, R. Voorakaranam, C. Christenson, W. Lin, T. Gu, D. Flores, P. Wang, W.-Y. Hsieh, M. Kathaperumal, B. Rachwal, O. Siddiqui, J. Thomas, R.A. Norwood, M. Yamamoto, N. Peyghambarian, Holographic three-dimensional telepresence using large-area photorefractive polymer, Nature 468 (7320) (2010) 80–83.

[15] Y. Ohsawa, K. Yamaguchi, T. Ichikawa, Y. Sakamoto, Computer-generated holograms using multiview images captured by a small number of sparsely arranged cameras, Applied Optics 52 (1) (2013) A167–A176.

[16] W.F. Schreiber, Fundamentals of Electronic Imaging Systems, Springer-Verlag, Berlin, 1986, pp. 15–17, 80-83.

[17] M. Yamaguchi, Light-field and holographic three-dimensional displays, Journal of the Optical Society of America A 33 (12) (Dec. 2016) 2348–2364.

[18] S.A. Benton, V.M. Bove Jr., Holographic Imaging, Wiley, Hoboken, NJ, 2008, pp. 59–61.

[19] L. Onural, F. Yaras, H. Kang, Digital holographic three-dimensional video displays, Proceedings of the IEEE 99 (4) (April 2011) 576–589.

[20] P. St-Hilaire, S.A. Benton, M. Lucente, M.L. Jepsen, J. Kollin, H. Yoshikawa, Electronic display system for computational holography, Proceedings of SPIE Practical Holography IV 1212 (1990) 174–182.

[21] C. Slinger, C. Cameron, M. Stanley, Computer-generated holography as a generic display technology, IEEE Computer 38 (8) (August 2005) 46–53.

[22] F. Yaras, H. Kang, L. Onural, State of the art in holographic displays: a survey, Journal of Display Technology 6 (10) (2010) 443–454.

[23] H.W. Lee, The Scophony television receiver, Nature 142 (3584) (1938) 59–62.

[24] P. St-Hilaire, S.A. Benton, M. Lucente, P.M. Hubel, Color images with the MIT holographic video display, Proceedings of SPIE Practical Holography VI 1667 (1992) 73–84.

[25] P. St-Hilaire, S.A. Benton, M. Lucente, J.D. Sutter, W.J. Plesniak, Advances in holographic video, Proceedings SPIE Practical Holography VII 1914 (1993) 188–196.

[26] T.C. Poon, B.W. Schilling, M.H. Wu, K. Shinoda, Y. Suzuki, Real-time two-dimensional holographic imaging by using an electron-beam-addressed spatial light modulator, Optics Letters 18 (1) (January, 1993) 63–65.

[27] R. Tudela, E. Martin-Badosa, I. Labastida, S. Vallmitjana, A. Carnicer, Wavefront reconstruction by adding modulation capabilities of two liquid crystal devices, Optical Engineering 43 (11) (November 2004) 2650–2657.

[28] J.-P. Liu, W.-Y. Hsieh, T.-C. Poon, P. Tsang, Complex Fresnel hologram display using a single SLM, Applied Optics 50 (34) (2011) H128–H135.

[29] P. Ambs, L. Bigue, Y. Fainman, Dynamic computer generated holograms displayed on a digital micromirror device, in: R. Magnusson (Ed.), Diffractive Optics and Micro-optics, OSA Trends in Optics and Photonics Series, vol. 75, 2002.

[30] M. Huebschman, B. Munjuluri, H. Garner, Dynamic holographic 3-D image projection, Optics Express 11 (5) (2003) 437–445.

[31] K. Sato, A. Sugita, M. Morimoto, K. Fujii, Reconstruction of color images at high quality by a holographic display, Proceedings SPIE Practical Holography XX 6136 (2006) 61360V.

[32] J. Hahn, H. Kim, Y. Lim, G. Park, B. Lee, Wide viewing angle dynamic holographic stereogram with a curved array of spatial light modulators, Optics Express 16 (16) (2008) 12372–12386.

[33] F. Yaraş, H. Kang, L. Onural, Circular holographic video display system, Optics Express 19 (10) (2011) 9147–9156.

[34] N. Okada, Y. Takaki, Horizontally scanning holography to enlarge both image size and viewing zone angle, Proceedings SPIE Practical Holography XXIII 7233 (2009) 723309.

[35] T. Inoue, Y. Takaki, Table screen 360-degree holographic display using circular viewing-zone scanning, Optics Express 23 (5) (2015) 6533–6542.

[36] Y. Lim, K. Hong, H. Kim, H.-E. Kim, E.-Y. Chang, S. Lee, T. Kim, J. Nam, H.-G. Choo, J. Kim, J. Hahn, 360-degree tabletop electronic holographic display, Optics Express 24 (22) (2016) 24999–25009.

[37] G. Li, J. Jeong, D. Lee, J. Yeom, C. Jang, S. Lee, B. Lee, Space bandwidth product enhancement of holographic display using high-order diffraction guided by holographic optical element, Optics Express 23 (No. 26) (2015) 33170–33183.

[38] J.S. Chen, Q.Y.J. Smithwick, D.P. Chu, Coarse integral holography approach for real 3D color video displays, Optics Express 24 (6) (2016) 6705–6718.

[39] S. Tay, P.A. Blanche, R. Voorakaranam, A.V. Tunc, W. Lin, S. Rokutanda, T. Gu, D. Flores, P. Wang, G. Li, P. St-Hilaire, J. Thomas, R.A. Norwood, M. Yamamoto, N. Peyghambarian, An updatable holographic three-dimensional display, Nature 451 (7179) (February, 2008) 694–698.

[40] R. Haussler, S. Reichelt, N. Leister, E. Zschau, R. Missbach, S. Schwerdtner, Large real-time holographic displays: from prototypes to a consumer product, Proceedings SPIE Stereoscopic Displays and Applications XX 7237 (2009) 72370S.

[41] D.E. Smalley, Q.Y.J. Smithwick, V.M. Bove, J. Barabas, S. Jolly, Anisotropic leaky-mode modulator for holographic video displays, Nature 498 (June, 2013) 313–317.
[42] S. Jolly, N. Savidis, B. Datta, D. Smalley, V.M. Bove Jr., Progress in transparent, flat-panel holographic displays enabled by guided-wave acousto-optics, Proceedings SPIE Practical Holography XXXII: Displays, Materials, and Applications 10558 (2018).
[43] H. Mukawa, A. Funanami, Head-mounted Display Apparatus, 2013. United States patent application pub. no. 2013/0069850A1.
[44] B.C. Kress, W.J. Cummings, Towards the ultimate mixed reality experience: HoloLens display architecture choices, SID Symposium Digest of Technical Papers 48 (1) (2017) 127–131.
[45] S. Lee, C. Jang, S. Moon, J. Cho, B. Lee, Additive light field displays: realization of augmented reality with holographic optical elements, ACM Transactions on Graphics 35 (4) (2016) paper 60.
[46] A. Putilin, V. Druzhin, E. Malinovskaya, A. Morozov, I. Bovsunovsky, Holographic Imaging Optical Device, 2014. United States patent application pub. no. 2014/0160543A1.
[47] A. Maimone, A. Georgiou, J.S. Kollin, Holographic near-eye displays for virtual and augmented reality, ACM Transactions on Graphics 36 (4) (2017) paper 85.
[48] S. Jolly, N. Savidis, B. Datta, D. Smalley, V.M. Bove Jr., Near-to-eye electroholography via guided-wave acousto-optics for augmented reality, Proceedings of SPIE Practical Holography XXXI: Materials and Applications 10127 (2017).
[49] B. Mullins, P. Greenhalgh, J. Christmas, The holographic future of head up displays, SID Symposium Digest of Technical Papers 48 (1) (May 2017) 886–889.

CHAPTER 5

Digital Holography

PASCAL PICART, PHD • SILVIO MONTRESOR, PHD

INTRODUCTION

The idea of digitally reconstructing the optical wavefront was born in the 1960s. The oldest study on the subject dates back to 1967 with the article published by J.W. Goodman in *Applied Physics Letters* [1]. The goal was to replace the "analogue" recording/decoding of the object by a "digital" recording/decoding simulating diffraction from a digital interferogram consisting of the recorded image. Holography thus became "digital," replacing the silvered material with a matrix of the discrete values of the hologram. Then, in 1971, Huang discussed the computer analysis of optical wavefronts and introduced for the first time the concept of "digital holography" [2]. The works presented in 1972 by Kronrod [3] historically constitute the first attempts at the reconstruction of an object coded in a hologram by the calculation. At the time, 6 h of calculation were required for the reconstruction of 512×512 pixels with the Minsk-22 computer, the discrete values being obtained from a holographic plate by 64-bit digitization with a scanner. However, it took until the 1990s for array detector-based digital holography to become effective [4]. Indeed, at the time the required technology had reached the sufficient maturity to make possible digital holographic recording and reconstruction. Two important progresses were at the origin of this development: the size reduction of the pixels, and the significant improvement in microprocessor performance, in particular their processing units as well as their storage capacities.

Basically, fundamentals of digital holography are the same as holography with nondigital materials. The main difference relies in the spatial resolution both during the recording of the hologram and throughout the course of the numerical reconstruction. This chapter aims at presenting digital holography and describing the properties of the digitally reconstructed images. The first part is devoted to the basic fundamentals of diffraction and Fourier optics, then we discuss the digital holographic process to reconstruct images; the properties of such images are discussed. The last part of the chapter constitutes an insight on the speckle noise reduction in holographic images.

BASIC FUNDAMENTALS OF FOURIER OPTICS

Digital holography and its related numerical image reconstruction techniques are based on the scalar wave diffraction approaches. To provide to the reader the few basic fundamentals that are necessarily required to understand and to carry out digital holographic experiments, this section proposes a brief summary of the basics of the scalar diffraction of light. Algorithms used to numerically reconstruct digital holograms are based on discrete versions of the described theoretical approaches.

Monochromatic Spherical and Plane Waves

Note that in all what follows, the time dependence of the optical wavefront is implicit and is not included in equations. In a space described by a Cartesian coordinate system $Oxyz$, a point P is described by a set of three coordinates (x,y,z) and we will use the modulus of the distance **OP** by $r = |\mathbf{OP}| = \sqrt{x^2+y^2+z^2}$. We also use $k = 2\pi/\lambda$, the modulus of the wave vector, here λ is the wavelength of the light. We assume that the complex optical field is written in the form:

$$U(P,t) = U(P)\exp(-2i\pi\nu t). \quad (5.1)$$

Here $U(P)$ is the complex amplitude at the observation point $P(x,y,z)$, and ν is the frequency of the light wave. In Fourier optics, two kinds of waves are of great interest to describe propagation of light: spherical waves and plane waves. The complex amplitude of a spherical wave can be expressed by Refs. [5–7]:

$$U(x,y,z) = \left\{ \begin{array}{ll} \dfrac{A_0}{r}\exp(ikr) & (divergent) \\ \dfrac{A_0}{r}\exp(-ikr) & (convergent) \end{array} \right\} \quad (5.2)$$

Optical Holography-Materials, Theory and Applications. https://doi.org/10.1016/B978-0-12-815467-0.00005-0

In Eq. (5.2), the point source of the spherical wave is at the origin of the Cartesian coordinate system, and A_0 is the modulus of the amplitude of the wave. When the center of the spherical wave is at the point (x_c, y_c, z_c), instead of the origin, the expressions are identical, with r substituted for:

$$r = \sqrt{(x - x_c)^2 + (y - y_c)^2 + (z - z_c)^2} \tag{5.3}$$

For a plane wave propagating in a homogeneous medium, the wavefront is perpendicular to the propagation direction. The plane wave can be written as:

$$U(x, y, z) = A_0 \exp[ik(x\cos\alpha + y\cos\beta + z\cos\gamma)] \tag{5.4}$$

The propagation direction is defined by the direction of the cosines $\cos\alpha$, $\cos\beta$, $\cos\gamma$ of Eq. (5.4). This relation shows that for a real number C, the expression $x\cos\alpha + y\cos\beta + z\cos\gamma = C$ describes a phase plane whose normal is in the direction given by the cosines $\cos\alpha$, $\cos\beta$, $\cos\gamma$. Because different values of C correspond to different parallel planes, Eq. (5.4) represents a wave propagating in the direction normal to these planes. For example, for a plane wave propagating in the direction of positive z (that is perpendicular to x,y plane), one can write

$$U(x, y, z) = A_0 \exp(ikz) \tag{5.5}$$

Fig. 5.1 illustrates spherical and plane waves. In Fig. 5.1A, a spherical wavefront with center A emits a divergent spherical wavefront Σ. In a homogenous medium, rays are perpendicular to Σ. So, the wave is deformed when propagating to the right. When the point source tends to infinity, the spherical wave tends to a plane wave, as illustrated in Fig. 5.1B. In this case, the rays become parallel and the beam propagates without any deformation.

Propagation Equation

The wave aspect of light is described by the classical theory of electromagnetism, that is, by the Maxwell's equations [5,8,9]. For the sake of simplicity, we will consider the case of a homogeneous medium. Given this simplification, the Maxwell's equations can be reduced to this propagation equation:

$$\nabla^2 E - \frac{1}{c^2}\frac{\partial^2 E}{\partial t^2} = 0 \tag{5.6}$$

with **E** being the electric field and c the velocity of light in the medium. The operator $\nabla^2 = \partial^2/\partial x^2 + \partial^2/\partial y^2 + \partial^2/\partial z^2$ is the Laplacian operator. Eq. (5.6) is valid also for the magnetic field **B**. From Eq. (5.6), the angular spectrum transfer function can be derived. This is discussed in the next section.

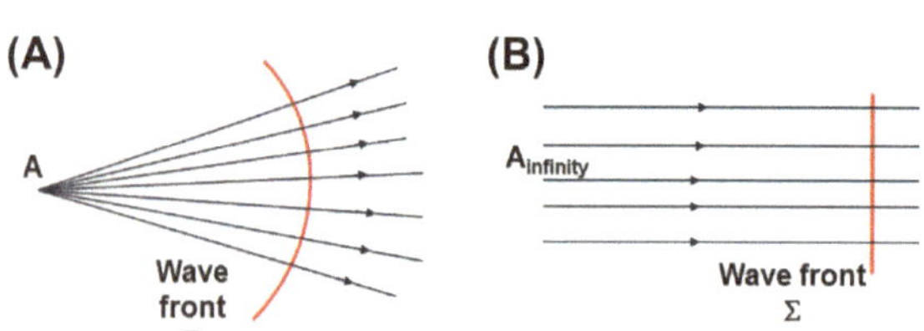

FIG. 5.1 **(A)** Spherical wave and **(B)** plane wave.

Angular Spectrum Transfer Function

By substituting Eq. (5.1), Eq. 8.2 into Eq. (5.6), the Helmholtz equation is obtained and does not depend on time t. We have

$$(\nabla^2 + k^2)U(P) = 0 \tag{5.7}$$

Fourier analysis can be advantageously used to solve this equation. If z is the distance between the initial field plane and the diffracted plane, we note $U(x,y,0)$ and $U(x,y,z)$ the respective complex amplitudes at these two planes. In addition, in the frequency space (Fourier space), the respective spectral functions are noted $G_0(u,v)$ and $G_z(u,v)$. Coordinates (u,v) are the spatial frequencies associated to the spatial coordinates (x,y). The two frequency functions are defined by the Fourier integrals such as:

$$G_0(u, v) = \int_{-\infty}^{\infty}\int_{-\infty}^{\infty} U(x, y, 0)\exp[-2i\pi(ux + vy)]dxdy \tag{5.8}$$

$$G_z(u, v) = \int_{-\infty}^{\infty}\int_{-\infty}^{\infty} U(x, y, z)\exp[-2i\pi(ux + vy)]dxdy \tag{5.9}$$

A general solution to the differential Eq. (5.7) can be obtained with the Fourier components of $U(x,y,0)$ and $U(x,y,z)$ according to (demonstration not provided, refer to Refs. [10–12]):

$$G_z(u, v) = G_0(u, v)\exp\left[\frac{2i\pi}{\lambda} z\sqrt{1 - (\lambda u)^2 - (\lambda v)^2}\right] \tag{5.10}$$

The complex field at distance z can be obtained by the superposition theorem and the inverse Fourier transform integral:

$$U(x, y, z) = \int_{-\infty}^{\infty}\int_{-\infty}^{\infty} G_z(u, v)\exp[2i\pi(ux + vy)]dudv \tag{5.11}$$

It follows that we get a relationship between the spectrum of the wave in the initial plane and that obtained in the diffraction plane. So, in the frequency space, the spectral variation in complex amplitude caused by the propagation of light over the distance z is represented by its multiplication by a phase-delay factor:

$$G(u,v) = \exp\left[\frac{2i\pi}{\lambda} z\sqrt{1-(\lambda u)^2-(\lambda v)^2}\right] \quad (5.12)$$

According to the theory of the linear systems, the diffraction process is a transformation of the light field across an optical system (*free space* in the present case), as the pure phase-delay factor of Eq. (5.12) can be interpreted as a transfer function in the frequency space. This interpretation of the propagation of light is called *the propagation of the angular spectrum*, and the associated transfer function Eq. (5.12) is called *the angular spectrum transfer function*. Fig. 5.2 illustrates this approach.

Fig. 5.2 shows that the field $U(x,y,z)$ can be considered as the superposition of multiple plane waves of amplitude $G_z(u,v)dudv$ propagating in the direction whose cosines are $\{\cos\alpha, \cos\beta, \cos\gamma\} = \left\{\sqrt{\lambda u, \lambda v, 1-(\lambda u)^2-(\lambda v)^2}\right\}$. From the diffraction of the angular spectrum, Eq. (5.12) means that the elementary waves satisfying $1-(\lambda u)^2-(\lambda v)^2 < 0$ are attenuated by the propagation, that is, all the components satisfying this relation only exist in a zone very close to the initial plane (evanescent waves).

Because the components of the observation plane must satisfy the relation $1-(\lambda u)^2-(\lambda v)^2>0$, that is, $u^2+v^2<1/\lambda^2$, propagation in free space can be considered as an ideal low-pass filter of radius $1/\lambda$ in the frequency space. Consequently, on the condition that we can obtain the spectrum of $U(x,y,0)$, the spectrum in the observation plane, $U(x,y,z)$ can be expressed by Eq. (5.13). Using the direct and inverse Fourier transforms (FT and FT^{-1}), the diffraction calculation process can be described as:

$$U(x,y,z) = FT^{-1}\left\{FT\{U(x,y,0)\} \exp\left[\frac{2i\pi}{\lambda} z\sqrt{1-(\lambda u)^2-(\lambda v)^2}\right]\right\} \quad (5.13)$$

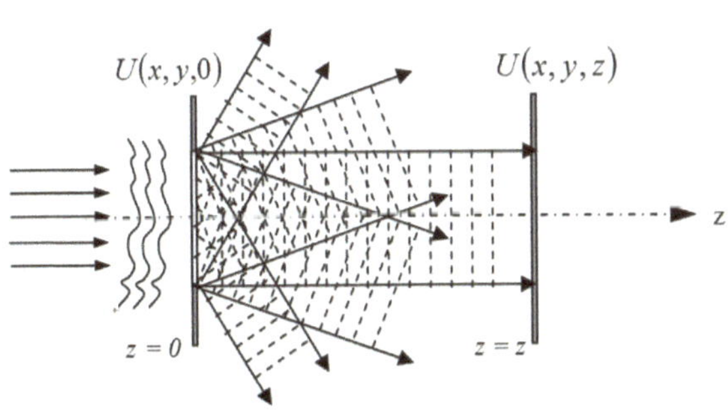

FIG. 5.2 Diffraction by the angular spectrum.

Kirchhoff and Rayleigh-Sommerfeld Formulas

The Kirchhoff and Rayleigh-Sommerfeld formulas are two more solutions to the Helmholtz equation. If we consider the scheme in Fig. 5.3, the two formulas can be written in the same mathematical expression [10–12]:

$$U(x,y,d_0) = \frac{1}{i\lambda}\int_{-\infty}^{\infty}\int_{-\infty}^{\infty} U(X,Y,0)\frac{\exp(ikr)}{r}K(\theta)dXdY \quad (5.14)$$

where

$$r = \sqrt{(x-X)^2+(y-Y)^2+d_0^2} \quad (5.15)$$

In Eq. (5.14), θ is the angle between the normal at the point $(X,Y,0)$, and the vector **MP** from the point $(X,Y,0)$ to the point (x,y,d_0) (refer to Fig. 5.3). Coefficient $K(\theta)$ is called the obliquity factor, and its three different expressions correspond to three different formulations [10–12].

- $K(\theta) = \frac{\cos\theta+1}{2}$ Kirchhoff's formula,
- $K(\theta) = \cos\theta$ First Rayleigh-Sommerfeld solution
- $K(\theta) = 1$ Second Rayleigh-Sommerfeld solution

The different Kirchhoff's formulas provide results in remarkable agreement with experiments. For that, they are widely used in practice. Furthermore, because the angle θ is often small in experimental configurations,

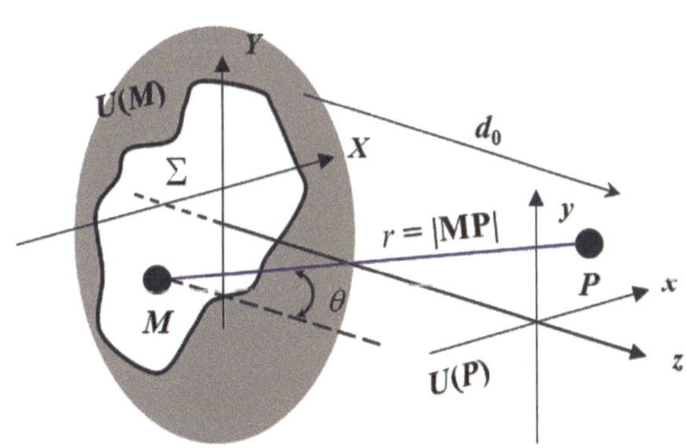

FIG. 5.3 Scheme of spatial coordinates for the initial plane and the diffraction plane.

the obliquity factors of the three formulations are roughly equal to unity. Even though there do exist certain inconsistencies [8,10], the Kirchhoff, Rayleigh-Sommerfeld, and angular spectrum formulas are considered as equivalent representations of the diffraction of light. The derivations of the Kirchhoff and Rayleigh-Sommerfeld approaches are presented in detail in the book of J.W. Goodman [10]. Readers who wish to go into these aspects in more detail are invited to familiarize themselves with this work.

Fresnel Approximation and Fresnel Diffraction Integral

The presence of the square root in the complex exponentials in Eq. (5.14) makes this approach difficult to use and laborious to mathematically manipulate. Practically however, diffraction conditions in digital holography, are often constrained to the paraxial propagation. To simplify the theoretical analysis, the Fresnel approximation is generally considered. Note d_0 the diffraction distance. By expanding the square root in Eq. (5.12) to the first order, one gets

$$G(u,v) \cong \exp[ikd_0]\exp\left[-i\pi\lambda d_0\left(u^2+v^2\right)\right] \tag{5.16}$$

Considering that Eq. (5.13) can be written in the form of a convolution (* means convolution), we have

$$U(x,y,d_0) = U(x,y,0) * FT^{-1}\{G(u,v)\}. \tag{5.17}$$

Considering that the inverse Fourier transform of $G(u,v)$ is an analytic function and putting Eq. (5.16) into Eq. (5.17), we have

$$U(x,y,d_0) = U(x,y,0) * \frac{\exp(ikd_0)}{i\lambda d_0}\exp\left[\frac{i\pi}{\lambda d_0}\left(x^2+y^2\right)\right], \tag{5.18}$$

Eq. (5.18) is the convolution of $U(x,y,0)$ with the impulse response of the free space propagation that will be noted $h(x,y,d_0)$:

$$h(x,y,d_0) = -\frac{i}{\lambda d_0}\exp[ikd_0]\exp\left[\frac{i\pi}{\lambda d_0}\left(x^2+y^2\right)\right] \tag{5.19}$$

Eq. (5.18) can be also written:

$$U(x,y,d_0) = \frac{\exp(ikd_0)}{i\lambda d_0}\int_{-\infty}^{\infty}\int_{-\infty}^{\infty} U(X,Y,0)\exp\left\{\frac{i\pi}{\lambda d_0}\left[(x-X)^2 + (y-Y)^2\right]\right\}dXdY \tag{5.20}$$

Eq. (5.20) is known as the *Fresnel's diffraction integral* and is equivalent to a parabolic approximation of the propagation of spherical waves. The expansions of the quadratic terms in the exponential leads to

$$U(x,y,d_0) = \frac{\exp(ikd_0)}{i\lambda d_0}\exp\left[\frac{i\pi}{\lambda d_0}\left(x^2+y^2\right)\right] \times \int_{-\infty}^{\infty}\int_{-\infty}^{\infty} U(X,Y,0)\exp\left[\frac{i\pi}{\lambda d_0}\left(X^2+Y^2\right)\right] \exp\left[-2i\pi\left(\frac{x}{\lambda d_0}X+\frac{y}{\lambda d_0}Y\right)\right]dXdY \tag{5.21}$$

Eq. (5.21) shows that diffraction is a Fourier transform of $U(x,y,0)$ multiplied by the quadratic phase function $\exp(i\pi(x^2+y^2)/\lambda d_0)$ that is calculated at the frequencies $(u,v)=(x/\lambda d_0, y/\lambda d_0)$. The calculation of the Fresnel diffraction integrals is relatively simple compared with the rigorous formulas satisfying the Helmholtz equation. In the paraxial approximation, this equation provides good precision. From above, the Fresnel transfer function [10] is defined as:

$$G_F(u,v) = \exp\left[\frac{2i\pi z}{\lambda}\right]\exp\left[-i\lambda z\left(u^2+v^2\right)\right] \tag{5.22}$$

and the Fresnel approximation can be expressed by:

$$U(x,y,z) = FT^{-1}\{FT\{U(x,y,0)\}G_F(u,v)\} \tag{5.23}$$

This expression is analogous to the angular spectrum formulation of Eq. (5.12), but the difference is related to the different transfer functions present in the two formulations.

In digital holography, the large majority of the numerical reconstruction algorithms are based on the Fresnel transform and the angular spectrum transfer function. In the following, we will consider applications of this theory to the reconstruction of digital holograms.

The next section discusses on recording of digital holograms.

RECORDING DIGITAL HOLOGRAMS

As also discussed in the previous chapters of this book, digital holography found its background in the coherent properties of light, and especially interferences. A hologram is an interferometric mixing between a reference wave and a wave from the object of interest. A digital hologram is a hologram recorded with a digital image sensor. It follows that digital holography is in close connection with the digital word and that an infinite space of possibility is open for processing the data.

Notations

Considering the scheme depicted in Fig. 5.4 where an extended object is illuminated with a monochromatic wave, the object surface diffracts a wave to the observation plane localized at a distance $d_O = |z_O|$.

The wavefront is noted:

$$A(x,y) = A_0(x,y)\exp(i\psi_0(x,y)). \quad (5.24)$$

The amplitude A_0 describes the reflectivity/transmission of the object, and phase ψ_0 is related to its surface and shape. Because of the natural roughness of the object, ψ_0 is a random variable, uniformly distributed over $[-\pi,+\pi]$. The diffracted field U_O at distance d_O, and at spatial coordinates (X,Y) of the observation plane is given by the propagation of the object wave to the recording plane. In the observation plane, the diffracted wave can be simply written as:

$$U_O(X,Y,d_O) = a_O(X,Y)\exp(i\varphi_O(X,Y)), \quad (5.25)$$

with a_O the modulus of the complex amplitude and φ_O its optical phase.

The complex amplitude of the reference wavefront, noted U_r, at the recording plane is

$$U_r(X,Y) = a_r(X,Y)\exp(i\varphi_r(X,Y)) \quad (5.26)$$

with a_r the modulus and φ_r the optical phase. Considering (x_s,y_s,z_s) the coordinates of the reference source point in the hologram reference frame ($z_s < 0$), the optical phase of the reference wave can be written in the paraxial approximations by Refs. [5,10–12]:

$$\varphi_r(X,Y) \cong -\frac{\pi}{\lambda z_s}\left((X-x_s)^2 + (Y-y_s)^2\right). \quad (5.27)$$

Expanding Eq. (5.27) and regrouping terms, leads to

$$\varphi_r(X,Y) = 2\pi(u_0X + v_0Y) - \frac{\pi}{\lambda z_s}(X^2+Y^2) + \varphi_s. \quad (5.28)$$

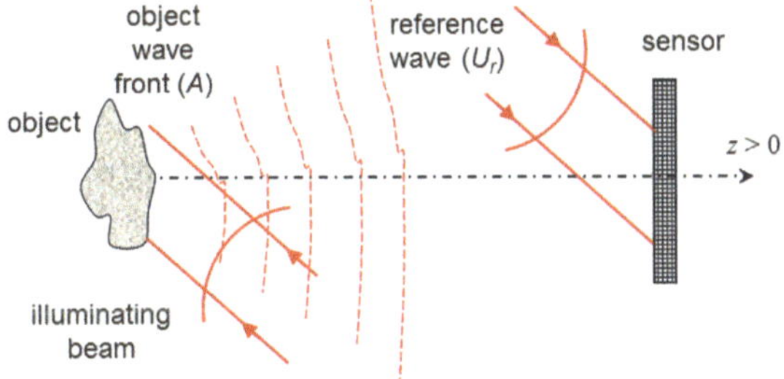

FIG. 5.4 Scheme for digital holographic recording.

In Eq. (5.28) $(u_0,v_0) = (-x_s/\lambda z_s, -y_s/\lambda z_s)$ are the spatial carrier frequencies of the hologram, $\varphi_s = 2\pi(x_s^2+y_s^2)/\lambda z_s$ is a constant term that can be omitted. This latter term can be useful in case of taking into account any phase shifting in the digital holographic recording process.

The total illumination at the sensor plane, noted H, is then written [13–15]:

$$H = |U_r + U_O|^2 = |U_r|^2 + |U_O|^2 + U_r^*U_O + U_rU_O^*. \quad (5.29)$$

In Eq. (5.29), we can identify three terms corresponding to the well-known three diffraction orders of holography:

- $|U_r|^2 + |U_O|^2$ the 0-order,
- U_rU_O the -1-order,
- U_rU_O the $+1$-order.

The $+1$ order is the one of interest because most of the time it is the orthoscopic image of the object that is sought-after.

Note: Eq. (5.29) can also be written according to the well-known equation of interferences:

$$H = a_r^2 + a_O^2 + 2a_ra_O\cos(\varphi_r - \varphi_O). \quad (5.30)$$

This equation is often useful to better understand the properties and constraints of digital holographic recording.

Fourier Analysis of the Digital Hologram

According to Eqs. (5.28) and (5.29), the Fourier transform of the hologram can be written (FT and FT^{-1} means, respectively, Fourier Transform and inverse Fourier Transform):

$$\begin{aligned} FT[H](u,v) &= C_0(u,v) + C_1(u-u_0, v-v_0) \\ &\quad + C_1^*(-u-u_0, -v-v_0), \end{aligned} \quad (5.31)$$

where C_0 is the Fourier transform of the zero-order and C_1 is the Fourier transform of $U_r^*U_O$ (related to $+1$ order). If the three orders are well separated in the Fourier plane, the $+1$ order can be extracted from the Fourier spectrum. From Eq. (5.31), the spatial frequencies (u_0,v_0) localize the useful information ($+1$ or -1 orders). So, they have to be adjusted to minimize the overlapping of the three diffraction orders. It follows that the structure of the numerical diffracted field depends on the (u_0,v_0) of the hologram. One can distinguish two cases described here after.

On-Axis Digital Holography and Phase Shifting

If we have $(u_O,v_O)=(0,0)$, that is $(x_s,y_s)=(0,0)$, the reference point source is localized on the z axis, and there does not exist any tip/tilt between the two waves. That is the case of an "on-axis" hologram. It follows that the three orders are overlapping as illustrated in Fig. 5.5.

In this case, there is no possibility to extract directly the +1 order from one single digital hologram due to the cross-mixing of the different orders. The only option is to apply a phase shifting approach, which was first described by Yamaguchi in 1997 [16–18]. This approach leads to the reconstruction of an image free from the zero order and of the −1 order.

Consider the hologram Eq. (5.30), there are basically three unknowns: the offset term $a_r^2+a_O^2$, the modulation term $2a_ra_O$, and the phase of the cosine function, $\varphi_O-\varphi_r$. So, with at least three values for H, one should be able to solve these three unknowns. This can be done by shifting the phase in the cosine function, by adding in the holographic interferometer a phase modulator. Practically, a piezo-electric transducer (PZT) is attached to a mirror, and by applying a small voltage, the mirror is slightly moved, and thus shifting the optical phase (although other methods do exist) [19,20]. With at least three positions of the mirror, the object wave field can be recovered.

The robustness of the method increases when increasing the number of phase-shifted holograms. Consider a phase-shifted hologram with a phase shift being an integer division of 2π, that is, $2\pi/P$, with P an integer. We have

$$H_p = a_r^2 + a_O^2 + 2a_ra_O\cos(\varphi_O - \varphi_r + 2(p-1)\pi/P) \quad (5.32)$$

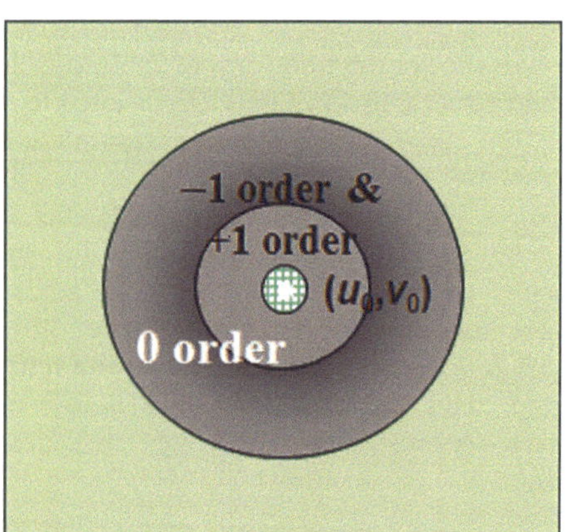

FIG. 5.5 Structure of the diffracted field with on-axis digital holography.

with $P = 1, 2, \ldots, P$. For $P \geq 3$, the phase of the object wave in the detector plane may be calculated by Ref. [21]:

$$a_O = a_r + \arctan\left\{\frac{\sum_{n=1}^{P} H_p\sin(2\pi(p-1)/P)}{\sum_{n=1}^{P} H_p\cos(2\pi(p-1)/P)}\right\} \quad (5.33)$$

and the amplitude may be calculated by:

$$a_O = \frac{1}{2a_r}\sqrt{\left(\sum_{p=1}^{P} H_p\sin(2\pi(p-1)/P)\right)^2 + \left(\sum_{p=1}^{P} H_p\cos(2\pi(p-1)/P)\right)^2} \quad (5.34)$$

If the reference wave is plane or spherical, that is, free from aberrations, the phase $\varphi_O(x,y)$ may be determined without ambiguity and compensated. With $P=4$, the most widely used method is obtained, which was proposed by Ref. [16], using four $\pi/2$ phase-shifted holograms [19,22]. In this case, we have

$$\varphi_O = \varphi_r + \arctan\left\{\frac{H_4 - H_2}{H_1 - H_3}\right\}$$
$$a_O = \frac{1}{4a_r}\sqrt{[H_1 - H_3]^2 + [H_4 - H_2]^2} \quad (5.35)$$

Thus, the complex-valued object field $a_O\exp(i\varphi_O)$ at the hologram plane can be obtained.

Off-Axis Digital Holography and Fourier Filtering

In the case where $(u_O,v_O)\neq(0,0)$, that is $(x_s,y_s)\neq(0,0)$, holography is said "off-axis." There does exist a slight tip/tilt between the two interfering waves. As a general rule, the reference wave has uniform amplitude, that is, $a_r(X,Y) = C^{te}$. It follows that the three orders can be separated as illustrated in Fig. 5.6, and the +1 order can be isolated and filtered (refer to the white dashed square in Fig. 5.6).

The spatial frequencies (u_O,v_O) localize the useful information, and they must be adjusted to minimize the overlapping of the three diffraction orders. By applying a bandwidth-limited filter ($\Delta u\times\Delta v$ width) around the spatial frequency (u_O,v_O), and after filtering and inverse 2D Fourier transform, one gets the object complex amplitude. Consider the Fourier filtering applied to Eq. (5.31), we can write the filtered +1 order as:

$$O_{+1}(x,y) = FT^{-1}[C_1(u-u_O, v-v_O)]$$
$$a_r\{a_O(x,y)\exp[\varphi_O(x,y)]\exp[2i\pi(u_Ox+v_Oy)]\} * h_f(x,y), \quad (5.36)$$

where * means convolution and $h_f(x,y)$ is the impulse response corresponding to the filtering applied in the

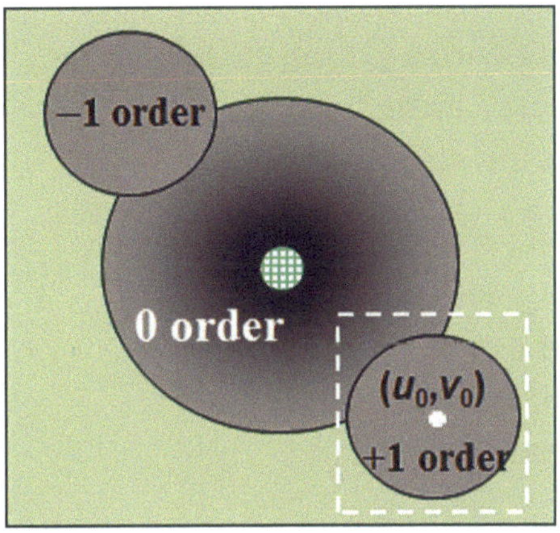

FIG. 5.6 Structure of the diffracted field with off-axis digital holography.

Fourier domain. The impulse response of the filter is such that:

$$h_f(x,y) = \Delta u \Delta v \exp[2i\pi(u_0x + v_0y)]\mathrm{sinc}(\pi\Delta ux)\mathrm{sinc}(\pi\Delta vy), \tag{5.37}$$

The spatial resolution is then related to $1/\Delta u$ and $1/\Delta v$, respectively, in the *x-y* axis. In addition, the phase recovered with Eq. (5.36) includes the spatial carrier modulation that has to be removed. This may be achieved by multiplying O_{+1} by $\exp[-2i\pi(u_0x + v_0y)]$. Then, the optical object phase at the hologram plane can be estimated from relation:

$$\varphi_O(x,y) = \tan^{-1}\left(\frac{\Im_m[O_{+1}(x,y)]}{\Re_e[O_{+1}(x,y)]}\right), \tag{5.38}$$

and the object amplitude by:

$$a_O(x,y) = \sqrt{\Im_m^2[O_{+1}(x,y)] + \Re_e^2[O_{+1}(x,y)]} \tag{5.39}$$

In Eqs. (5.38) and (5.39), $\Im_m[...]$ and $\Re_e[...]$ mean imaginary and real part of complex-valued amplitude. As for the on-line configuration, the complex-valued object field $a_O\exp(i\varphi_O)$ at the hologram plane can be obtained.

The main advantage of this approach is its capability of recovering the complex object wave through only one acquisition. Thus, there is no time to spend for any heterodyning or moving mirrors, and the influence of vibrations is thus greatly reduced. However, as the diffraction terms are spatially encoded in the hologram, this one-shot capability potentially comes at the cost of useable spatial bandwidth (filter with width $\Delta u \times \Delta v$). In addition, the frequency modulation, induced by the angle between the reference and the object wave, have to guarantee the separation of the information contained in the different diffraction terms that are encoded in the hologram while carrying a frequency compatible with the sampling capacity of digital detectors.

Case of Digital Color Holograms

There do exist different approaches for recording digital color holograms, in particular for *simultaneously* recording the three colors. Fig. 5.7 illustrates the different recording strategies. The first possibility consists of using a chromatic filter organized in a Bayer mosaic (Fig. 5.7A). However, in such a detector, half of the pixels detect green, and only a quarter detect red or blue [23,24]. The spatial color filter creates holes in the mesh and therefore a loss of information, which translates into a loss of resolution.

The second possibility consists of using three detectors organized as a "tri-CCD," the spectral selection being carried out by a prism with dichroic coatings (Fig. 5.7B). Such a detector guarantees a high spatial resolution and a spectral selectivity compatible with the constraints of digital color holography.

A third possibility consists of using a color detector based on a stack of photodiodes [24–27], refer to Fig. 5.7C. The spectral selectivity is relative to the mean penetration depth of the photons in the silicon: blue photons at 425 nm penetrate to around 0.2 μm, green photons at 532 nm to around 2 μm and red photons at 630 nm to around 3 μm. Thus, the construction of junctions at depths at around 0.2, 0.8, and 3.0 μm give the correct spectral selectivity for color imaging. However, the spectral selectivity is not perfect, as green photons may be detected in the blue and red bands, but the architecture guarantees a maximum spectral resolution because the number of effective pixels for each wavelength is that of the entire matrix.

The last possibility consists of using a monochromatic detector combined with spatial chromatic multiplexing (Fig. 5.7D). Each reference wave must have different separately adjusted spatial frequencies according to their wavelengths. The complexity of the experimental apparatus increases with the number of colors. Demonstrations of this approach are provided in Refs. [28–31].

Few Comments on the Role of the Image Sensor

The sampling of the hologram has to fulfill the Shannon conditions. Considering the maximum angle θ_{max} (consider Fig. 5.4) between the two waves, then the microfringes locally produced by the two wavefronts must be sampled so that the sampling pitch is at least

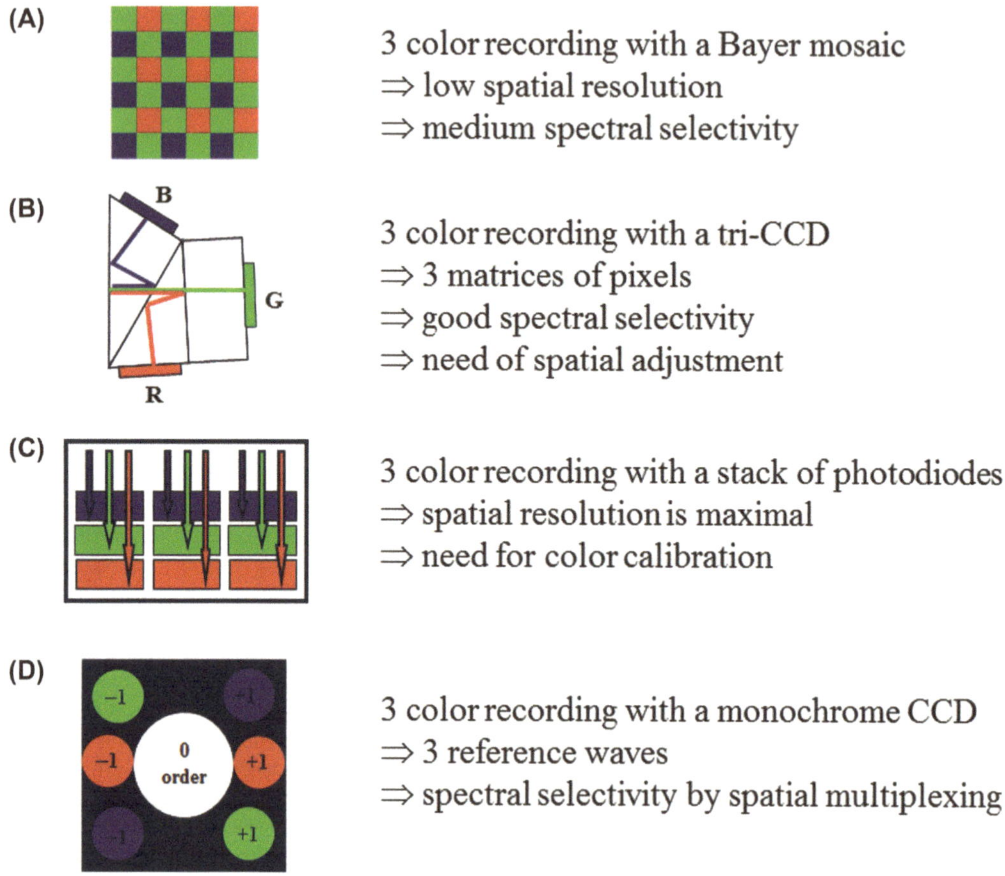

FIG. 5.7 Approaches for recording digital color holograms.

equal to two times the fringe period. The maximum acceptable angle for the setup is given by Ref. [4]:

$$\theta_{\max} < 2\sin^{-1}\left(\frac{\lambda}{4p_x}\right) \cong \frac{\lambda}{2p_x}. \tag{5.40}$$

In Eq. (5.40), p_x is the pixel pitch of the image sensor and λ is the wavelength of light. For example, with $\lambda = 532$ nm and $p_x = p_y = 4$ μm, the maximum acceptable angle is smaller than 3.8 degrees. This means that the setup must be precisely adjusted so as to fulfill the Shannon conditions. In addition, the active surface of pixels has a role of local spatial integration because pixels are spatially extended. The basic effect can be understood: because the pixel provides local integration of the micro fringes, the consequence is a blurring of theses fringes. Qualitatively, this means that the spatial resolution will be deteriorated and that the pixel induces a low-pass filtering to the digital hologram.

The last comment is related to the exposure time during which the pixels accumulate light to produce photoelectrons and then a digital image. When the hologram has no temporal dependence (case of static and nonvarying objects), the time integration can be omitted. However, in the case where the object exhibits time dependence, that is, sinusoidal oscillation, the exposure time influences significantly the recorded hologram. According to the time-average regime, for which the oscillation period is much smaller than the exposure time [32,33], the reconstructed image will be amplitude modulated by a Bessel function. This point will be discussed in chapter 6 devoted to interferometry and nondestructive testing.

Case of Long-Wave Infrared Holography

The recording constraints in digital holography are related to the ratio between the wavelength of light and the pixel pitch. For the visible range of light, this ratio is clearly inferior to 1 (typ. 0.1 for $\lambda = 0.5$ μm and pitch at 5 μm), meaning the pixel pitch is too large compared with the wavelength. To get a more advantageous ratio, one has to move to long wavelengths. For example, with a wavelength at about 10 μm and a pixel pitch at 8 μm, the ratio is now 1.25. So, relaxing the recoding constraints permit to record and reconstruct larger fields of view. To record digital holograms in the long-wave infrared (LWIR) regime, the most common coherent source is the CO_2 laser. Pulse and continuous mode systems can be found with power ranging from GigaWatts to milliWatts. The most common emission lines of this laser are 10.6 μm, 10.3 μm, 9.6 μm, and 9.3 μm, depending on the manufacturer and models. In the case of holography, the CO_2 lasers that are currently used have power from a few Watts to a few tens of Watts. The coherence lengths are quite long (>100 m), so there is no problem related to optical path difference. However, if one wishes to apply LWIR holography for holographic interferometry purposes, one needs CO_2 laser with optional mode stabilization, which is proposed by some manufacturers. Indeed, CO_2 lasers usually suffer from mode hop, which consists in change of longitudinal mode, which has the effect of changing the wavelength.

Alternatively, semiconductor quantum cascade laser technology offers laser lines between 3 and 25 μm, with the added benefit of the possibility to tune the wavelength [34]. Recently, applications of such lasers in various digital holography applications were demonstrated [35], among which two-wavelength digital holography, which can be useable for contouring [36].

For the recording of the LWIR digital holograms, the two important classes of infrared focal plane arrays (FPAs) that can be used in LWIR imaging are based on photon detection or on thermal detection [37]. The photon detection class is mostly represented in LWIR by two technologies: quantum well infrared photodetectors (QWIPs), and mercury cadmium telluride (MCT).

QWIPs are made of III-V semiconductor materials containing one or more quantum wells formed by layers of different materials with different bandgaps. Generally, the spectral sensitivity of QWIPs is in the range of 8–9 μm with a bandwidth of 1 μm, but one can find examples in literature of sensitivities centered at 10-μm wavelengths [38], making them compatible with some lines of the CO_2 laser.

MCT is made of alloys of II-VI semiconductor materials HgTe and CdTe with various stoichiometries [39]. The MCT technology can cover short-WIR, middle-WIR, and long-WIR ranges, or even higher, depending on the alloy $Hg_{1-x}Cd_xTe$ chosen. In the LWIR range, the spectral sensitivity generally extends between 8 and 10 μm with strong cutoff at the highest wavelength, making the MCTs compatible with some CO_2 laser lines. State-of-art LWIR FPAs can be found in 640 × 512 pixels format with 15–20 μm pitch (depending on manufacturer) while in the mid-WIR range, 1280 × 1024 pixels FPAs are available. Recently, 5-μm pitches were demonstrated [40]. To get optimum detection performances, QWIP and MCT technologies require cryogenic cooling operation.

The thermal detection class of FPAs is the main competitor of direct photon detection technology. Because thermal detection does not require cooling, it is often referred as the uncooled infrared detectors. The main uncooled technologies used in LWIR imaging are the pyroelectric and the microbolometer arrays [37]. One of the most used effects for infrared detection is the pyroelectric effect. The earliest technology used for LWIR imaging was that of pyroelectric vidicon (PEV) [41]. Depending on the window material sealing the tube, PEVs could image between 2 and 400 μm, hence covering the LWIR range. PEVs were used in the very first commercial 10.6 μm interferometer, which required varying the infrared light intensity by introducing a phase shift in one arm of the interferometer [42]. PEV also allowed the very first implementation of electronic hologram recording in the LWIR [43], but the author limited their investigation to objects varying due to vibrations for which the speckle irradiance was naturally varying. State-of-art pyroelectric cameras, such as Pyrocam by company Ophir, include typically 320 × 320 pixels with 75 μm pitch and they are sensitive from 1 to 3000 μm wavelengths. Pyroelectric cameras are currently used as infrared laser beam profiler. Such a camera, although with a smaller resolution of 124 × 124 pixels, was used for the very first demonstration of LWIR lensless digital holography [44]. The other main, and nowadays more prominent uncooled technology, is the one of microbolometer FPAs.

A bolometer exhibits electrical resistance variation when its temperature changes due to absorption of light energy [37]. Microbolometer-based FPAs are made of silicon, which is micromachined to obtain a pixel structure. The very first use of a microbolometer FPA in holography was demonstrated in 2007 in an in-plane electronic speckle pattern interferometry with a 320 × 240 pixels camera [45] and in lensless digital holography with similar resolution [46]. The rapidly growing market of uncooled thermal camera with increasing performances at affordable prices (compared with cooled cameras) make this technology the most attractive for LWIR digital holography [47–49].

NUMERICAL RECONSTRUCTION OF DIGITAL HOLOGRAMS

General Considerations

There are mainly two ways to perform the numerical reconstruction of digitally recorded holograms. These methods are based on Basic fundamentals of Fourier optics section and the different diffraction formulas. At any reconstruction distance d_r, which may be different from d_0, the numerically reconstructed hologram corresponds to the discrete version of the following equation:

$$A_r(x,y,d_r) = (H*h)(x,y). \quad (5.41)$$

Eq. (5.41) can be interpreted as the exact convolution formula as given in Eq. (5.14). It can also be interpreted as the two-dimensional (2D) discrete Fresnel transform in Eq. (5.20).

The last way to calculate the numerical image from the hologram plane consists of using the convolution theorem: in this case, the reconstructed object can also be written as follows (*FT* means Fourier Transform):

$$A_r = FT^{-1}[FT[H] \times FT[h]], \quad (5.42)$$

or,

$$A_r = FT^{-1}[FT[H] \times G]. \quad (5.43)$$

Eqs. (5.42) and Eq. (5.43) are two other possibilities to perform the reconstruction of the object in the spectral domain. Eq. (5.42) uses the transfer function of the free-space propagation, which is the Fourier transform of the impulse response (exact or Fresnel, consider Basic Fundamentals of Fourier Optics Section). Eq. (5.43) uses the angular spectrum transfer function. This transfer function is analytically defined in the spatial frequency domain according to Eq. (5.12).

The next sections discuss on the different algorithms, such as the discrete Fresnel transform and the family of convolution algorithms.

The Discrete Fresnel Transform

The numerical reconstruction with the discrete Fresnel transform is based on the discrete version of expanded Eq. (5.20) and considering the input plane as the hologram plane. The spatial sampling in the hologram plane $(X,Y)=(np_x,mp_y)$ where $(m;n)\in(-M/2,+M/2-1;\ -N/2,+N/2-1)$ has to be taken into account, with N, M the number of pixels along, respectively, the x and y direction and p_x, p_y the pixel pitch of the sensor. At any distance d_r from the recording plane, the numerically diffracted field can be calculated according to Eq. (5.44):

$$A_r(x,y,d_r) = -\frac{i}{\lambda d_r}\exp\left(\frac{2i\pi d_r}{\lambda}\right)\exp\left[\frac{i\pi}{\lambda d_r}(x^2+y^2)\right] \times \sum_{l=-L/2}^{l=+L/2}\sum_{k=-K/2}^{k=+K/2} H(lp_x,kp_y)\exp\left[\frac{i\pi}{\lambda d_r}\left(l^2p_x^2+k^2p_y^2\right)\right] \times \exp\left[-\frac{2i\pi}{\lambda d_r}(lp_xx+kp_yy)\right] \quad (5.44)$$

Practically, Eq. (5.44) is implemented using fast Fourier transform (FFT) algorithms. In Eq. (5.44), the double discrete summation is performed over a set of $K\times L$ data points. Note that this set can be larger than the initial size $M\times N$ of the image sensor. In this case, one has to add zeros around the hologram, and this is called zero-padding. This point will be discussed further in next sections. The spatial coordinates of $A_r(x,y,d_r)$ must be sampled because one cannot compute data over an infinite number of values. From Eq. (5.44), and because the Fourier transform is performed with the FFT algorithm, the sampling pitches in the reconstructed plane depend on the physical parameters of diffraction and on the number of sampling points from the discrete Fresnel transform, respectively, in the horizontal and vertical directions, according to

$$\Delta\eta = \frac{\lambda d_r}{Lp_x} \qquad \Delta\xi = \frac{\lambda d_r}{Kp_y} \quad (5.45)$$

Thus, it follows that the spatial extents of the reconstructed field are equal to $\lambda d_r/p_x \times \lambda d_r/p_y$. So, the width of the reconstructed field is wavelength and distance dependent.

The sampling of the quadric phase that is multiplied by H in Eq. (5.44) must fulfill the Shannon condition. This means that the minimal distance $d_r^{\min}$ that can be put into the algorithm must fulfill this relation [50–52]:

$$d_r^{\min} \geq \max\left\{\frac{Np_x^2}{\lambda},\frac{Mp_y^2}{\lambda}\right\}. \quad (5.46)$$

Thus, the discrete Fresnel transform cannot be calculated for distance shorter than $d_r^{\min}$.

Note that in Eq. (5.44) we considered diffraction calculation from the hologram H. In case where the object field U_O is known at the hologram plane, then it can be inserted in Eq. (5.44) in place of H.

Alternative approaches to overcome the drawbacks introduced by the wavelength dependence and the minimum distance are provided by the convolution algorithms discussed in the next section.

The Convolution Algorithms

Reconstructed horizon

Consider Eqs. (5.42) and (5.43): the double FFT included in the reconstruction algorithm has for consequence that the reconstructed horizon (that is width and height of the reconstructed field) is related to that of the initial hologram (with or without any padding). So, the pixel pitch in the reconstructed image is simply equal to that of the sensor, $\Delta\eta = p_x$, and is not wavelength and distance dependent as for the discrete Fresnel transform (see Eq. 5.45) [53].

Spatial frequency localization

From the spatial analysis of the digital hologram, the +1 order is localized at spatial frequencies (u_0,v_0). Thus, to reconstruct from H, by convolution, the +1 order localized at spatial frequencies (u_0,v_0), one needs to check the suitable frequency localization of the convolution kernel and the object-adapted spatial bandwidth in Eqs. (5.42) and (5.43). Note that, basically, the transfer function of the convolution in Eqs. (5.42) and (5.43) is a band-pass filter centered at (0,0) spatial frequency [53]. In case where the object field is extracted at the hologram plane (see 5.3.4) and compensated for the spatial carrier frequency, then the transfer function does not require to be shifted at (u_0,v_0) in the Fourier spectrum.

Spatial bandwidth requirements

To reconstruct the digital hologram according to Eq. (5.42) or Eq. (5.43), the analytical functions require to be spatially (versions of h) or spectrally (versions of G) sampled. The analysis of the sampling requirements is related to the oscillation properties of both functions. From the convolution Eq. (5.42) and from Eq. (5.20), the impulse response of free-space propagation is a bandwidth-limited function whose spatial bandwidth is related to the finite extent on which it is calculated. Thus, the spatial bandwidth of this convolution kernel is given by Ref. [53]:

$$\Delta u_{kernel} \times \Delta v_{kernel} = \left(\frac{Np_x}{\lambda d_r}\right) \times \left(\frac{Mp_y}{\lambda d_r}\right) \tag{5.47}$$

So, to fully reconstruct the object, the spatial frequency extents of the convolution kernel (G or the Fourier transform of h) must be at least that of the object, giving:

$$\Delta u_{kernel} \times \Delta v_{kernel} \geq \Delta u_{object} \times \Delta v_{object} = \left(\frac{\Delta A_x}{\lambda d_r}\right) \times \left(\frac{\Delta A_y}{\lambda d_r}\right), \tag{5.48}$$

where ΔA_x, ΔA_y are, respectively, the object width along the x and y directions. Thus, from Eqs. (5.47) and (5.48), it is not possible to reconstruct directly an object which size is larger than the width on which is calculated the impulse response (i.e., $Np_x \times Mp_y$).

So, the requirement is that the spatial frequency bandwidth of the convolution kernel needs to be extended to fit that of the object. The possible strategies are describes here after.

Zero-padding of the impulse response

A simple way to increase the spatial frequency bandwidth consists of using the zero-padding of the impulse response or of the angular spectrum transfer function, to get:

$$\Delta u_{kernel} = \frac{Lp_x}{\lambda d_r} = \Delta u_{object} = \frac{\Delta A_x}{\lambda d_r} \tag{5.49}$$

simply leading to $L = \Delta A_x / p_x$. So, the number of required data points is the ratio between the object size and the pixel pitch of the sensor.

The zero-padding of the digital hologram with $(K,L)>(M,N)$ leads to extending the horizon for the calculation of the impulse response of the free-space propagation, or to an oversampling of the angular transfer function.

As an example, let us consider the numerical reconstruction of the digital hologram of an object sized $\Delta A_x = \Delta A_y = \Delta A = 60$ mm that is recorded with a pixel pitch at $p_x = p_y = 4.65$ μm. Given $L = \Delta A_x / p_x$, one needs at least 12,000 × 12,000 data points to reconstruct the full object, which is greater than the calculation capacity of standard PC computers. So, the approach based on the zero-padding remains limited to objects having a "moderate" size, that is, not too large compared with the sensor width [53].

Adjustable magnification

Another method to extend the spatial frequency bandwidth is based on modifying the reconstruction distance. This approach is quite useful when the object size is much larger than the sensor width, and when the zero-padding approach leads to prohibitive matrix sizes. For a given value of $\{K,L\}$, if d_r decreases, then the spatial bandwidth of Eq. (5.49) increases. The modification of the reconstruction distance can be obtained by using a spherical reconstruction wave, instead of a plane wave as a numerical wave impacting the hologram. This means that one needs to consider H multiplied by a spherical wave (see Eq. 5.2) in Eqs. (5.42) and (5.43).

The notion of "adjustable magnification" in digital holography is related to the field of view (sized $Lp_x \times Kp_y$) that is reconstructed by the convolution. The reconstruction of an extended object, whose physical size $\Delta A_x \times \Delta A_y$ is larger than the field of view, is expected. In order for the object to fully appear in the reconstruction horizon, a transversal magnification $\gamma = \min(Lp_x/\Delta A_x; Kp_y/\Delta A_y)$ must be applied; it is the ratio between the horizon and the object widths. So, one may conceive that the adjustable magnification method results in adjusting the reconstructed object size to that of the field of view, which is imposed by the calculation capacities or the computation speed.

Note that several authors proposed algorithms qualified "with adjustable magnification." In 2004 [54], proposed an algorithm based on a double Fresnel transform resulting in the adjustment of the side length of the field of view. In 2010, Restrepo and Garcia-Sucerquia [55], discussed the adjustable magnification using the Fresnel-Bluestein transform. Both methods are based on the Fresnel transform (single or double) and the notion of magnification is linked to the ratio between the reconstructed pixel pitch and that of the sensor. The transverse magnification is here related to the ratio between reconstruction and recording distances [56]:

$$\gamma = -\frac{d_r}{d_0}. \tag{5.50}$$

Thus, the focus on the orthoscopic image (+1 order) is not obtained for $d_r = -d_0$, but for a different distance that depends on the curvature of the recording reference wave as well as the curvature of the numerical reconstruction wave. The reconstruction with adjustable magnification is obtained as follows: a spherical wave, either a reference or a reconstruction wave [25,56,57], that is multiplied to the digital hologram, will modify the reconstruction distance, which will modify the spatial frequency bandwidth of convolution, leading to an adaptation to the object bandwidth. The consequence is that the reconstructed object does get a size, which is now compatible with that of the reconstruction field of view. For further details, the reader in invited to consider [53,56].

Case of Digital Color Holograms

The reconstruction of images from the digitally recorded three-color holograms requires that the size of each individual image is the same and does not depend on the wavelength, or on any other wavelength-dependent parameter. When calculating with the discrete Fresnel transform, the pixel pitch in the reconstructed image is given by $\Delta\eta = \lambda d_r/Lp_x$ (see Eq. 5.45). So, the pixel pitch is wavelength-dependent, and the consequence is that the reconstructed image size is also wavelength-dependent. There are several approaches to reconstruct color holograms while preventing the image to vary with the wavelength: wavelength-dependent zero-padding Fresnel transform algorithm [31,58], convolution [53], convolution with adjustable magnification [56] or also spectral scanning [59].

The basic principle of the wavelength-dependent zero-padding Fresnel transform algorithm is to ensure that the pixel pitch in each reconstructed R-G-B plane is invariant that is [31]:

$$\Delta\eta_\lambda = \frac{\lambda d_r^\lambda}{L_\lambda p_x} = cste \quad \forall \lambda. \tag{5.51}$$

The adaptation is based on slightly varying the reconstruction distance with the wavelength and choosing specific zero-padding of the hologram for each wavelength.

In the case of the use of lenses between the object plane and the sensor plane, chromatic aberrations are introduced. Then, this modifies both the sensor-to-object distance and the virtual object size along each wavelength. Several actions were taken in the past years to correct optical aberrations in digital holography. In 2000, Stadelmaier et al. proposed a technique using a quadratic phase term to get rid of the spherical aberration [60]. In 2006, Colomb extended this approach and proposed a method for full compensation of geometrical aberrations using a numerical parametric lens, whose parameters were based on a polynomial decomposition [61]. For chromatic aberrations, De Nicola proposed a method to equalize the pixel pitch in reconstructed color images, to retrieve the corrected superposed phase images. This method used a zero-padding approach that depends on the wavelengths used in the experiment [62,63].

The opportunity provided by digital color holography is that chromatic aberrations can be corrected in a pure numerical way. An all-numerical robust and simple method to compensate for chromatic aberrations in a digital three-color holographic scheme can be as follows: the method is based on a full numerical process and combines both a modified version of the zero-padding algorithm and the convolution with adjustable magnification. For the calibration of the algorithm, the reference test object that has to be used is a rectangular reference grid with a good contrast, but any object containing parallel lines can be also acceptable. The first step is to reconstruct, with the Fresnel transform, the three-color image with the same pixel pitch using the zero-padding algorithm at the different distances given by the chromatic aberration. Thus, the reconstructed images have got the same physical reconstruction horizon (i.e., the reconstructed area does not depend on λ). However, they are slightly laterally shifted and exhibit different sizes. The lateral shift can be estimated by measuring the center of the grid for each wavelength and will be compensated at the next step with the convolution algorithm.

Using the Hough transform, one has access to the equations of any straight line in the test target [64]. So, one gets an average value of the distances between each vertical and horizontal line of the grid, thus providing the measurement of the grid period. If one considers the blue color to be the reference image, then one has to compensate for the red and green color. The ratio $G(\lambda)$ of the values of the estimated periods for the red and green images in regard of the blue, one determines the transverse magnification that has to be applied to the green and red image to get the red, green, blue images with the same size.

The last step is to use the estimated transverse magnification and lateral shift position as inputs in the convolution algorithm with adjustable magnification [56]. Finally, the magnification $G(\lambda)$ between colors is used to get the same image sizes from the convolution algorithm by retrieving the new reconstruction distance d'_r.

$$d_r'^{\lambda} = G(\lambda) d_0^{\lambda}. \tag{5.52}$$

To illustrate the reconstruction algorithm fully described in Ref. [64], Fig. 5.8A shows the superposition of the three R-G-B individual images from the reference test object (grid). Fig. 5.8B shows that a lateral shift clearly exists. Fig. 5.8C and D shows the result of the correction of the lateral shift obtained by making the center of the grating the new center of the reconstructed images: one can also observe that the size of the color images depends on the wavelength, as the interspaces between each line differ from one color to the other. There is a chromatic error because the images are different in size. Fig. 5.8E and F shows the results obtained at the final step of the compensation approach, using the convolution algorithm with adjustable magnification. The three images are then perfectly superposed, giving the opportunity to reconstruct color images without any chromatic error.

The reconstruction of digital color holograms from three-color objects is illustrated in Fig. 5.9 with the colored Chinese porcelain masks (Beijing Opera) around 50 mm high and containing different colors. Fig. 5.9A shows a snapshot of the mask using a classical color camera with an imaging lens. The superposition of the three-color images reconstructed simply using the Fresnel transform is shown in Fig. 5.9B. The superposition of the three-color images corrected from the chromatic aberrations with the proposed method is shown in Fig. 5.9C.

Fig. 5.9 clearly demonstrates the improvement of the full compensated image because details combining different colors are perfectly restituted.

PROPERTIES OF IMAGES RECONSTRUCTED FROM DIGITAL HOLOGRAMS

Speckle in Digital Holograms

When using light sources having a large coherence length, such as single longitudinal mode lasers, the nature of the surface (or thickness) of the illuminated sample may generate a speckle pattern. This effect is easy to comprehend: each "point" at the surface of the object emits a wavelet onto the sensor, and this wavelet has a random phase because of the surface roughness (refer to Fig. 5.4). At the detector plane, all emitted wavelets are coherently mixed, and they produce a phenomenon of multiple random beam interferences. So, the object wave at the detector plane is a speckle wave. It follows that digital holograms of a naturally rough object are constituted of a speckle pattern modulated by microfringes, originated from the coherent interference between the speckle field and the reference wave (supposed to be free of speckle).

Statistics of speckle in the hologram plane can be determined by considering that the speckle field generated by the object surface is a random Gaussian process with zero mean. In the off-axis configuration, the speckle grain covers several pixels in the recording plane. Typically, a single grain covers 3 to 4 pixels (or maybe more), so that speckles can be considered as "resolved." In the on-axis configuration, the speckle grains are about the same size of the pixel so, the speckle is considered "unresolved" [65].

In the case of resolved speckles, the statistics of the +1 order in the hologram plane follow a negative exponential for the intensity, $I = |U_r U_O|^2$, and the

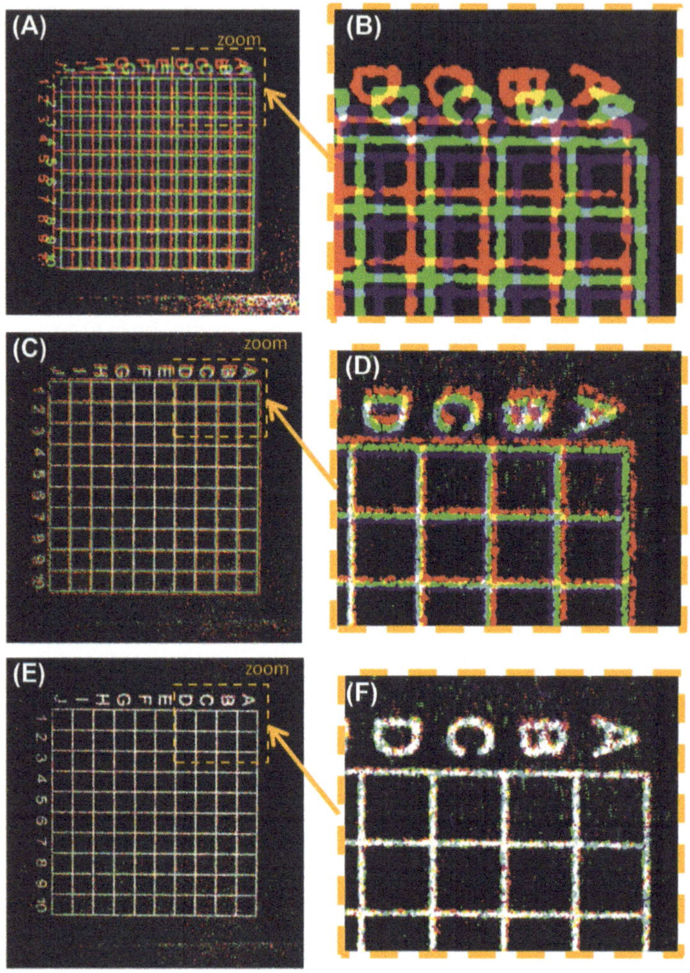

FIG. 5.8 **(A)** Three reconstructed images superposed without any correction, **(B)** zoom of **(A)**, **(C)** three reconstructed images superposed using the convolution algorithm with correction of the lateral shift only, **(D)** zoom of **(C)**, **(E)** three images superposed after the adjustable magnification, and **(F)** zoom of **(E)**. (From M. Leclercq, P. Picart, Method for chromatic errors compensation in digital color holographic imaging, Optics Express 21 (22) (2013) 26456–26467.)

Rayleigh law for the amplitude, $a_O = |U_r U_O|$, according to Eqs. (5.53) and (5.54):

$$p_I(I) = \left\{ \frac{1}{2\sigma^2} \exp\left(-\frac{I}{2\sigma^2} \right) \quad I \geq 0 \right\} \tag{5.53}$$

$$p_a(a_O) = \frac{2a_O}{\langle I \rangle} e^{-\frac{a_O^2}{\langle I \rangle}} \quad a_O 0. \tag{5.54}$$

In Eq. (5.53), we have $\sigma^2 = <I>/2$ and $< \ldots >$ means ensemble average. It follows that the speckle contrast, which can be estimated by calculating $C = \sigma/<I>$, is equal to 1. The phase of the +1 order, $\varphi_O = \arg\{U_r^* U_O\}$, is random and uniformly distributed over $[-\pi, +\pi]$ (uniform value at $1/2\pi$).

In the case of unresolved speckles, the active surface of the pixels plays a role in the speckle statistics in the hologram plane due to its low-pass filtering action. The probability density function of the intensity is modified and depends on the number of grains per pixel, according to Eq. (5.55):

$$P_I(I) = \left(\frac{n_0}{I} \right)^{n_0} \frac{1}{\Gamma(n_0)} I^{n_0 - 1} \exp\left(-\frac{n_0}{I} I \right). \tag{5.55}$$

FIG. 5.9 **(A)** Color photography, **(B)** superposition of the noncorrected reconstructed images, and **(C)** superposition of the corrected reconstructed images of a green and purple Chinese porcelain mask. (From M. Leclercq, P. Picart, Method for chromatic errors compensation in digital color holographic imaging, Optics Express 21 (22) (2013) 26456–26467.)

where Γ is the Gamma function and $n_0 = \langle I \rangle^2/\sigma^2 = C^{-2}$ can be interpreted as the "number of speckle grains per pixel."

To illustrate the speckle nature of a digital hologram, Fig. 5.10A shows a digitally recorded off-axis hologram of a rough object. Fig. 5.10B shows a zoom exhibiting the microfringes in the hologram, which are due to the spatial carrier frequency of the reference wave (off-axis configuration). Fig. 5.10C shows the +1 order extracted from the hologram after Fourier transform and filtering (see 5.3.4), and Fig. 5.10D shows the phase of the +1 order. As it can be observed, both data are speckled. The inset in Fig. 5.10D is a zoom showing the phase dislocations in the speckle phase (combination of random vortices).

Speckle in Holographic Images

The speckle pattern in the digital hologram is quite useful and necessary to encode any natural rough object. The drawback is that speckle also becomes a strong enemy in the reconstructed image, both for amplitude and phase images, because it affects the quality of the reconstructed field, and appears as parasitic noise. Practically, the reconstructed amplitude image from objects having a rough surface is not uniform and exhibits dark and bright gains due to the speckle phenomenon (even if the initial object surface exhibits uniform brightness). Then, the speckle noise is a multiplicative noise on the intensity image reconstructed from the digital hologram. Illustration is provided in Fig. 5.11. Fig. 5.11A shows the reconstructed intensity image of the digital hologram of a loudspeaker. A zoom exhibits the speckle pattern as the fine structure of the image. Fig. 5.11B shows the reconstructed phase image of the loudspeaker, and the zoom exhibits its random nature.

The statistics in the digitally reconstructed image follows the same laws as the statistics of the speckle pattern in the hologram plane. Fig. 5.12 shows the probability densities evaluated from the +1 order in Fig. 5.10 and from the reconstructed images from Fig. 5.11. Fig. 5.12A and B exhibits the probability density of the intensity and phase of the +1 order in the hologram plane. Comparison with the theoretical equation is also provided and the agreement with the experimental curve is quite good. Fig. 5.12C and D show the probability density of the intensity and phase of the image in the reconstructed plane. Experimental and theoretical curves exhibit a very good agreement.

In Section Reduction of Noise in Digital Holographic Images, discussion on how to deal with the speckle noise in digital holographic images will be provided.

Spatial Resolution in the Reconstructed Plane

The spatial resolution in digitally reconstructed image depends on the initial hologram width and physical parameters of diffraction. The limiting factor is related to the computation of the reconstructed field using a finite number of pixels, which induces a truncation effect. Especially, the filtering function of the 2D discrete

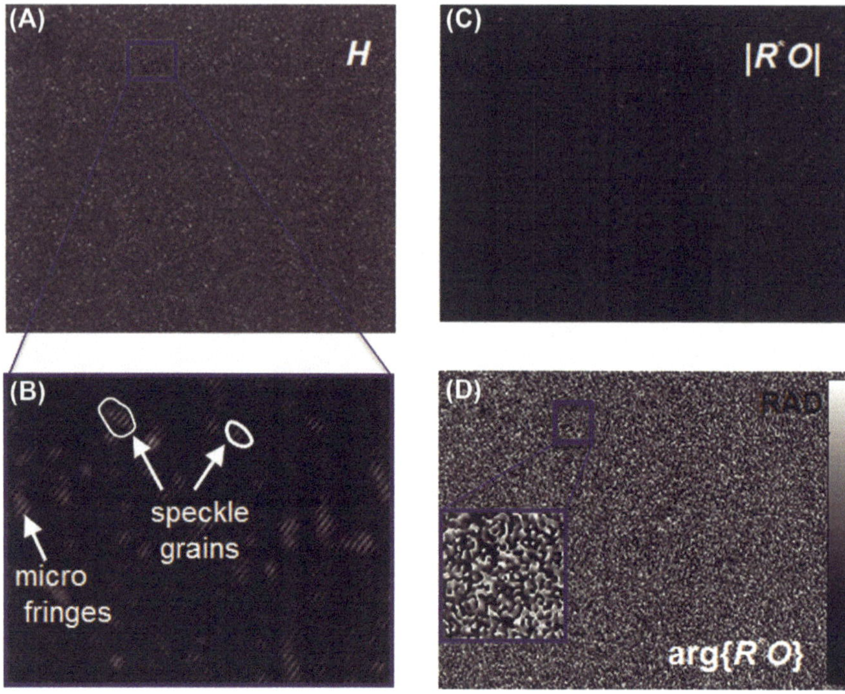

FIG. 5.10 **(A)** Digital hologram of a rough object, **(B)** zoom of a part of the hologram showing the microfringes and the speckle grains, **(C)** +1 order at the sensor plane extracted from the hologram after Fourier transform and filtering (refer to Off-axis Digital Holography and Fourier Filtering section), and **(D)** phase of the +1 order, insert: zoom showing phase dislocations.

Fourier transform is of interest, which limits the achievable spatial resolution in the reconstructed plane. This function can be interpreted to be the numerical diffraction pattern of the rectangular aperture constituted by the recorded hologram. The width of this function yields the *intrinsic spatial resolutions* in the reconstructed plane. The spatial resolutions in the horizontal and vertical directions are given by Ref. [67]:

$$\rho_x = \frac{\lambda d_r}{N p_x} \qquad \rho_y = \frac{\lambda d_r}{M p_y} \tag{5.56}$$

So, the spatial resolution depends on the wavelength, number of useful sampling pixels in the hologram ($M \times N$) and on the reconstruction distance. Note that when padding the digital hologram, the sampling pitches in the image are given by Eq. (5.45). If $K \times L > M \times N$, the pixel pitch is smaller than the spatial resolution, and the image definition is increased. But if $K \times L < M \times N$ both the resolution and the image definition are decreased. In this case, the image is pixelated, and one can observe that the image quality is strongly decreased.

Computation of the discrete Fresnel transform with $(K,L)=(N,M)$ uses the data from the raw hologram (H or U_O). If $(K,L) \geq (N,M)$, this case is called "zero-padding" and it consists of adding $(L-N,K-M)$ zeros to the hologram matrix. Fundamentally, these zeros do not add any pertinent information. Zero-padding of holograms has for consequence to make the speckles of the image appearing finely sampled but without decreasing their size. This is illustrated in Fig. 5.13 where the object is a 2€ coin 25 mm in diameter illuminated at $\lambda = 632.8$ nm which was located at a distance $d_0 = 660$ mm from the sensor ($M \times N = 1024 \times 1360$ pixels, $p_x = p_y = 4.65$ µm). The number of reconstructed points is chosen to be (512, 1024, 2048, 4096). The +1 order reconstructed with the raw hologram (without padding) is displayed in Fig. 5.13A.

In Fig. 5.13B, when $K = L = 512$, there is a strong reduction of the width of the hologram because the number of data points used for the computation is smaller than the initial matrix. In this case, the

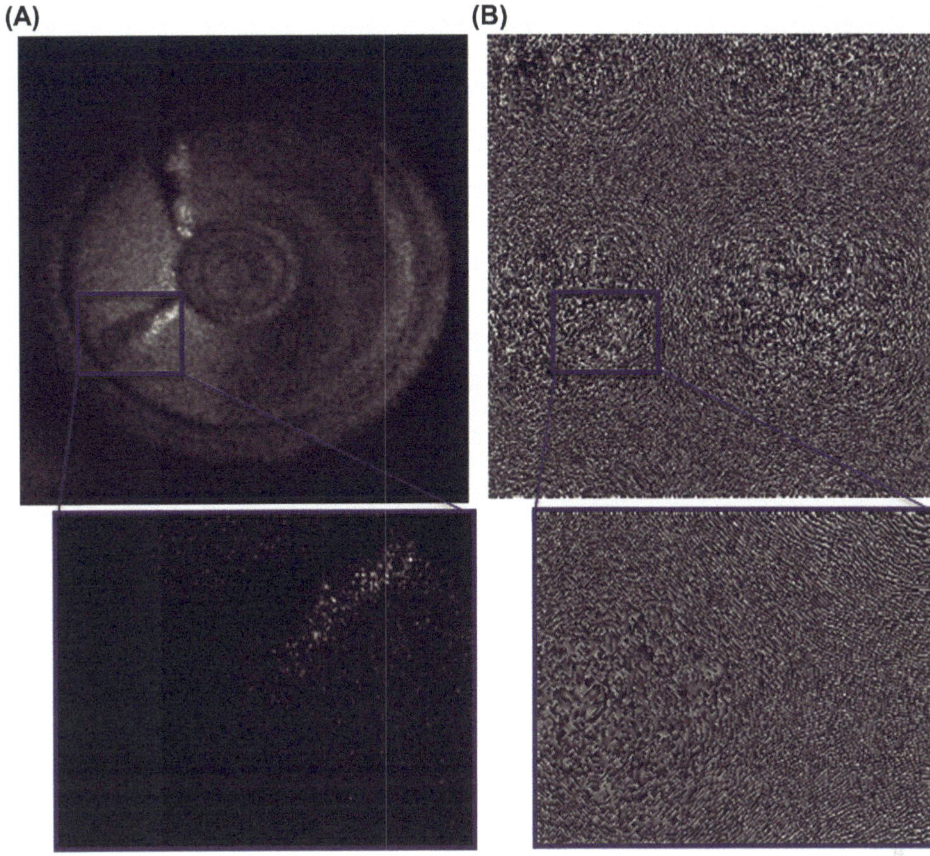

FIG. 5.11 **(A)** Reconstructed intensity image of the digital hologram of a loudspeaker and zoom exhibiting the speckle pattern, and **(B)** reconstructed phase image of the loudspeaker and zoom. (From V. Bianco, P. Memmolo, M. Leo, S. Montresor, C. Distante, M. Paturzo, P. Picart, B. Javidi, P. Ferraro, Strategies for reducing speckle noise in digital holography, Nature-Light Science & Applications 7 (2018) 45.)

resolution decreases, and the reconstructed image does look quite bad.

In Fig. 5.13C, when $K=L=1024$, the number of data points is approximately equal to that given by the sensor (1024 against 1360 in horizontal direction). The image sampling corresponds also approximately to the resolution. Thus, the image appears pixelated. In Fig. 5.13D, for $K=L=2048$, zero-padding is effective and image sampling (Eq. 5.45) is twice smaller than the resolution (Eq. 5.56). So, point spread function [67] is sampled with a better definition, and this allows the observation of the fine texture of the image, particularly the speckle structure.

In Fig. 5.13E, for $K=L=4096$, the image sampling is four times smaller than the resolution. The definition of the image plane is again increased, but the speckle does not change its size since it is imposed by Eq. (5.56) (i.e., ρ_x and ρ_y) [67].

Depth of Focus

Although digital holography is not a conventional imaging method, it exhibits some similarities to classical imaging. Especially, the reconstructed images include a depth of focus. The perfect focus is obtained if the spatial resolution reaches its theoretical limits. A criterion to determine the focal depth of the reconstructed image can be obtained by setting that the width of the defocusing function must be at most equal to ρ_x. If the perfect image distance is d_i, noting $\Delta z=|d_r-d_i|$, the full depth of focus on both sides of the perfect image plane is given by:

$$2\Delta z \cong \frac{2\lambda d_i^2}{N^2 p_x^2} \tag{5.57}$$

Thus, the focal depth in digital holography is proportional to the square of the angular aperture of the sensor as seen from the object [17].

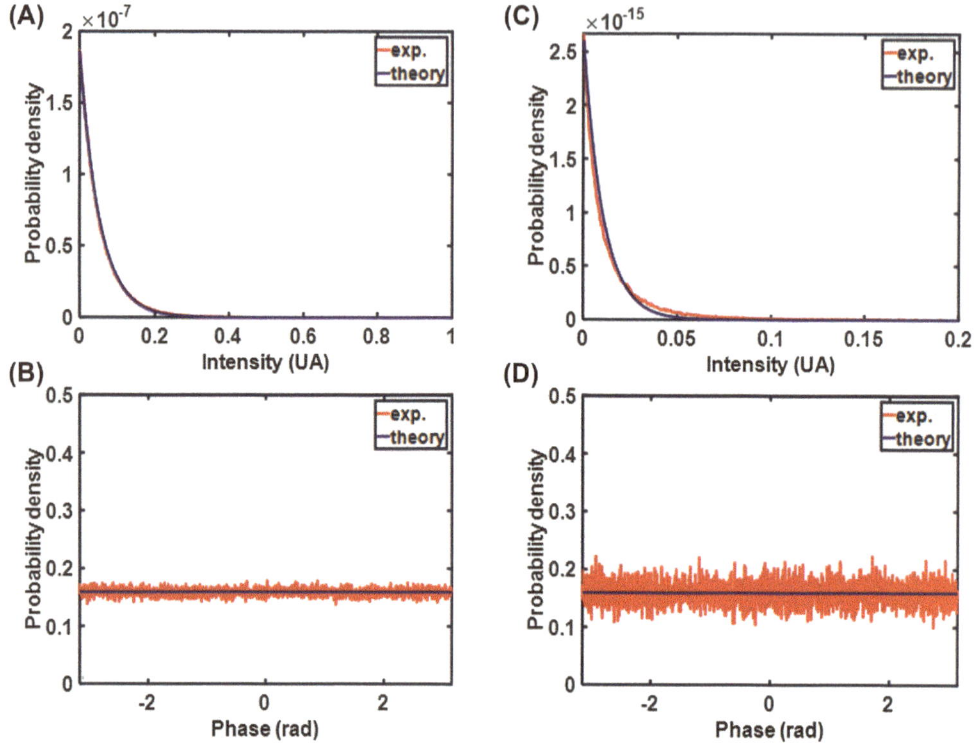

FIG. 5.12 **(A)** Probability density of the intensity of +1 order in the hologram plane, **(B)** probability density of the phase of +1 order in the hologram plane, **(C)** probability density of the intensity of the image in the reconstructed plane, and **(D)** probability density of the phase of the image in the reconstructed plane; experimental (red) and theoretical (blue) curves are provided. (From V. Bianco, P. Memmolo, M. Leo, S. Montresor, C. Distante, M. Paturzo, P. Picart, B. Javidi, P. Ferraro, Strategies for reducing speckle noise in digital holography, Nature-Light Science & Applications 7 (2018) 45.)

An illustration is provided with the digital hologram of the 2€ coin (see Speckle in Holographic Images section). The theoretical ideal focused image is obtained with $d_i = -660$ mm, and the spatial resolution is $\rho_x = 87.7$ μm in the reconstructed image plane. Then, the depth of focus is estimated at $\Delta z \approx 12.15$ mm. Fig. 5.14 provides a set of reconstructed images. Fig. 5.14A corresponds to the best focus image, and Fig. 5.14B corresponds to a 50 mm out of focus image with $d_r = d_i - 50$ mm. The image is highly blurred. Fig. 5.14C and D correspond to reconstructions symmetrically included in the depth of focus: the images are seen focused even if there does exist slight difference between the two images. Fig. 5.14E and F correspond to reconstructions slightly outside the depth of focus: the images become strongly blurred by the defocusing [67], whose width increases with the increase of $|d_r - d_i|$. The main conclusion from Fig. 5.14 is that if the reconstruction distance is slightly different than the ideal one, there is no significant difference between reconstructed images.

Decorrelation Noise in Doppler Phase Images from Holographic Interferometry

When using digital holography for quantitative measurements, a subtraction between the current object phase and a "reference" (or original) object phase is carried out. The obtained phase difference is given modulo 2π, and phase unwrapping is then required [68]. Further details on the method are provided in Chapter 6 in this book. It follows that a speckle phase decorrelation occurs between the two states of the object (current and original) and induces a noise in the phase variation between these two states. The speckle phase decorrelation noise is related to the correlation property of the two speckle fields given at the two object states. This correlation property is related to the second-order probability density function of the phase [69–71]. The

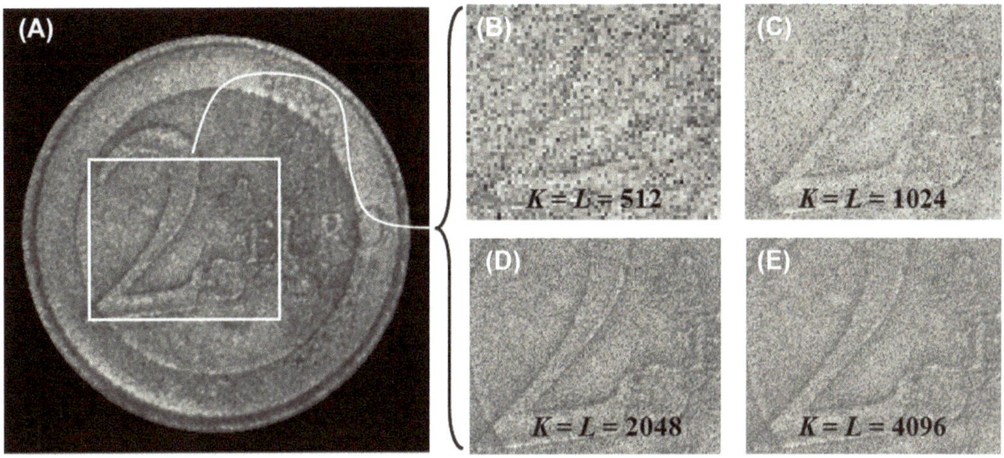

FIG. 5.13 **(A)** Reconstructed +1 order of the digital hologram of a 2€ coin using the raw hologram, **(B)** reconstruction obtained when the initial hologram is truncated to 512×512 pixels, **(C)** reconstruction obtained when the initial hologram is truncated to 1024×1024, **(D)** reconstruction obtained when the initial hologram is padded to 2048×2048, and **(E)** reconstruction obtained when the initial hologram is padded to 4096×4095. (From P. Picart, J. Leval, General theoretical formulation of image formation in digital Fresnel holography, Journal of the Optical Society of America A 25 (2008) 1744–1761.)

probability density function of the phase noise ε depends on the modulus of the complex coherence factor $|\mu|$ between the two speckle fields. Parameter $|\mu|$ indicates the correlation between the two speckle fields. With $\beta = |\mu|\cos(\varepsilon)$, the second-order probability density of the phase noise ε is given by Refs. [69–71]:

$$p(\varepsilon) = \frac{1-|\mu|^2}{2\pi}\left(1-\beta^2\right)^{-3/2}\left(\beta\sin^{-1}\beta + \frac{\pi\beta}{2} + \sqrt{1-\beta^2}\right). \tag{5.58}$$

Fig. 5.15 shows curves of Eq. (5.58) for different values of the coherence factor. If $|\mu| = 0$, the two fields are not correlated, and their phase difference is uniformly distributed over interval $[-\pi,+\pi]$, thus noise level in the phase measurement is high. In the case where $|\mu| = 1$, the fields are fully correlated, and the phase difference is close to 0, thus noise is weak, and the probability density is narrower but is not completely similar to a Gaussian distribution. It is important to note that the decorrelation noise is not a Gaussian noise [72–74], and, in addition, this is not a stationary noise. This means that the local standard deviation of noise (and also the coherence factor), in any phase map, depends on the fringe density.

The decorrelation noise may have several origins: modifications of the object surface (due to loading such as mechanical, vibrations, heating, pneumatic, …) [75–77], laser wavelength change between exposures (e.g., surface shape measurement) [78], reduced number of recording pixels (low resolution of the sensor) [74], defocusing of the reconstructed image (the reconstruction distance is "not good") [79], nonlinearity in the recorded hologram [80], quantization effect due to a low number of useful bits or shot-noise if the number of photoelectrons is not high [73].

REDUCTION OF NOISE IN DIGITAL HOLOGRAPHIC IMAGES

General Approach

As was discussed in the previous sections, the holographic image quality is degraded by undesired artifacts due to the coherent nature of the illuminating light. This encouraged many research groups to work on the topic of noise reduction in holographic images, and a large number of solutions, both numerical and optical, were proposed to overcome this problem [72,73,80–100], [101–121].

Considering the wide variety of methods developed to tackle the problem of denoising in coherent imaging systems, the different solutions could be classified according to the proper attributes they share. Method to denoise holographic images can be divided in two main categories: optical methods involving the concept of "multilooks" (MLs) [81,82,84,95–104], and

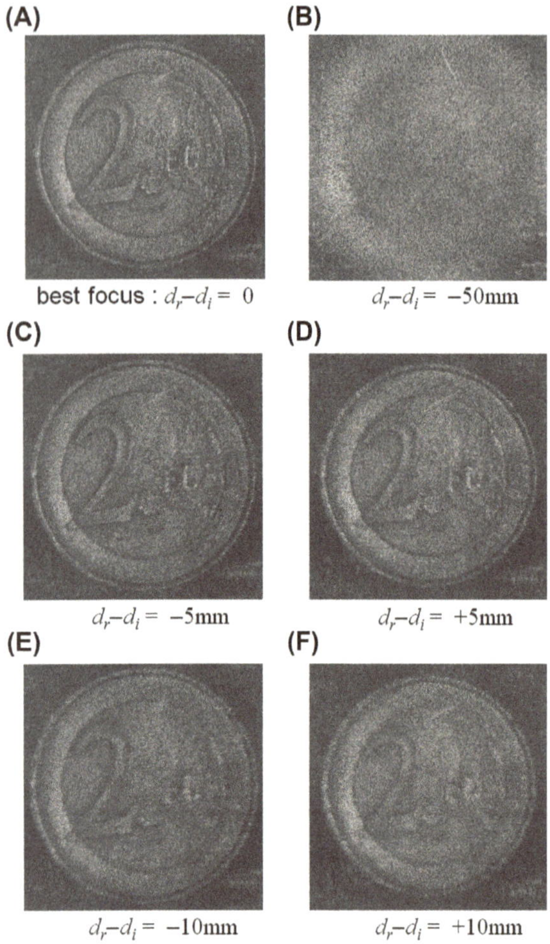

FIG. 5.14 **(A)** Best focus image, **(B)** out-of-focus image with $d_r = d_i - 50$ mm, **(C)** and **(D)** reconstructions symmetrically included in the depth of focus, and **(E)** and **(F)** reconstructions slightly outside the depth of focus.

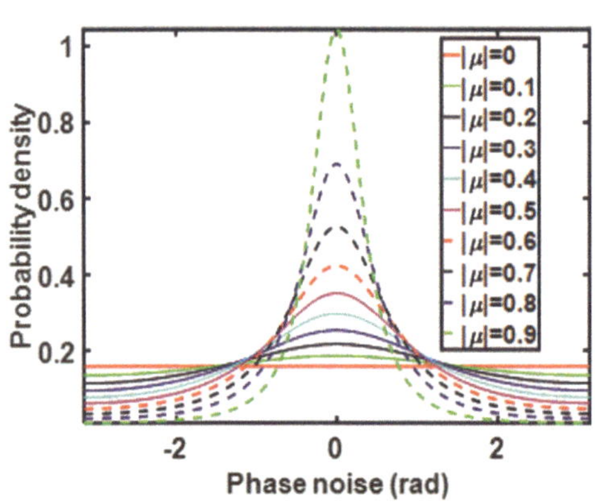

FIG. 5.15 Plots of curves from Eq. (5.58) for different values of the coherence factor μ.

numerical methods mainly based on advanced digital filters [85,90,105–109,122].

The $1/\sqrt{N}$ Law

Consider Eq. (5.55) describing the probability density function of the intensity in the speckle field when the speckles are unresolved, and in which n_0 can be interpreted as the "number of speckle grains per pixel." In Eq. (5.55), the relation $C = 1/\sqrt{n_0}$ holds and means that the speckle contrast decreases when averaging several speckle grains according to what is called the "$1/\sqrt{N}$ rate," where N is number of averaged speckle grains.

Fig. 5.16 shows the probability density of intensity from Eq. (5.55) versus several values of the speckle contrast. When the speckle contrast is equal to 1, the probability density of intensity follows the negative exponential from Eq. (5.53). When the speckle contrast decreases, meaning speckles are averaged, then the probability density moves progressively from the negative exponential to a Gaussian-type curve (e.g., case $C = 0.1$). This means that when averaging the speckles in any image, the residual noise tends to be a Gaussian noise, whereas this is not the case for the raw speckles.

As a general rule, the decreasing rate is considered as a reference rule to evaluate how efficient speckle reduction processes are. Indeed, when averaging several non-correlated speckle patterns, the contrast decreases and, as an ideal way, follows the "$1/\sqrt{N}$ rate." So, an efficient denoising process should be able to follow this rate.

The next section provides an overview of advanced filtering algorithms.

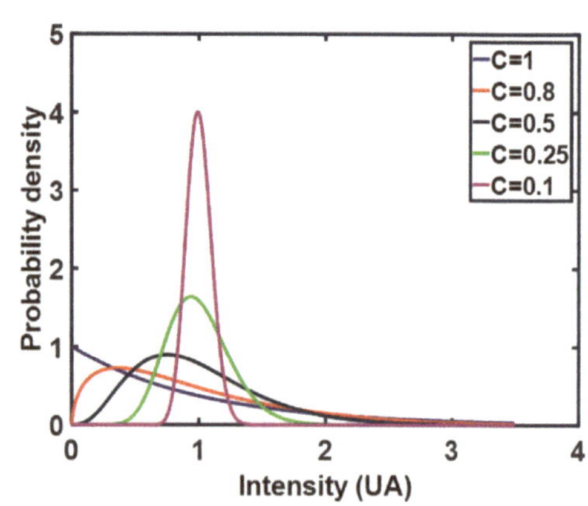

FIG. 5.16 Plots of several probability density functions of intensity from Eq. (5.55) versus values of the speckle contrast (0.1, 0.25, 0.5, 0.8, 1).

Overview of Advanced Filtering Algorithms

Spatial filters

Spatial filtering is the most commonly applied filtering in digital interferometry [123,124]. To preserve the phase jumps, filtering is performed on the sine and cosine of the phase. Several types of kernels can be considered: the moving average filter, the median filter (efficient for impulsive noise), or the Gaussian filter. The size of the kernel may vary from 3×3 to 21×21 for images at 512×512 or 1024×1024 pixels. These filters are generally very easy to implement and they constitute reference methods for comparing other more complex algorithms.

Wiener filter

The Wiener deconvolution is a mathematical operation applying a filter to eliminate the noise in any signal. This filtering operates in the frequency domain by trying to minimize the impact of noise where the signal-to-noise ratio (SNR) is bad. From a mathematical point of view, this is the solution, in the least square sense, of the minimization of the mean square error computed between the original image $s(x,y)$ and the filtered image $d(x,y)$, in the case of linear spreading model with additive noise $b(x,y)$. According to Ref. [125], the transfer function of the Wiener filter, $W(u,v)$, is given by Eq. (5.59):

$$W(u,v) = \frac{F(u,v)\Phi_s(u,v)}{|F(u,v)|^2\Phi_s(u,v) + \Phi_b(u,v)}. \tag{5.59}$$

In Eq. (5.59) the spreading function $F(u,v)$ can be chosen to be unitary, Φ_s is the power spectrum density of the original image and Φ_b is the power spectrum density of noise. The power spectrum densities of the original image and the noise are not known a priori. The median filtering is applied on the noisy image to get estimations of both the noise and the original image. The original image is simply estimated by applying the median filtering on the noisy image. Noise is estimated by calculating the difference between the raw noisy image and the median-filtered noisy image.

Synthetic aperture radar filtering

Synthetic aperture radar (SAR) filtering can be applied to speckle images because intrinsically SAR images include the same properties as that of speckle images, the noise model being considered as multiplicative. Lee proposed an adaptive filter [126] in which the parameters are adapted according to the local statistical properties of the image to be processed. These parameters are estimated in a window centered on the pixel for which one expects restoring the correct values. Computation of the filtering can be described by Eq. (5.60):

$$d(x,y) = \alpha\ s(x,y) + (1 - \alpha\mu_s). \tag{5.60}$$

The parameter $\alpha = 1 - C_b^2/C_s^2$ depends on the ratio of the squares of the local variation coefficients of the image, C_s, and noise, C_b. This local variation coefficient is defined as the ratio of the variance between the mean square of the considered pixel values. In a region with homogeneous values, the local variation coefficient is low (≈ 0), and the filter does not modify the image. In the opposite case, if this variation is larger than that of the noise, the current value of the pixel is considered.

An alternative approach was proposed by Frost [105]. The filter consists of an adaptive Gaussian filter kernel. The kernel parameters are adjusted according to the local statistics of the image by considering the local variation coefficient C_s^2. In the presence of an edge or a discontinuity, the coefficient increases and the Gaussian kernel is concentrated around the current pixel, ultimately its original value is retained. In a region with homogeneous values, the local variation coefficient is low (≈ 0), and the filter behaves as a simple averaging. If the variation in the observed image is less than that of the noise, α is equal to 0, and the average of the window is affected at the value of the pixel.

Wavelet thresholding approaches

Image enhancement can also be considered in the sense of decomposition onto a wavelet basis associated with a threshold function [127,128]. The basic idea is to apply a threshold operator on the wavelet coefficients computed from the image to be restored. If the wavelet basis is adapted to the signal contents, its representation in the wavelet coefficients space is sparse. This means that the highest coefficients concentrate the information about the image. Then, applying a threshold leads to leave off the least significant coefficients, which are supposed to be related to the noise.

In such an approach, the model considers the noise to have an additive property. Using the wavelet transformation, the same process is applied to the coefficients. When the noise is not additive but multiplicative, additive signal and noise are obtained by computing the logarithm. This was studied in the case of speckle noise processing in SAR images [129]. Multiple variants of solutions, not detailed here, are available by adjusting many parameters of this method: choice of the wavelet basis, number of decomposition levels, profile of the thresholding operator, threshold values, and their different associated estimator. Also separable wavelets can be considered such as *Daubechies* and *symlets* built from the tensor product of their one-dimensional release, *curvelets* and *contourlets* [130,131] built in a circular paving plane, thus offering a wide range of orientation waveforms. Note that such an approach was applied to the denoising of noisy speckle interferograms [132–134].

Nonlocal means

Nonlocal means methods (NL-means) were recently proposed [135,136]. It constitutes an efficient technique for the enhancement of images corrupted by additive noise. The principle is to consider the pixel value replacement by a weighted sum of values included in patches that are chosen in the neighborhood of the pixel to be processed. The weighting is obtained from correlations estimated between the local patch of the pixel to be processed and the patch of the neighborhood patches of pixels that are taken into account in the algorithm. Thus, when a patch is strongly correlated with the local patch of the pixel to be processed, the pixel is taken into account in the weighting with a coefficient evaluated from a Gaussian kernel which argument is the Euclidean distance between the two patches. Such method has been used with success for denoising speckle images from digital holography [106].

Block matching 3D

Since its introduction by Dabov [137,138], the transform-based block-matching 3D (BM3D) filter is recognized as state-of-the art in image denoising. The important point to address is that the BM3D filter combines various efficient techniques and synthesizes major advances from recent years [139]. Among them, the approaches based on the decomposition of the image in patches, such as NL-means method [135] and the use of shrinkage operators applied on image transform domain as wavelet representations.

The BM3D filter exploits a particular NL image modeling through a procedure termed grouping and collaborative filtering. Grouping finds mutually similar 2D image blocks and stacks them together in 3D arrays. Collaborative filtering produces individual estimates of all grouped blocks by filtering them jointly, using transform-domain shrinkage operator of the 3D arrays. BM3D relies both on local and NL characteristics of images. It follows the presence of numerous similar patches and that the image is locally highly correlated. When these assumptions are verified, a sparse representation of the group is obtained by applying a decorrelation 3D transform. The sparsity of the true image determines the success of the shrinkage operator to separate the noise from the observation, and it means that its energy is compactly represented in the 3D transform domain.

These initial operations constitute in fact the first stage of the overall method. The estimated true image is then used to perform a second stage filtering such as a Wiener-like operation onto the coefficients of the 3D transform of the similar stacked patch groups.

As a final stage, an aggregation procedure is performed on the estimated true patches to compute the resulting value of pixels by summing patches using a weighting in a similar way as uses in the NL-means method. The BM3D algorithm is based on the optimization of a very large number of parameters, at least a dozen. As such, it is difficult to measure the impact of various components separately.

2D windowed Fourier transform filtering

The 2D windowed Fourier transform filter (WFT2F) method [140,141] is based on a local Fourier transform, which can takes into account the nonstationary characteristics of the speckle noise in Doppler phase images (see Decorrelation Noise in Doppler Phase Images From Holographic Interferometry section). So, the frequency components of a fringe pattern can be extracted from the noise with more efficiency than with an FT computed over the entire phase maps, such as, for example, the Wiener filter method. This operation is similar to the wavelet transforms, but more adapted to a complex representation. The filtering process consists of applying a threshold to the modulus of the 2D Fourier coefficients, letting the phase unchanged. Finally, an inverse 2D windowed Fourier transform of the filtered frequency domain is computed to get the modulo 2π phase of the fringe pattern.

The WFT2F can be seen as a projection of the phase map onto 2D oscillating functions, which are localized in the spatial and the frequency domain. The discrete version of this algorithm leads to setting eight parameters, defining the window size, the threshold value, and six parameters for sampling of the frequency axis [142,143].

SPADEDH

SPADEDH algorithm was proposed by Memmolo [89], and means SPArsity DEnoising of Digital Holograms. This approach considers L_1 norm minimization algorithm and is able to suppress the noise in digital holographic images without any prior knowledge or estimation about the noise statistics [87]. During the last decade, compressive sensing (CS) approaches have brought major improvements in many fields of applications, including signal processing. CS appears as a new paradigm of sampling theory in which samples are not anymore resulting from the acquisition along a regular grid of the quantity to record but from a projection onto an operator in which the signal is sparsely represented. To select the coefficients from the representation, a sparsity measure associated to a selection algorithm must be applied. L_0 and L_1 norms have been shown to be the appropriate measures. In certain conditions, L_0 and L_1 norms lead to the same solutions.

There exist two main families of selection algorithms: greedy and minimization algorithms. The SPADEDH algorithm is based on a greedy approach, which is a stage-wise orthogonal matching pursuit. This algorithm is able to suppress the noise components on digital holograms without any prior knowledge or estimation about the noise statistics [81].

Denoising with Multilook Approaches

Optical multilook methods

As mentioned above, the ML methods rely on merging multiple recordings of the object with different realizations of the noise process. There are several ways discussed in the literature. In this section, optical methods are reviewed. Optical methods mean that the setup is adapted and organized to permit to record several digital holograms in which the speckle patterns are independent.

A first approach consists of illuminating the object surface with a moving diffuser [80,92,102,114]. At each position of the diffuser, the speckle in the illuminating beam changes significantly so that it is fully decorrelated from that of the previous position. One digital hologram is recorded at each position of the diffuser. When reconstructing the images with the appropriated algorithm, the amplitudes exhibit decorrelated speckle patterns. Averaging these images yields a strong decrease of the speckle noise, according to the "$1/\sqrt{N}$ rate."

In the same idea, the possibility of getting independent speckles can be obtained when slightly shifting the object illumination direction [101]. For example, a series of holograms can be obtained at different illuminating angles, which vary within small range as 1.5 degrees. This allowed getting speckle reduction at "$1/\sqrt{N}$ rate" in the final reconstructions. From shifting the object illuminations, several authors proposed to shift the object around its initial position [100,144]. In Ref. [144], it was demonstrated that shifting the object distance similar to the pixel size of the camera yielded the "$1/\sqrt{N}$ rate."

Instead of shifting the object, several authors proposed to laterally shift the sensor [99,103,104]. Thus, this is equivalent to SAR imaging in which a collection of multiple views of the scene are processed to increase the spatial resolution. In Refs. [99,103,104], averaging of the collection of the different views from the different sensor positions yields to highly contrasted and reduced speckle images obeying the "$1/\sqrt{N}$ rate."

Finally, the last way to get experimental ML is related to engineering the light source, from the point of view of coherence or polarization [96–98,118,145,146]. Especially, the use of multimode lasers or wavelength-tunable coherent light source is adapted to get the "$1/\sqrt{N}$ rate" in the averaged reconstructed images.

The speckle noise reduction depends on the number of available recordings and speckle decorrelation among them. This implies that an intrinsic trade-off exists between the setup complexity (as well as the recording time) and the achievable denoising gain. To break this trade-off, numerical approaches to ML aim at simulating noise diversity instead of acquiring it, from a single recorded hologram. This is discussed in the next section.

Numerical multilook methods

The numerical approach consists in computing the set of uncorrelated speckle patterns from one single recording. In what follows, the approaches are demonstrated to fulfill the "$1/\sqrt{N}$ rate." A first algorithm was proposed for reconstructing the captured hologram using different filtering masks in the Fourier domain, to introduce the noise diversity [90]. The basic idea is similar to that from shifting the sensor, but from the Fourier point of view. The set of laterally shifted Fourier masks yields a set of independent speckle patterns in the reconstructed amplitudes. It follows that such approach is a good emulation of the moving apertures typically used in classical holography setups to reduce the correlated noise [119].

An alternative possibility is to obtain a set of subholograms from parts of the acquired hologram [107,121,147]. The hologram can be divided in several smaller parts, with or without any spatial overlapping, each one containing a reduced aperture version of the original hologram. Reconstructions obtained from each part can be incoherently averaged. The speckle realizations are then different from one computation to another. The drawback of such approach is related to the lost in spatial resolution due to the reduction of the horizon (i.e., the width) of the original hologram.

Compared with optical approaches, the spatial jittering method [109] has similarities to object shifting [100,144], but, in this case, shifting the object is performed by several numerical reconstructions in which the Fresnel kernel is slightly shifted to an amount of the speckle grain in the reconstruction plane. In Ref. [109], the spatial resolutions of several methods [80,92,102,114] are compared when considering an USAF target. The spatial jittering approach [109] is demonstrated to be better than the resampling mask [121], but it exhibits loss in spatial resolution. To overcome this problem, random resampling masks were introduced [88]. These masks are binary random patterns that generate multiple uncorrelated holograms. If the stack of binary masks is properly chosen, the spatial resolution is preserved while reducing the noise.

To illustrate a few optical and numerical ML approaches, Fig. 5.17 summarizes three methods applied to an USAF target. The two first approaches are numerical MLs with spatial jittering [109] and subholograms [107,121,147], and the last one is the optical moving scatterer [80,82,102,109,114]. Fig. 5.17A shows the averaged image with 16 diffuser positions and is considered as the reference image. Fig. 5.17B shows the image obtained with 16 jitters at an amount of the speckle size. Fig. 5.17C exhibits the image obtained with a set of 16 masks sized 256 × 256 pixels [107,121,147]. As can be seen, Fig. 5.17C exhibits a lower resolution than Fig. 5.17B. A qualitative appraisal is shown in Fig. 5.17D, which provides profiles along the red line in Fig. 5.17A. The red profile from subholograms (Fig. 5.17C) is strongly distorted compared with the blue (Fig. 5.17A) and green ones (Fig. 5.17B).

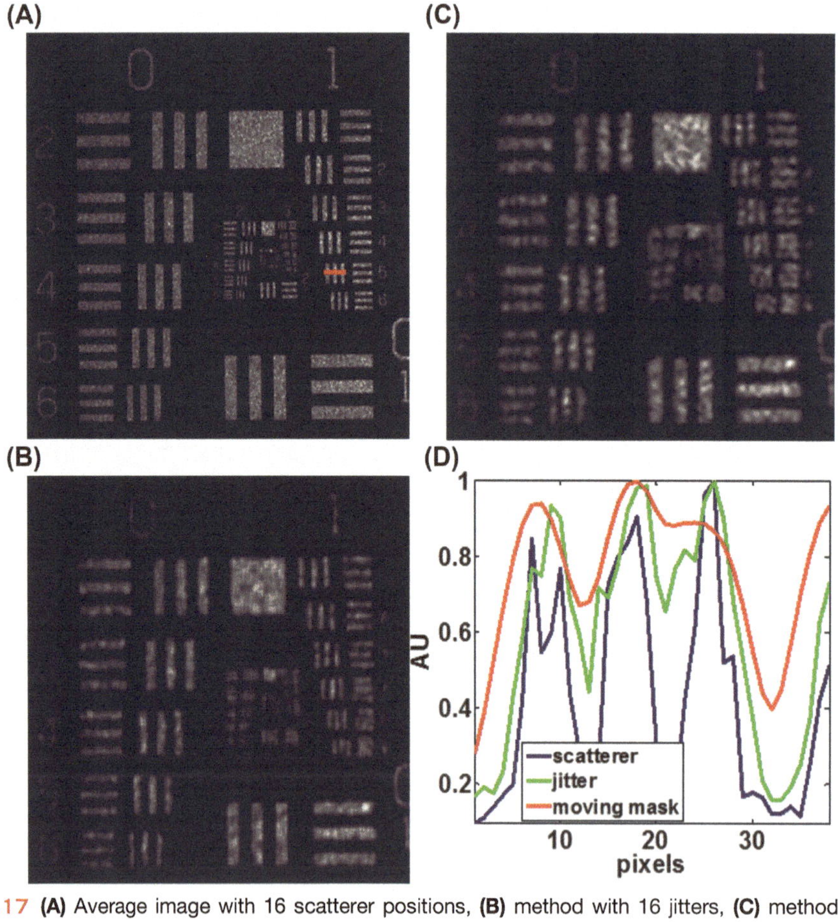

FIG. 5.17 **(A)** Average image with 16 scatterer positions, **(B)** method with 16 jitters, **(C)** method with 16 subholograms, and **(D)** three profiles along the red line in Fig. 5.17A.

Denoising with Advanced Combined Approaches

In this section, we provide an overview of the speckle denoising schemes for holographic image when considering combined digital filtering. We will also report combined approach of MLs and BM3D leading to quasi-noise-free digital holography.

A first approach was proposed by Cai [95,148] with mixing Wiener filtering and homomorphic transform. Considering that the speckle noise is not additive but multiplicative on the amplitude image, the noise is made additive by calculating the logarithm of the hologram. Noting I the image amplitude, I_0 the noise-free image and S_n the speckle noise, then the homomorphic transform leads to Eq. (5.61):

$$\ln(I) = \ln(I_0 \times S_n) = \ln(I_0) + \ln(S_n) \qquad (5.61)$$

The proposed hybrid method first sets an appropriate aperture function matching the recording parameter and the aperture size of the hologram. Then, deconvolution between the reconstructed image of the hologram and the aperture function is applied. Next, the multiplicative noise is converted into additive noise by means of homomorphic transformation according to Eq. (5.61). The applied Wiener filtering permits to reduce the additive noise (refer to Eq. 5.59). Finally, the inverse transformation yielded an effective noise reduction in digital holographic images.

In 2014, Leo proposed a denoising scheme based on combining the 2D-Empirical Mode Decomposition and the Frost filter [85]. This technique performs multiscale analysis that provides different denoising images and automatically identifies the best among these. The effectiveness of the method was simulated on the Lena image, on which speckle noise was artificially added. Numerical evaluations and quantitative information of the denoising capabilities were obtained when comparing ideal and denoised images. The authors also applied the method to digital holographic images obtained from LWIR digital holography [82,149].

More recently, Bianco proposed to combine the ML approach with the BM3D filtering [81]. They demonstrated the robustness of the approach by considering both MLs from the numerical random mask resampling [88] and from optical looks with a moving scatterer [80,92,102,114]. They concluded that the proposed approach leads to "quasi-noise-free" digital holographic images. An illustration is provided in Fig. 5.18 for one single wavelength and for three-wavelength holographic images. Fig. 5.18A shows a single-wavelength speckled image, and Fig. 5.18B shows the image obtained with the approach when applying ML and BM3D filtering. In Fig. 5.18C, the case of the raw color image with three-wavelength holography is considered. In Fig. 5.18D, the image was obtained from MLs and BM3D filtering. In both cases, Fig. 5.18B and D, the images have high quality, and the speckle noise is nearly totally eliminated.

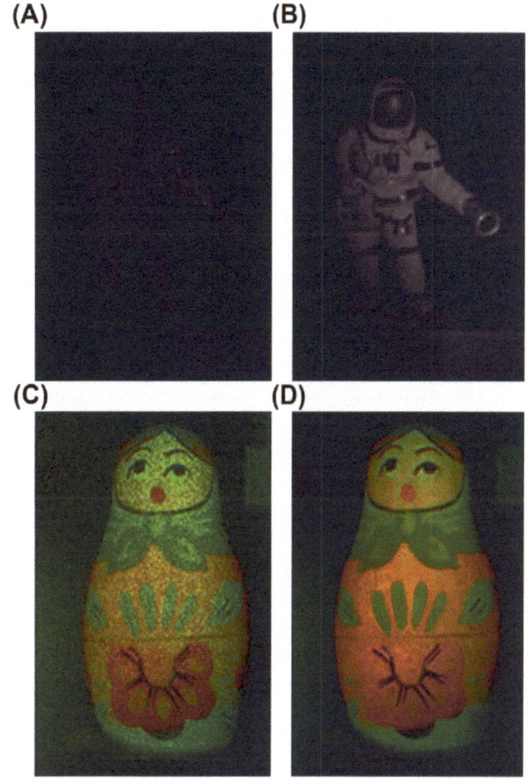

FIGURE 5.18 **(A)** Single wavelength raw holographic image, **(B)** image obtained after applying the combined approach with multilooks and BM3D, **(C)** three-wavelength raw holographic image, and **(D)** color image obtained after applying the combined approach with multilooks and BM3D. (From V. Bianco, P. Memmolo, M. Leo, S. Montresor, C. Distante, M. Paturzo, P. Picart, B. Javidi, P. Ferraro, Strategies for reducing speckle noise in digital holography, Nature-Light Science and Applications 7 (2018) 45; V. Bianco, P. Memmolo, M. Paturzo, A. Finizio, B. Javidi, P. Ferraro, Quasi noise-free digital holography, Light: Science and Applications 5 (2016) e16142.)

Quantitative Appraisal of Denoising Filters for Amplitude Images and Doppler Phases

Denoising of amplitude images

We previously discussed that the reconstructed amplitude images suffer from speckle noise. The speckle noise is a very particular noise because it is a multiplicative and non-Gaussian noise. However, denoising algorithms (see Depth of Focus section) are generally designed on the assumption of an additive Gaussian noise. So, they are not specifically dedicated to eliminate the speckle noise. Consequently, their robustness and efficiency to deal with the speckle noise has to be evaluated. The best way to deal with such an evaluation is to consider a benchmark carried out with a database composed of experimental images extracted from several holograms acquired with a random illumination [80,92,102,114]. For example, using 20 amplitude images, 20 daughter images can be computed to get images with increasing SNR.

There do exist a lot of evaluation criteria for the quantification of the denoising of amplitude images [150]. The most used criteria in image processing are the root-mean-square error (*RMSE*), the peak-signal-to-noise-ratio (P_{SNR}), and the quality index (Q_{index}). The standard deviation of the phase error (noted here σ_φ) is the most relevant of all available criteria from the literature for Doppler phase data [150,151].

The *RMSE* is based on the estimation of the standard deviation of the difference between the noise-free image ($s(i,j)$) and the denoised image ($d(i,j)$). The *RMSE* is mathematically defined as:

$$RMSE = \sqrt{\frac{1}{PQ}\sum_{i,j} r(i,j)^2 - \left(\frac{1}{PQ}\sum_{i,j} r(i,j)\right)^2}, \quad (5.62)$$

with $r(i,j) = s(i,j) - d(i,j)$ and P,Q are the number of data points in the region of interest. The P_{SNR} is given as (for images digitized with n bits):

$$P_{SNR} = 10\log 10\left[\frac{(2^n-1)^2}{\frac{1}{PQ}\sum_{i,j}(s(i,j)-d(i,j))^2}\right]. \quad (5.63)$$

The two previous criteria do not account for distortions, which may affect the structures in the image. Thus, the same amount of error can be observed between two images restored for an image having a very different perceived quality. To address this drawback, Wang and Bovik proposed what they called the "quality index" criterion [152]. This criterion has the property of better exhibiting the quality of the perceived image. Eq. (5.64) provides the useful expression to calculate the quality index:

$$Q_{index} = \frac{\sigma_{sd}}{\sigma_s\sigma_d}\frac{2\mu_s\mu_d}{\mu_s^2+\mu_d^2}\frac{2\sigma_s\sigma_d}{\sigma_s^2+\sigma_d^2}. \quad (5.64)$$

In Eq. (5.64), μ_s and μ_d are the mean values of the images $s(i,j)$ and $d(i,j)$, respectively, σ_s and σ_d are their standard deviations, whereas σ^2_{sd} is their covariance. The value of Q_{index} is included in the interval [−1,+1], the latter being reached when the two images are quite similar.

From Depth of Focus section, a set of 34 denoising algorithms were evaluated [153]. Denoising algorithms are ranked using *RMSE*, P_{SNR}, and Q_{index} metrics. Briefly, they are Gaussian filtering, median filtering, SAR filtering, Wiener filtering, wavelet thresholding approaches (such as Daubechies, symlets, curvelets, contourlets), NL-means, and BM3D. Note that the Wtfr2 is only applicable on Doppler phase images.

The database was constituted from experimental holograms acquired with the USAF1951 target as object. A moving scatterer generates a speckle pattern at the object surface. A collection of 20 holograms for 20 positions of the scatterer yields 20 noncorrelated speckle patterns. From the individual reconstructed image amplitudes from the target, one can get 20 other images by averaging 1, 2, 3, ..., 20 speckled images. By this way, the SNR increases, and the speckle statistics move from the negative exponential to the Gaussian law. From the 20 noisy images, the estimated input SNR varies from 0.4 to 4.7 dB. Fig. 5.19A and Fig. 5.19B illustrate the amplitude images from the target, obtained with the raw speckle (SNR = 0.4 dB) and when averaging 20 noisy images (SNR = 4.7 dB), respectively.

As few examples from the benchmark, Fig. 5.19C exhibits the image from the BM3D algorithm applied to the raw speckled image, and Fig. 5.19C shows that obtained with the image from the averaging of 20 speckled images. Note that the BM3D approach is known to be the state of the art for the processing of natural images. Fig. 5.19E exhibits the image from the Daubechies8 wavelet algorithm applied to the raw speckled image, and Fig. 5.19F to the averaging of 20 speckled images. Finally, Fig. 5.19G and H consider the result obtained with the median 9×9 applied, respectively, to the raw speckled image and to the averaging of 20 speckled images. The difference in the quality of the restored images can be appreciated with the naked eye.

The different rankings obtained from the benchmark and the 34 denoising algorithms are shown in Fig. 5.20, Figs. 5.21 and 5.22. Fig. 5.21 shows the *RMSE*, Fig. 5.21 shows the Q_{index}, and Fig. 5.22 shows the P_{SNR}. The different rankings are found to be quiet similar, so

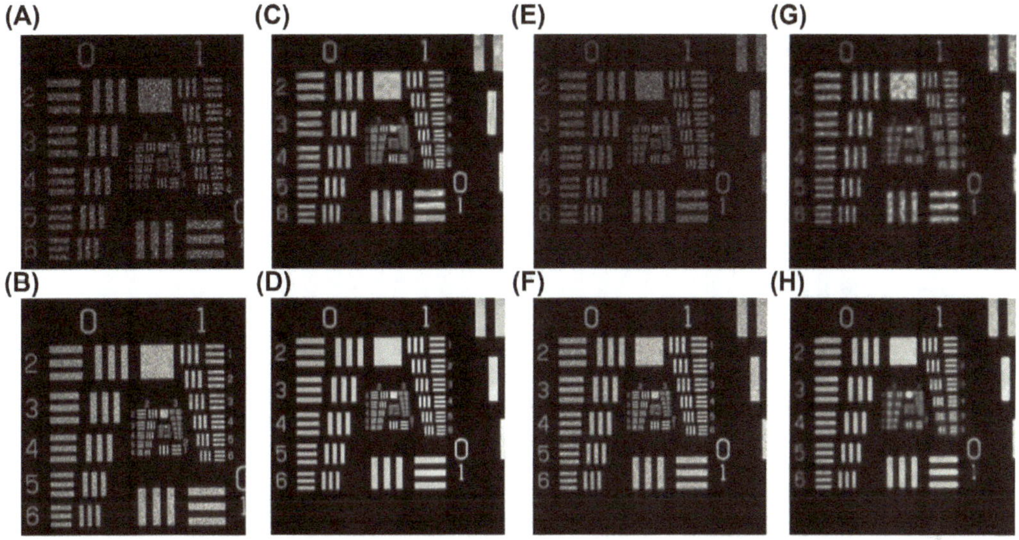

FIG. 5.19 **(A)** Amplitude image with the raw speckle noise, **(B)** amplitude image obtained after averaging 20 noncorrelated realizations of noisy images, **(C)** denoised raw image from BM3D, **(D)** denoised image from BM3D and from averaging 20 speckled images, **(E)** denoised raw image from Daubechies8, **(F)** denoised image from Daubechies8 and from averaging 20 speckled images, **(G)** denoised raw image from Median9, and **(H)** denoised image from Median9 and from averaging 20 speckled images.

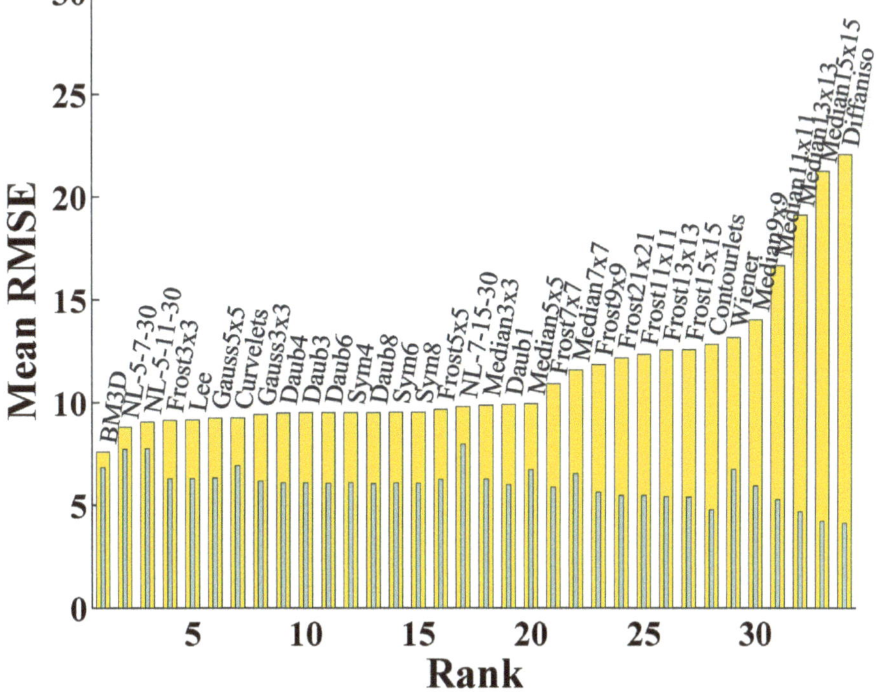

FIG. 5.20 Ranking with root-mean-square error (RMSE).

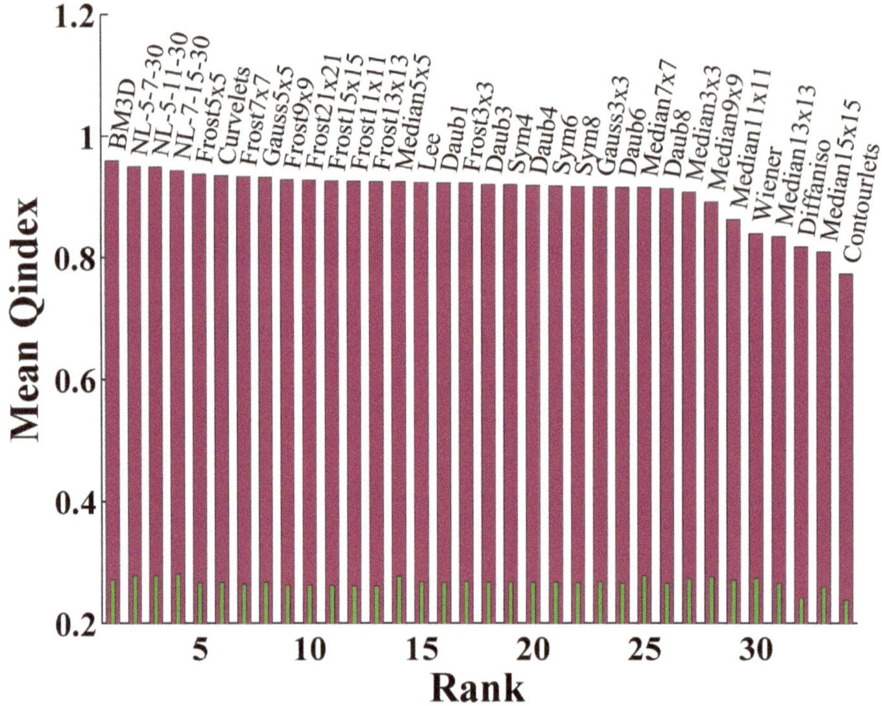

FIG. 5.21 Ranking with Q_{index}.

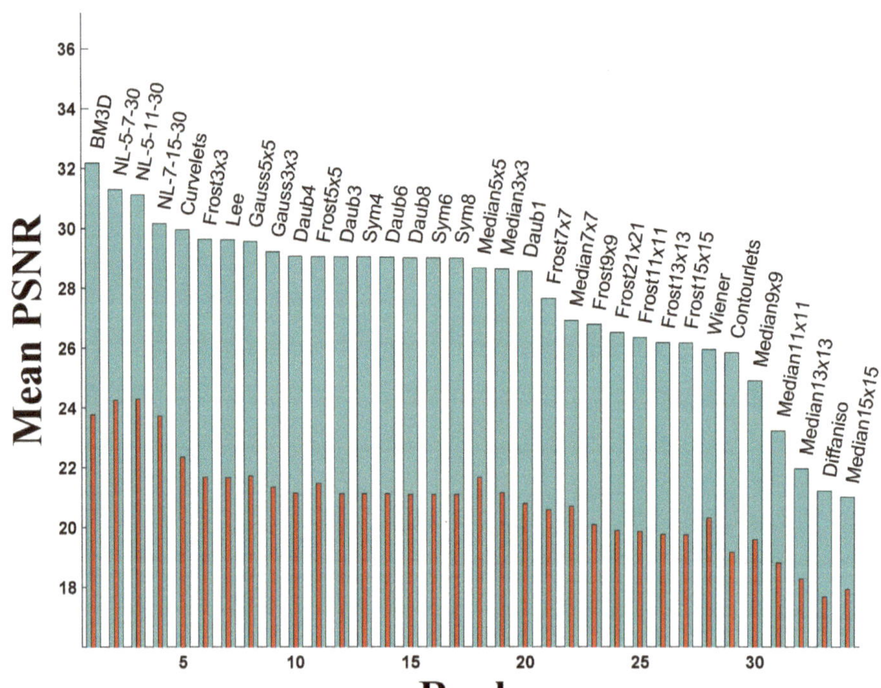

FIG. 5.22 Ranking of P_{SNR}.

that one can conclude that these criteria lead to the same rankings.

Fig. 5.23, Figs. 5.24 and 5.25 show the trends of the selected denoising algorithms versus input SNR for *RMSE*, Q_{index}, P_{SNR}, respectively. Note that the trends are similar from one criterion to another.

The set of Figs. 5.20–5.25 shows an important result: BM3D exhibits the best results and is systematically followed by NL-means and SAR filters. BM3D is known to be the state of the art of denoising for texture images (with Gaussian noise). The set of Figs. 5.20–5.25 shows that this is also the case for strongly speckled amplitude images.

Denoising of Doppler phase maps

This section discusses the quantitative comparison of the performances of the different advanced algorithms (see Depth of Focus section) for Doppler phase denoising. Remind that the Doppler phase is extracted from digital holographic interferometry and that it can be used to measure a lot of physical, mechanical, thermal, etc., parameters on a large variety of object surfaces (refer to Chapter 6). The quantification of the denoising for Doppler phase has to consider, as evaluation criteria, the standard deviation of the phase error (noted here σ_φ), which is the most relevant of all available criteria from the literature, for phase data [150,151] (see also Eq. 5.62). The methodology can be based on the use of a realistic simulation of noise-corrupted phase data. For example, a database including 25 fringe patterns divided into five patterns and five different SNRs was generated to evaluate a collection of denoising algorithms [72]. Quantitative appraisal is based on a benchmark with ranking of 34 algorithms (divided into different families) within the considered criterion.

The data base for the benchmark consists of simulated fringe images in which the type of the fringe pattern and the noise level are controlled. The use of several fringe pattern permits to consider a certain "fringe diversity" so as to establish statistics on the obtained results [72]. As an example, Fig. 5.26 shows five fringe patterns of the set of data for one SNR value. Fig. 5.26A to Fig. 5.26E show the cosine of the

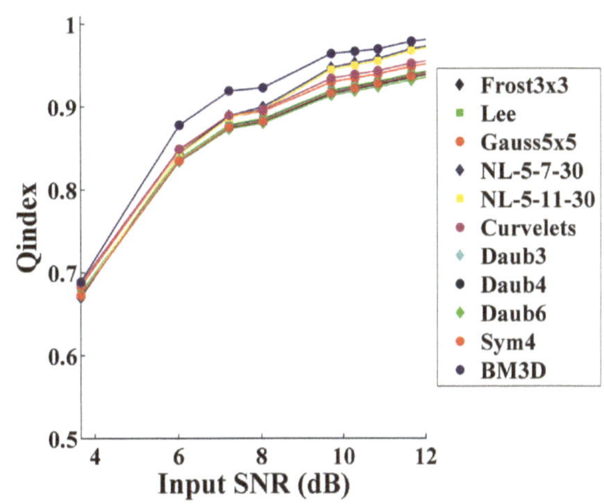

FIG. 5.24 Trends of Q_{index} versus signal-to-noise ratio (SNR).

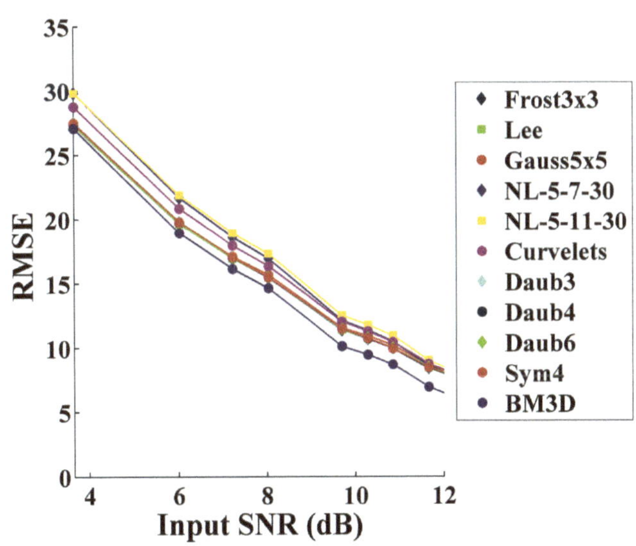

FIG. 5.23 Trends of root-mean-square error (RMSE) versus signal-to-noise ratio (SNR).

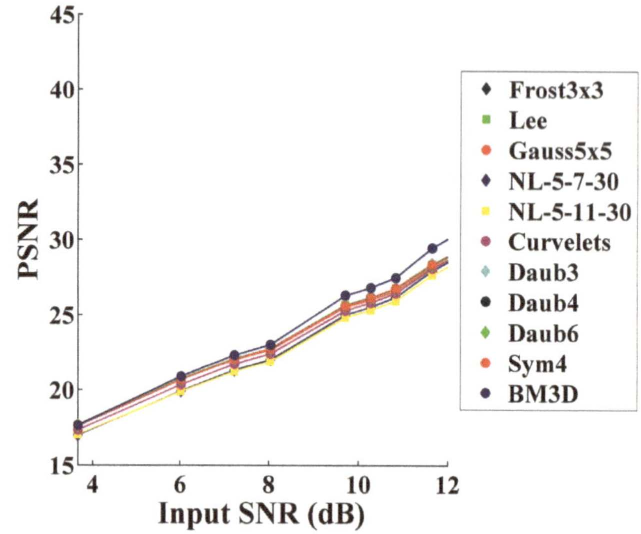

FIG. 5.25 Trends of P_{SNR} versus signal-to-noise ratio (SNR).

simulated deformation phase. Fig. 5.26F–J shows the cosine of the noisy phase deformation for the SNR at, respectively, 3.10, 3.59, 3.34, 3.66, and 3.65 dB (average value of 3.46 dB). Fig. 5.26K to 5.26(o) shows the simulated modulo 2π phase maps. In Fig. 5.26P–T, the modulo 2π noisy Doppler phase maps are exhibited. Last, Fig. 5.26U–Y shows the noise maps extracted from the simulation.

A total of 34 algorithms, differentiated by the used methods and their parameters, were evaluated. These methods are linear filter method with a Gaussian kernel sized 3×3 and 5×5, median filter with kernel from 3×3 to 21×21 (7 odd mask sizes), Wiener filter, Lee filter, Frost filter the same sizes as the median filter, NL-means method with three sets of parameters (5,7,30), (5,11,40), and (7,15,30) (which, respectively, represent the size of the patch, the size of the neighborhood to search for similar patches, and the declining value of the Gaussian kernel used for computation of distances between patches), stationary wavelets such as Daubechies and symlets with five and three values of the regularity parameter (1,3,4,6,8) and (4,6,8), and SPADEDH. Finally, decompositions with curvelet, contourlets, BM3D, and Wtfr2 were considered with the parameters provided by their respective authors.

Fig. 5.27 shows the results obtained for the average value of σ_φ. The main important result is that the WFT2F algorithm provides the best performance with an average value at 0.035 rad, followed by curvelets at 0.07 rad, with in addition, the lowest variance with the curvelets over the whole database. Note that median filters 9×9 and 11×11 at 0.09 rad are followed by SPADEDH near 0.1 rad and that their ranking exhibits good performance. Above the average value of 0.11 rad, ranking is similar for all stationary wavelet methods with close similar variance values, excepted for the regularity 1 which is the Haar wavelet. Around the value of 0.13 rad are the two Frost filters. Finally, beyond 0.2 rad are the three less efficient methods, namely Gauss, Lee filter, and NL-means, which also have the largest variances over the database.

The NL-means, which was presented as close to the state of the art before the emergence of BM3D, is at the end of ranking. This fact can be explained by considering that the fringe patterns contain areas of continuous curves, which affect the search for similar patches in the near vicinity. Thus, the algorithm calculates averaged patches that include major distortions compared with the original image.

It can be observed that these results are different compared with those obtained from the benchmark on denoising of amplitude images (Fig. 5.21).

Emergence of Deep Learning

Since 10 years now, deep learning has emerged as a very efficient tool in signal and image processing with various applications mainly in speech and image recognition, as well as natural language understanding [154,155]. At the heart of this new tool is the convolutional neural network (CNN) [156]. In the field of image processing, CNN integrates three fundamental advances obtained in the past decades: wavelet and

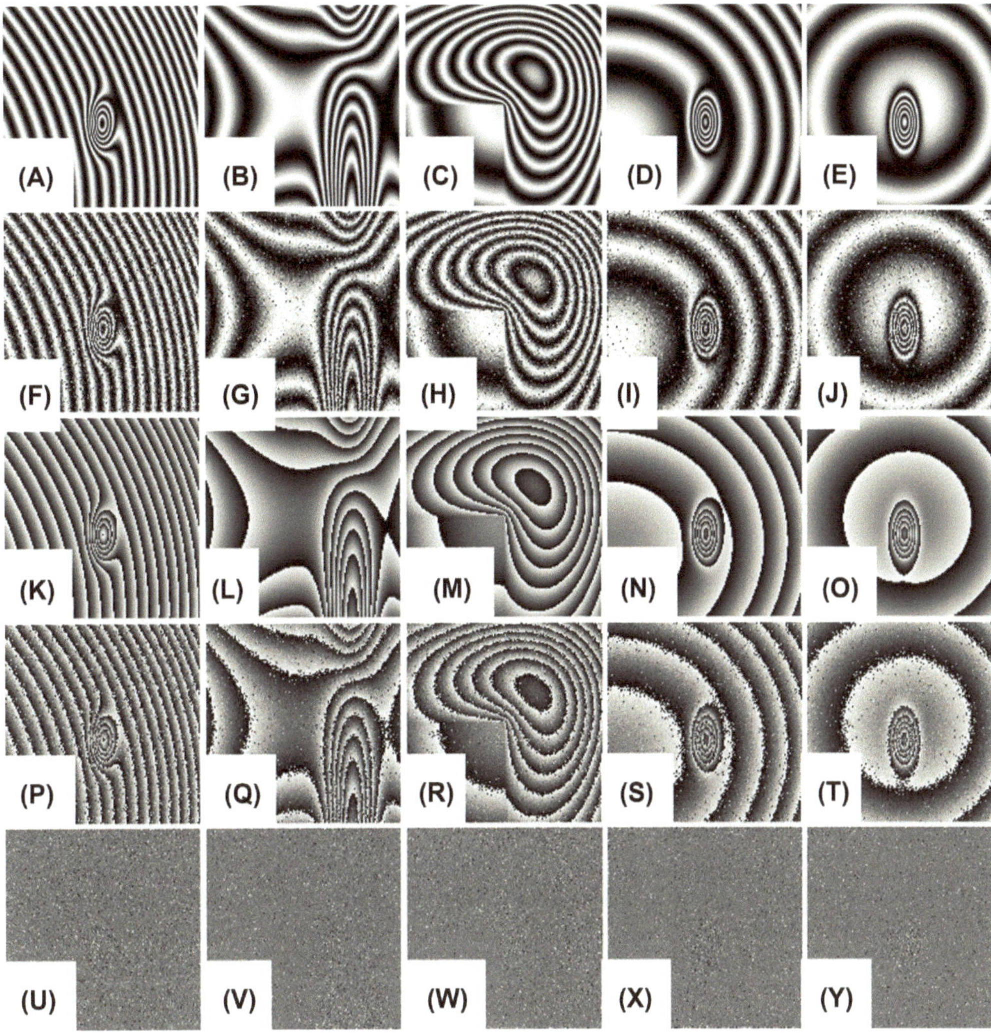

FIG. 5.26 Outputs from the numerical simulations for five different fringe patterns, **(A)** to **(E)**: cosine of the simulated phase, **(F)** to **(J)**: cosine of the noisy phase for signal-to-noise ratio (SNR) at respectively 3.10, 3.59, 3.34, 3.66, and 3.65 dB (average value at 3.46 dB), **(K)** to **(O)**: simulated modulo 2π deformation phase maps, **(P)** to **(T)**: modulo 2π noisy deformation phase maps, **(U)** to **(Y)**: noise maps.

multiresolution analysis, shrinking and thresholding algorithms, and sparse representations [157].

In recent years, several applications of deep learning to optics have emerged. For example, several papers discussed on superresolution [158–160], aberration compensation [161], autofocus [162,163], imaging reconstruction [164,165], and phase recovery [166,167]. Recently, it has also been demonstrated that a deep neural network may significantly improve optical microscopy, by enhancing its spatial resolution over a large field of view and depth of field [168]. A method for computer-generated holography based on deep learning was presented in Ref. [169]. Concerning noise processing, several approaches were proposed in the field of optical coherence tomography [170–172]. Deep learning–based algorithms applied to the noise reduction in SAR imaging were proposed in Refs. [173–175]. Finally, the problem of speckle decorrelation is approached with a deep learning–based solution in Ref. [176], which took into account scattering effects.

To the best of our knowledge, at the date of the writing of this chapter, no realizations and evaluations

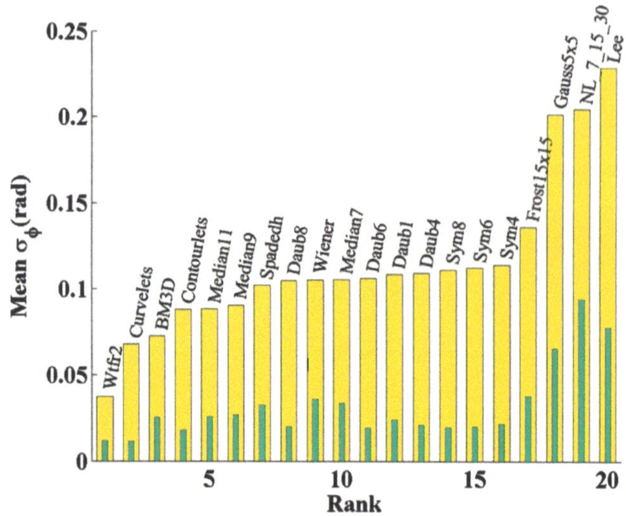

FIG. 5.27 Ranking of denoising algorithms obtained for the average value of σ_φ, overall the database of realistic fringe patterns.

of deep learning–based algorithms applied to the reduction of speckle noise in digital holography, whether for phase or amplitude imaging, has been proposed. This is therefore an essential step to climb in the near future.

CONCLUSION

In this chapter, we have discussed the principles of digital holography, as well as the associated reconstruction methods. Some demodulation techniques of eliminating the zero order and twin image have been discussed. The discrete Fresnel transform is the simplest and most direct calculation method, on the condition that the sampling theorem is respected in the plane of the hologram. This method is certainly the most widely used, even though it does not guarantee the invariance of the reconstructed horizon with the wavelength. We have presented the family of the methods based on convolution, especially the adjustable-magnification approach, which leads to the adaptation of the reconstructed image size to the size of the field of view.

We have introduced the properties of digitally reconstructed images from holograms. Especially, we pointed out their speckle property both for the amplitude and the phase. The most important point is that holographic images include multiplicative and non-Gaussian noise. We have also discussed on spatial resolution and depth of focus.

A large part of this chapter is devoted to denoising of holographic images. This was discussed through denoising of speckled amplitude images and noisy Doppler phase maps. The most advanced denoising algorithms from literature were presented. Quantitative appraisal of their efficiency was discussed through comparison and evaluation criteria. It was demonstrated that the BM3D algorithm is the best approach to denoise amplitude images and that the 2D windowed Fourier transform is the best one to deal with noisy Doppler phases.

As perspectives, we discussed the emergence of deep learning in image processing, which will probably strongly impact digital holography and its related post-processing approaches.

REFERENCES

[1] J.W. Goodman, R.W. Lawrence, Digital image formation from electronically detected holograms, Applied Physics Letters 11 (1967) 77–79.

[2] T.S. Huang, Digital holography, Proceedings of the IEEE 159 (1971) 1335–1345.

[3] M.A. Kronrod, N.S. Merzlyakov, L.P. Yaroslavskii, Reconstruction of a hologram with a computer, Soviet Physics – Technical Physics 17 (1972) 333–334.

[4] U. Schnars, W. Jüptner, Direct recording of holograms by a CCD target and numerical reconstruction, Applied Optics 33 (1984) 179–181.

[5] J.W. Goodman, Introduction à l'optique de Fourier et à l'holographie (Traduction Française par Christian Durou et José-Philippe Perez), Masson, Paris, 1972.

[6] S.A. Collins, Laser-system diffraction integral written in terms of matrix optics, Journal of the Optical Society of America 60 (1970) 11681177.

[7] A. Yariv, Optical Electronics, Holt, Rinehart and Winston, New York, 1985.
[8] M. Born, E. Wolf, Principles of Optics, seventh ed., Cambridge University Press, 1999.
[9] W. Lauterborn, T. Kurz, Coherent Optics: Fundamentals and Applications, Springer-Verlag, Berlin, 2010.
[10] J.W. Goodman, Introduction to Fourier Optics, third ed., Greenwood Village, Roberts & Company Publishers, 2005.
[11] J.C. Li, Y. Wu, Diffraction Calculation and Digital Holography-I, Science Press, Beijing, 2017.
[12] J.C. Li, Y. Wu, Diffraction Calculation and Digital Holography-II, Science Press, Beijing, 2017.
[13] T. Kreis, Holographic Interferometry—Principles and Methods, Akademie Verlag Gmbh, Berlin, 1995.
[14] P. Hariharan, Basics of Holography, Cambridge University Press, New York, 2002.
[15] T. Kreis, Handbook of Holographic Interferometry Optical and Digital Methods, Wiley-VCH, Weinheim, 2004.
[16] I. Yamaguchi, T. Zhang, Phase-shifting digital holography, Optics Letters 22 (1997) 1268—1270.
[17] I. Yamaguchi, J. Kato, S. Ohta, J. Mizuno, Image formation in phase shifting digital holography and application to microscopy, Applied Optics 40 (2001) 6177—6185.
[18] I. Yamaguchi, J. Kato, S. Ohta, Surface shape measurement by phase shifting digital holography, Optical Review 8 (2001) 85—89.
[19] K. Creath, in: E. Wolf (Ed.), Phase Measurement Interferometry Techniques, Progress in Optics 26, North-Holland publishing Company, 1988, pp. 349—393.
[20] B.V. Dorrio, J.L. Fernandez, Phase evaluation methods in whole-field optical measurement techniques, Measurement Science and Technology 10 (1999) 33—55.
[21] J.E. Greivenkamp, Generalized data reduction for heterodyne interferometry, Optical Engineering 23 (4) (1984) 350—352.
[22] J.C. Wyant, Use of an ac heterodyne lateral shear interferometer with real-time wavefront correction systems, Applied Optics 14 (1975) 2622—2625.
[23] I. Yamaguchi, T. Matsumura, J. Kato, Phase shifting color digital holography, Optics Letters 27 (2002) 1108—1110.
[24] J.M. Desse, P. Picart, P. Tankam, Sensor influence in digital 3λ holographic interferometry, Measurement Science and Technology 22 (2011) 064005.
[25] P. Tankam, Q. Song, M. Karray, J.C. Li, J.M. Desse, P. Picart, Real-time three-sensitivity measurements based on three-color digital Fresnel holographic interferometry, Optics Letters 35 (2010) 2055—2057.
[26] P. Tankam, P. Picart, D. Mounier, J.M. Desse, J.C. Li, Method of digital holographic recording and reconstruction using a stacked color image sensor, Applied Optics 49 (2010) 320—325.
[27] J.M. Desse, P. Picart, P. Tankam, Digital three-color holographic interferometry for flow analysis, Optics Express 16 (2008) 5471—5480.
[28] P. Picart, P. Tankam, D. Mounier, Z. Peng, J.C. Li, Spatial bandwidth extended reconstruction for digital color Fresnel holograms, Optics Express 17 (2009) 9145—9155.
[29] C.J. Mann, P.R. Bingham, V.C. Paquit, K.W. Tobin, Quantitative phase imaging by three wavelength digital holography, Optics Express 16 (2008) 9753—9764.
[30] J. Kuhn, T. Colomb, F. Montfort, F. Charriere, Y. Emery, E. Cuch, P. Marquet, C. Depeursinge, Real-time dual-wavelength digital holographic microscopy with a single hologram acquisition, Optics Express 15 (2007) 7231—7242.
[31] P. Tankam, P. Picart, Use of digital color holography for crack investigation in electronic components, Optics and Lasers in Engineering 49 (2011) 1335—1342.
[32] P. Picart, J. Leval, D. Mounier, S. Gougeon, Time-averaged digital holography, Optics Letters 28 (2003) 1900—1902.
[33] P. Picart, J. Leval, D. Mounier, S. Gougeon, Some opportunities for vibration analysis with time averaging in digital Fresnel holography, Applied Optics 44 (2005) 337—343.
[34] Y. Yao, A.J. Hoffman, C.F. Gmachl, Mid-infrared quantum cascade lasers, Nature Photonics 6 (2012) 432—439.
[35] M. Ravaro, M. Locatelli, E. Pugliese, I. Di Leo, M. Siciliani de Cumis, F. d'Amato, P. Poggi, R. Meucci, P. Ferraro, P. de Natale, Mid-infrared digital holography and holographic interferometry with a tunable quantum cascade laser, Optics Letters 39 (16) (2014) 4843—4845.
[36] G. Pedrini, P. Froning, H.J. Tiziani, M.E. Gusev, Pulsed digital holography for high-speed contouring that uses a two-wavelength method, Applied Optics 38 (16) (1999) 3460—3467.
[37] P.W. Kruse, Uncooled Thermal Imaging Arrays, Systems and Applications, Tutorial Text Series, vol. TT51, SPIE, Bellingham, 2001.
[38] N. Diakides, J. Bronzino, Medical Thermal Imaging, CRC Press, 2007.
[39] A. Rogalski, HgCgTe infrared detector material: history, status and outlook, Reports on Progress in Physics 68 (2005) 2267—2335.
[40] J.M. Armstrong, M.R. Skokan, M.A. Kinch, J.D. Luttmer, HDVIP five-micron pitch HgCdTe focal plane arrays, Proceedings of SPIE 9070 (2014) 907033.
[41] T. Conklin, E.H. Stupp, Applications of the pyroelectric vidicon, Optical Engineering 15 (6) (1976) 510—515.
[42] O. Kwon, J.C. Wyant, C.R. Hayslett, Rough surface interferometry at 10.6 μm, Applied Optics 19 (11) (1980) 1862—1869.
[43] O.J. LØkberg, O. Kwon, Electronic Speckle Pattern Interferometry Using a CO_2 Laser, Optics & Laser Technology, 1984, pp. 187—192.
[44] E. Allaria, S. Brugioni, S. de Nicola, P. Ferraro, S. Grillis, R. Meucci, Digital holography at 10.6 μm, Optics Communications 215 (2003) 257—262.

[45] J.-F. Vandenrijt, M. Georges, Infrared electronic speckle pattern interferometry at 10 μm, Proceedings of SPIE 6616 (2007) 66162Q.

[46] N. George, K. Khare, W. Chi, Infrared holography using a microbolometer array, Applied Optics 47 (54) (2008) A7–A12.

[47] M.P. Georges, J.-F. Vandenrijt, C. Thizy, Y. Stockman, P. Queeckers, F. Dubois, D. Doyle, Digital holographic interferometry with CO_2 lasers and diffuse illumination applied to large space reflector metrology, Applied Optics 52 (1) (2013) A102–A115.

[48] M.P. Georges, J.-F. Vandenrijt, C. Thizy, I. Alexeenko, G. Pedrini, B. Vollheim, I. Lopez, I. Jorge, J. Rochet, W. Osten, Combined holography and thermography in a single sensor through image-plane holography at thermal infrared wavelengths, Optics Express 22 (21) (2014) 25517–25529.

[49] F. Languy, J.-F. Vandenrijt, P. Saint-Georges, S. Paquay, P. De Vincenzo, M.P. Georges, Space mirror deformation: from thermo-mechanical measurements by speckle interferometry to optical comparison with multiphysics simulation, Applied Optics 57 (24) (2018) 6982–6989.

[50] D. Mas, J. Perez, C. Hernandez, C. Vazquez, J.J. Miret, C. Illueca, Fast numerical calculation of Fresnel patterns in convergent systems, Optics Communications 227 (2003) 245–255.

[51] D. Mas, J. Garcia, C. Ferreira, L.M. Bernardo, F. Marinho, Fast algorithms for free-space diffraction patterns calculation, Optics Communications 164 (1999) 233–245.

[52] J.C. Li, Z. Peng, Y. Fu, Diffraction transfer function and its calculation of classic diffraction formula, Optics Communications 280 (2007) 243–245.

[53] P. Picart, P. Tankam, Analysis and adaptation of convolution algorithms to reconstruct extended objects in digital holography, Applied Optics 52 (2013) A240–A253.

[54] F. Zhang, I. Yamaguchi, Algorithm for reconstruction of digital holograms with adjustable magnification, Optics Letters 29 (2004) 1668–1670.

[55] J.F. Restrepo, J. Garcia-Sucerquia, Magnified reconstruction of digitally recorded holograms by Fresnel-Bluestein transform, Applied Optics 49 (2010) 6430–6435.

[56] J.C. Li, P. Tankam, Z. Peng, P. Picart, Digital holographic reconstruction of large objects using a convolution approach and adjustable magnification, Optics Letters 34 (2009) 572–574.

[57] J.C. Li, Z. Peng, P. Tankam, Q. Song, P. Picart, Digital holographic reconstruction of a local object field using an adjustable magnification, Journal of the Optical Society of America A 28 (6) (2011) 1291–1295.

[58] P. Ferraro, S. De Nicola, G. Coppola, A. Finizio, D. Alfieri, G. Pierattini, Controlling image size as a function of distance and wavelength in Fresnel-transform reconstruction of digital holograms, Optics Letters 29 (2004) 854–855.

[59] P. Picart, D. Mounier, J.M. Desse, High resolution digital two-color holographic metrology, Optics Letters 33 (3) (2008), 276-275.

[60] A. Stadelmaier, J.H. Massig, Compensation of lens aberrations in digital holography, Optics Letters 25 (2000) 1630–1632.

[61] T. Colomb, F. Montfort, J. Kühn, N. Aspert, E. Cuche, A. Marian, F. Charrière, S. Bourquin, P. Marquet, C. Depeursinge, Numerical parametric lens for shifting, magnification, and complete aberration compensation in digital holographic microscopy, Journal of the Optical Society of America A 23 (2006) 3177–3190.

[62] S. De Nicola, A. Finizio, G. Pierattini, D. Alfieri, S. Grilli, L. Sansone, P. Ferraro, Recovering correct phase information in multiwavelength digital holographic microscopy by compensation for chromatic aberrations, Optics Letters 30 (2005), 2706-2705.

[63] P. Ferraro, S. Grilli, L. Miccio, D. Alfieri, S. De Nicola, A. Finizio, B. Javidi, Full color 3-D imaging by digital holography and removal of chromatic aberrations, Journal of Display Technology 4 (2008) 97–100.

[64] M. Leclercq, P. Picart, Method for chromatic errors compensation in digital color holographic imaging, Optics Express 21 (22) (2013) 26456–26467.

[65] P. Picart, Speckle noise in digital holographic images: genesis and reduction methods, in: Digital Holography and Three-Dimensional Imaging, OSA Technical Digest (online) (Optical Society of America, 2017 paper Th1A.1.

[66] V. Bianco, P. Memmolo, M. Leo, S. Montresor, C. Distante, M. Paturzo, P. Picart, B. Javidi, P. Ferraro, Strategies for reducing speckle noise in digital holography, Nature-Light Science & Applications 7 (2018) 45.

[67] P. Picart, J. Leval, General theoretical formulation of image formation in digital Fresnel holography, Journal of the Optical Society of America A 25 (2008) 1744–1761.

[68] D.C. Ghiglia, M.D. Pritt, Two-dimensional Phase Unwrapping: Theory, Algorithms and Software, Wiley-Interscience, 1998.

[69] J.W. Goodman, Speckle Phenomena in Optics, Roberts and Company Publishers, Greenwood village, 2005.

[70] J.C. Dainty, Laser Speckle and Related Phenomena, Springer-Verlag Berlin, 1984.

[71] J.W. Goodman, Statistical Optics, Wiley, New York, 1985.

[72] S. Montresor, P. Picart, Quantitative appraisal for noise reduction in digital holographic phase imaging, Optics Express 24 (2016) 14322–14343.

[73] J. Poittevin, P. Picart, F. Gautier, C. Pezerat, Quality assessment of combined quantization-shot-noise-induced decorrelation noise in high-speed digital holographic metrology, Optics Express 23 (2015) 30917–30932.

[74] J. Poittevin, F. Gautier, C. Pezerat, P. Picart, High-speed holographic metrology: principle, limitations, and application to vibroacoustics of structures, Optical Engineering 55 (12) (2016), 121717-1-13.

[75] U. Schnars, W. Jüptner, Digital Holography: Digital Hologram Recording, Numerical Reconstruction, and Related Techniques, Springer-Verlag, Berlin and Heidelberg, 2004.
[76] P. Picart, J.C. Li, Digital Holography, ISTE-Wiley, London, 2012.
[77] P. Picart, New Techniques in Digital Holography, ISTE-Wiley, London, 2015.
[78] I. Yamaguchi, A. Yamamoto, S. Kuwamura, Speckle decorrelation in surface profilometry by wavelength scanning interferometry, Applied Optics 37 (1998) 6721–6725.
[79] P. Picart, S. Montresor, O. Sakharuk, L. Muravsky, Refocus criterion based on maximization of the coherence factor in digital three-wavelength holographic interferometry, Optics Letters 42 (2017) 275.
[80] P. Picart, P. Tankam, Q. Song, Experimental and theoretical investigation of the pixel saturation effect in digital holography, Journal of the Optical Society of America A 28 (2011) 1262–1275.
[81] V. Bianco, P. Memmolo, M. Paturzo, A. Finizio, B. Javidi, P. Ferraro, Quasi noise-free digital holography, Light: Science and Applications 5 (2016) e16142.
[82] V. Bianco, P. Memmolo, M. Paturzo, P. Ferraro, On-speckle suppression in IR digital holography, Optics Letters 41 (2016) 5226–5229.
[83] M. Matrecano, P. Memmolo, L. Miccio, A. Persano, F. Quaranta, P. Siciliano, P. Ferraro, Improving holographic reconstruction by automatic Butterworth filtering for microelectromechanical systems characterization, Applied Optics 54 (2015) 3428–3432.
[84] P. Memmolo, V. Bianco, M. Paturzo, B. Javidi, P.A. Netti, P. Ferraro, Encoding multiple holograms for speckle-noise reduction in optical display, Optics Express 22 (2014) 25768–25775.
[85] M. Leo, R. Piccolo, C. Distante, P. Memmolo, M. Paturzo, P. Ferraro, Multilevel bidimensional empirical mode decomposition: a new speckle reduction method in digital holography, Optical Engineering 53 (11) (2014) 112314.
[86] M. Leo, C. Distante, M. Paturzo, P. Memmolo, M. Locatelli, E. Pugliese, R. Meucci, P. Ferraro, Automatic digital hologram denoising by spatiotemporal analysis of pixel-wise statistics, Journal of Display Technology 9 (11) (2013) 904–909.
[87] P. Memmolo, M. Iannone, M. Ventre, P.A. Netti, A. Finizio, M. Paturzo, P. Ferraro, Quantitative phase maps denoising of long holographic sequences by using SPADEDH algorithm, Applied Optics 52 (2013) 1453–1460.
[88] V. Bianco, M. Paturzo, P. Memmolo, A. Finizio, P. Ferraro, B. Javidi, Random resampling masks: a non-Bayesian one-shot strategy for noise reduction in digital holography, Optics Letters 38 (2013) 619–621.
[89] P. Memmolo, I. Esnaola, A. Finizio, M. Paturzo, P. Ferraro, A.M. Tulino, SPADEDH: a sparsity-based denoising method of digital holograms without knowing the noise statistics, Optics Express 20 (15) (2012) 17250–17257.
[90] J. Maycock, B.M. Hennelly, J.B. McDonald, Y. Frauel, A. Castro, B. Javidi, T.J. Naughton, Reduction of speckle in digital holography by discrete Fourier filtering, Journal of the Optical Society of America A 24 (2007) 1617–1622.
[91] H.Z. Jiang, J.L. Zhao, J.L. Di, Digital color holographic recording and reconstruction using synthetic aperture and multiple reference waves, Optics Communications 285 (2012) 3046–3049.
[92] Y. Kuratomi, K. Sekiya, H. Satoh, T. Tomiyama, T. Kawakami, B. Katagiri, Y. Suzuki, T. Uchida, Speckle reduction mechanism in laser rear projection displays using a small moving diffuser, Journal of the Optical Society of America A 27 (8) (2010) 1812–1817.
[93] S.-H. Shin, B. Javidi, Speckle-reduced three-dimensional volume holographic display by use of integral imaging, Applied Optics 41 (2002) 2644–2649.
[94] N. Bertaux, Y. Frauel, P. Réfrégier, B. Javidi, Speckle removal using a maximum-likelihood technique with isoline gray-level regularization, Journal of the Optical Society of America A 21 (2004) 2283–2291.
[95] X. Cai, Reduction of speckle noise in the reconstructed image of digital holography, Optik 121 (2010) 394–399.
[96] J. Garcia-Sucerquia, Noise reduction in digital lensless holographic microscopy by engineering the light from a light-emitting diode, Applied Optics 52 (2013) A232–A239.
[97] B. Redding, M.A. Choma, H. Cao, Speckle-free laser imaging using random laser illumination, Nature Photonics 6 (2012) 355–359.
[98] T. Nomura, M. Okamura, E. Nitanai, T. Numata, Image quality improvement of digital holography by superposition of reconstructed images obtained by multiple wavelengths, Applied Optics 47 (19) (2008) D38–D43.
[99] F. Pan, W. Xiao, S. Liu, L. Rong, Coherent noise reduction in digital holographic microscopy by laterally shifting camera, Optics Communications 292 (2013) 68–72.
[100] F. Pan, W. Xiao, S. Liu, F. Wang, L. Rong, R. Li, Coherent noise reduction in digital holographic phase contrast microscopy by slightly shifting object, Optics Express 19 (2011) 3862–3869.
[101] C. Quan, X. Kang, C.J. Tay, Speckle noise reduction in digital holography by multiple holograms, Optical Engineering 46 (2007) 115801.
[102] J. Garcia-Sucerquia, J.H. Ramirez, R. Castaneda, Incoherent recovering of the spatial resolution in digital holography, Optics Communications 260 (2006) 62–67.

[103] T. Baumbach, E. Kolenovi, V. Kebbel, W. Jüptner, Improvement of accuracy in digital holography by use of multiple holograms, Applied Optics 45 (2006) 6077–6085.

[104] D. Claus, D. Iliescu, B.H. Timmerman, P.J. Bryanston-Cross, Resolution improvement in digital holography: comparison between synthetic aperture method and spatial averaging method, Proceedings of SPIE 8001 (2011) 80010Z–80015Z.

[105] V.S. Frost, J.A. Stiles, K.S. Shanmugan, J.C. Horltzman, A model for radar images and its application to adaptive digital filtering of multiplicative noise, IEEE Transactions on Pattern Analysis and Machine Intelligence 4 (1982) 157–165.

[106] A. Uzan, Y. Rivenson, A. Stern, Speckle denoising in digital holography by nonlocal means filtering, Applied Optics 52 (2012) A195–A200.

[107] D. Hincapie, J. Herrera-Ramírez, J. Garcia-Sucerquia, Single-shot speckle reduction in numerical reconstruction of digitally recorded holograms, Optics Letters 40 (2015) 1623–1625.

[108] T. Fukuoka, Y. Mori, T. Nomura, Speckle reduction by spatial domain mask in digital holography, Journal of Display Technology 12 (4) (2016) 315–322.

[109] M. Haouat, J. Garcia-Sucerquia, A. Kellou, P. Picart, Reduction of speckle noise in holographic images using spatial jittering in numerical reconstructions, Optics Letters 42 (6) (2017) 1047–1050.

[110] E.Y. Lam, X. Zhang, H. Vo, T.-C. Poon, G. Indebetouw, Three-dimensional microscopy and sectional image reconstruction using optical scanning holography, Applied Optics 48 (2009) H113–H119.

[111] S. Sotthivirat, J.A. Fessler, Penalized likelihood image reconstruction for digital holography, Journal of the Optical Society of America A 21 (2004) 737–750.

[112] B. Katz, D. Wulich, J. Rosen, Optimal noise suppression in Fresnel incoherent correlation holography (FINCH) configured for maximum imaging resolution, Applied Optics 49 (2010) 5757–5763.

[113] G. Chen, Q. Li, Markov chain Monte Carlo sampling based terahertz holography image denoising, Applied Optics 54 (2015) 4345–4351.

[114] S. Kubota, J.W. Goodman, Very efficient speckle contrast reduction realized by moving diffuser device, Applied Optics 49 (2010) 4385–4391.

[115] I. Peled, M. Zenou, B. Greenberg, Z. Kotler, MEMS based speckle reduction obtain by angle diversity for fast imaging, in: CLEO/QELS, Conference on Quantum Electronics and Laser Science Conference, 2009.

[116] J. Amako, H. Miura, T. Sonehara, Speckle noise reduction on kinoform reconstruction using a phase only spatial light modulator, Applied Optics 34 (1995) 3165–3171.

[117] A. Brozeit, J. Burke, H. Helmers, H. Sagehorn, R. Schuh, Noise reduction in ESPI fringes by merging orthogonally polarised speckle fields, Optics & Laser Technology 30 (1998) 325–329.

[118] L. Rong, W. Xiao, F. Pan, S. Liu, R. Li, Speckle noise reduction in digital holography by use of multiple polarization holograms, Chinese Optics Letters 8 (2010) 653–655.

[119] F.T.S. Yu, E.Y. Wang, Speckle reduction in holography by means of random spatial sampling, Applied Optics 12 (1973) 1656–1659.

[120] M. Matsumura, Speckle noise reduction by random phase shifters, Applied Optics 14 (1975) 660–665.

[121] M. Abollasshani, Y. Rostami, Speckle noise reduction by division and digital processing of a hologram, Optik 123 (10) (2012) 937–939.

[122] Y. Rivenson, M.A. Shalev, Z. Zalevsky, Compressive Fresnel holography approach for high-resolution viewpoint inference, Optics Letters 40 (2015) 5606–5609.

[123] H.A. Aebischer, S. Waldner, A simple and effective method for filtering speckle-interferometric phase fringe patterns, Optics Communications 162 (1999) 205–210.

[124] Q. Kemao, S.H. Soon, A. Asundi, Smoothing filters in phase-shifting interferometry, Optics & Laser Technology 35 (8) (2003) 649–654.

[125] R.C. Gonzales, R.E. Woods, Digital Image Processing, third ed., Prentice Hall, Upper Saddle River, 2008.

[126] J.S. Lee, Digital image enhancement and noise filtering by using local statistics, IEEE Transactions on Pattern Analysis and Machine Intelligence 2 (1980) 165–1655.

[127] S. Mallat, A Wavelet Tour of Signal Processing, Academic Press, New York, 1999.

[128] D.L. Donoho, De-noising by soft-thresholding, IEEE Transactions on Information Theory 41 (1995) 613–627.

[129] H. Xie, L.E. Pierce, F.T. Ulaby, Sar speckle reduction using wavelet denoising and markov random field modeling, IEEE Transactions on Geoscience and Remote Sensing 40 (10) (2002) 2196–2212.

[130] J.-L. Starck, E.J. Candès, D.L. Donoho, The curvelet transform for image denoising, IEEE Transactions on Image Processing 11 (2002) 670–684.

[131] M.N. Do, M. Vetterli, The contourlet transform: an efficient directional multiresolution image representation, IEEE Transactions on Image Processing 14 (12) (2005) 2091–2105.

[132] A. Frederico, G.H. Kaufmann, Denoising in digital speckle pattern interferometry using wave atoms, Optics Letters 32 (2007) 1232–1234.

[133] G.H. Kaufmann, G.E. Galizzi, Speckle noise reduction in television holography fringes using wavelet thresholding, Optical Engineering 35 (1) (1996) 9–14.

[134] A.A. Shulev, A. Gotchev, A. Foi, I.R. Roussev, Threshold selection in transform-domain denoising of speckle pattern fringes, Proceedings of SPIE 6252 (2006) 625220–625221, 625220-7.

[135] A. Buades, B. Coll, J.M. Morel, A review of image denoising algorithms, with a new one, Multiscale Modeling and Simulation 4 (5) (2005) 490–530.

[136] A. Buades, B. Coll, J. Morel, A non-local algorithm for image denoising, Proceedings – IEEE Computer Society Conference on Computer Vision and Pattern Recognition 2 (2) (2005) 60–65.

[137] K. Dabov, A. Foi, V. Katkovnik, K. Egiazarian, Image denoising with block-matching and 3D filtering, Proceedings of SPIE 6064 (2006), 606414-606421-606414-12.

[138] K. Dabov, A. Foi, V. Katkovnik, K. Egiazarian, Image denoising by sparse 3D transform-domain collaborative filtering, IEEE Transactions on Image Processing 16 (8) (2007) 2080–2095.

[139] V. Katkovnik, A. Foi, K. Egiazarian, J. Astola, From local kernel to nonlocal multiple-model image denoising, International Journal of Computer Vision 86 (1) (2010) 1–32.

[140] Q. Kemao, Windowed Fourier transform for fringe pattern analysis, Applied Optics 43 (2004) 2695–2702.

[141] L. Huang, Q. Kemao, B. Pan, A. Asundi, Comparison of Fourier transform, windowed Fourier transform, and wavelet transform methods for phase extraction from a single fringe pattern in fringe projection profilometry, Optics and Lasers in Engineering 48 (2) (2010) 141–145.

[142] Q. Kemao, L. Nam, L. Feng, S. Soon, Comparative analysis on some filters for wrapped phase maps, Applied Optics 46 (2007) 7412–7415.

[143] Q. Kemao, On window size selection in the windowed Fourier ridges algorithm, Optics and Lasers in Engineering 45 (12) (2007) 1186–1192.

[144] S. Panezai, J. Zhao, Y. Wang, D. Wang, L. Rong, Speckle suppression in off-axis lensless Fourier transform digital holography, Optics Communications 397 (2017) 100–104.

[145] J.A. Rodrigo, T. Alieva, Illumination coherence engineering and quantitative phase imaging, Optics Letters 39 (2014) 5634–5637.

[146] F. Pan, L. Yang, W. Xiao, Coherent noise reduction in digital holographic microscopy by averaging multiple holograms recorded with a multimode laser, Optics Express 25 (2017) 21815–21825.

[147] W. Zhang, L. Cao, H. Zhang, H. Zhang, C. Han, G. Jin, Y. Sheng, Quantitative study on a resampling mask method for speckle reduction with amplitude superposition, Applied Optics 56 (2017) F205–F212.

[148] X. Cai, H. Wang, The influence of hologram aperture on speckle noise in the reconstructed image of digital holography and its reduction, Optics Communications 281 (2008) 232–237.

[149] M. Paturzo, A. Pelagotti, A. Finizio, L. Miccio, M. Locatelli, A. Gertrude, P. Poggi, R. Meucci, P. Ferraro, Optical reconstruction of digital holograms recorded at 10.6 μm: route for 3D imaging at long infrared wavelengths, Optics Letters 35 (12) (2010) 2112–2114.

[150] S. Montresor, P. Picart, M. Karray, Reference-free metric for quantitative noise appraisal in holographic phase measurements, Journal of the Optical Society of America A 35 (2018) A53–A60.

[151] S. Montresor, P. Picart, O. Sakharuk, L. Muravsky, Error analysis for noise reduction in 3D deformation measurement with digital color holography, Journal of the Optical Society of America B 34 (5) (2017) B9–B14.

[152] Z. Wang, A.C. Bovik, L. Lu, Why is image quality assessment so difficult? Proceedings IEEE ICASSP 4 (2002) 3313–3315.

[153] S. Montresor, P. Picart, Evaluation of de-noising algorithms for amplitude image restoration in digital holography, in: Imaging and Applied Optics, OSA Technical Digest, 2018 paper JTu4A.5.

[154] Y. Le Cun, Y. Bengio, G. Hinton, Deep learning, Nature 523 (2015) 436–444.

[155] J. Xie, L. Xu, E. Chen, Image denoising and inpainting with deep neural networks, Advances in Neural Information Processing Systems 25 (2012) 341–349.

[156] V.P. Lutsiv, Convolutional deep-learning artificial networks, Journal of Optical Technology 82 (8) (2015) 499–505.

[157] S. Mallat, Understanding deep convolution networks, Philosophical Transactions of the Royal Society A374 (2016) 20150203.

[158] L. Mukherjee, A. Keikhosravi, D. Bui, K. W Eliceiri, Convolutional neural networks for whole slide image superresolution, Biomedical Optics Express 9 (11) (2018) 5368–5385.

[159] S. Jonna, S. Satapathy, R.R. Sahay, Super resolution image de-fencing using a nonlocal nonconvex prior, Applied Optics 57 (2) (2018) 322–333.

[160] E. Nehme, L.E. Weiss, T. Michaeli, Y. Shechtman, Deep-STORM: super-resolution single-molecule microscopy by deep learning, Optica 5 (4) (2018) 458–464.

[161] T. Nguyen, V. Bui, D. Tran, G. Nehmetallah, Deep learning bi-telecentric digital holographic microscopy for aberration compensation applied to cancer cells, in: Imaging and Applied Optics, OSA Technical Digest, 2017 paper Tu2A.5.

[162] Y. Wu, Y. Rivenson, Y. Zhang, Z. Wei, H. Gunaydin, X. Lin, A. Ozcan, Extended depth-of-field in holographic imaging using deep-learning-based autofocusing and phase recovery, Optica 5 (6) (2018) 704–710.

[163] J. Lee, W. Jeong, K. Son, W. Jeon, H. Yang, Autofocusing using deep learning in off-axis digital holography, in: Imaging and Applied Optics, OSA Technical Digest, 2018 paper DTh1C.4.

[164] T. Zahavy, A. Dikopoltsev, D. Moss, G.I. Haham, O. Cohen, S. Mannor, M. Segev, Deep learning reconstruction of ultrashort pulse, Optica 5 (5) (2018) 666–673.

[165] Z. Göröcs, M. Tamamitsu, V. Bianco, P. Wolf, S. Roy, K. Shindo, K. Yanny, Y. Wu, H.C. Koydemir, Y. Rivenson, A. Ozcan, A deep learning-enabled portable imaging flow cytometer for cost-effective, high throughput, and label-free analysis of natural water samples, Light: Science and Applications 7 (66) (2018).

[166] Y. Rivenson, Y. Zhang, H. Gunaydin, D. Teng, A. Ozcan, Phase recovery and holographic image reconstruction using deep learning in neural networks, Light: Science and Applications 7 (2018) 17141.

[167] G. Zhang, T. Guan, Z. Shen, X. Wang, T. Hu, D. Wang, Y. He, N. Xie, Fast phase retrieval in off-axis digital holographic microscopy through deep learning, Optics Express 26 (15) (2018) 19388–19405.

[168] Y. Rivenson, Z. Göröcs, Gunaydin, Y. Zhang, H. Wang, A. Ozcan, Deep learning microscopy, Optica 4 (11) (2017) 1437–1443.

[169] R. Horisaki, R. Takagi, J. Tanida, Deep-learning-generated holography, Applied Optics 57 (14) (2018) 3859–3863.

[170] Y. Ma, X. Chen, W. Zhu, X. Cheng, D. Xiang, F. Shi, Speckle noise reduction in optical coherence tomography images based on edge-sensitive cGAN, Biomedical Optics Express 9 (11) (2018) 5129–5145.

[171] M.U. Ghani, W.C. Karl, CNN based sinogram denoising for low-dose CT, in: Imaging and Applied Optics, OSA Technical Digest, 2018 paper MM2D.5.

[172] H. Chen, Y. Zhang, W. Zhang, P. Liao, K. Li, J. Zhou, G. Wang, Low-dose CT via convolutional neural network, Biomedical Optics Express 8 (2) (2017) 679–694.

[173] T. Zeng, Z. Ren, E.Y. Lam, Speckle suppression using the convolutional neural network with an exponential linear unit, in: Imaging and Applied Optics, OSA Technical Digest, 2018 paper CW5B.3.

[174] G. Chierchia, D. Cozzolino, G. Poggi, L. Verdoliva, SAR image despeckling through convolutional neural networks, Proceedings of IEEE International Geoscience and Remote Sensing Symposium (2017) 5438–5441.

[175] P. Wang, H. Zhang, V.M. Patel, SAR image despeckling using a convolutional neural network, IEEE Signal Processing 24 (12) (2017) 1763–1767.

[176] Y. Li, Y. Xue, L. Tian, Deep speckle correlation: a deep learning approach toward scalable imaging through scattering media, Optica 5 (10) (2018) 1181–1190.

CHAPTER 6

Holographic Interferometry: From History to Modern Applications

MARC GEORGES, PHD

INTRODUCTION

Holography allows the recording and the reproduction of the wavefront scattered from an opaque object or transmitted through a transparent object. However, the phase and shape of that wavefront can change over time because of the deformation of the object. Holographic interferometry (HI) is the process of observing and recording the interference between two or more wavefronts and the quantitative interpretation of the interference pattern in terms of object modification.

Year 1965 marked the discovery of HI simultaneously by different groups. Some conceptually proposed interferometry through hologram or wavefront reconstruction [1,2], some just mentioned they observed it [3], and others showed experimental evidence through images [4–6]. The invention of HI is generally attributed to Powell and Stetson for having showed the first experimental evidence of time-averaged HI (TAHI) of vibrating objects during an OSA Spring Meeting [5]. Later that year, they submitted a manuscript in March to the Journal of Optical Society of America, which was published in December [6]. Meantime, Collier et al. showed real-time and double-exposure HI for the first time in a letter submitted in September 1965 and published in October the same year [4]; however, they attributed their observation to a Moiré effect.

The first utilization of HI came from Stetson and Powell [7]. For an expanded overview of the history of the discovery of HI, the reader should refer to several interesting and recent papers by Stetson [8,9]. This story lasts for more than 50 years now, and it would be difficult to summarize the many achievements in the field into a single chapter. Several excellent textbooks have been edited and provide a lot of detailed developments in all the directions taken by many groups since the HI invention [10–16]. This chapter will serve as an introduction to the HI technique.

Let us now recall that the phase of an object wave propagating in some material can be written as follows:

$$\phi = \frac{2\pi}{\lambda} nl, \qquad (6.1)$$

where λ is the wavelength in vacuum, n is the refractive index of the material, and l is the geometric path length traveled by the wave. In HI, the phase difference can be produced by any variation of the three quantities in Eq. (6.1). This leads to quite a large variety of HI configurations and applications, which have been demonstrated since almost the beginning of holography. We also can distinguish the different ways to record holograms. At the beginning of holography, this was achieved by photosensitive media, such as silver halides or others reviewed in Chapter 2 of this book. The wave coming from the object scene is recorded in a holographic physical support and readout later for object phase comparison. A key issue for the applicability of HI, especially in applications of industrial interest, is the performances of the hologram recording medium, not only in terms of usual figures of merit but also in terms of ease of use. Because of the tedious physico-chemical processing of traditional holographic recording media (such as photoplates or photothermoplastics), electronic recording of holograms directly on the imaging sensor has taken an important place in HI applications. Usually, one speaks about electronic speckle pattern interferometry (ESPI) and more recently digital holography (DH). The latter is covered in Chapter 5 of this book. Nowadays, it is usual to speak of "analog holography," when considering photosensitive plates, in comparison with DH.

In this chapter, we will review how these different hologram recording strategies can be applied in

Optical Holography-Materials, Theory and Applications. https://doi.org/10.1016/B978-0-12-815467-0.00006-2

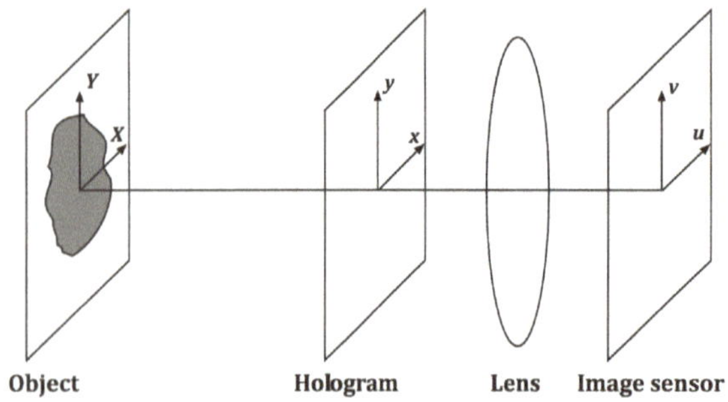

FIG. 6.1 Definition of reference systems.

metrology and nondestructive testing (NDT) techniques by interferometry. Before that, we will clarify the different reference axes systems that are used in the text and are consistent to other used in this book. As is shown in Fig. 6.1, the object is located in the plane (X,Y), the hologram recording plate in the plane (x,y), and the image sensor in the plane (u,v). Usually (but not mandatory, e.g., in DH), a lens is used in between the hologram plate and the sensor to form an image of the object on the sensor plane. For an ideal imaging system, and if the object image is well focused, the coordinates of the image plane are used instead of that in the object plane, taking into account the magnification of the lens.

We will see that, for the different hologram recording methods used, HI applications are globally identical. Before that, let us recall some basic hologram recording principles. In all cases, the interference of the reference and the object waves with electric fields, respectively $U_r(x,y)$ and $U_o(x,y)$, is recorded in the plane $(x,y,z=0)$. The resulting hologram irradiance is written as

$$H(x,y) = I_r(x,y) + I_o(x,y) + 2\sqrt{I_r(x,y)I_o(x,y)}\cos(\Delta\varphi(x,y)), \tag{6.2}$$

where $\Delta\varphi(x,y)$ is the difference of phases of reference and object waves

$$\Delta\varphi(x,y) = \varphi_r(x,y) - \varphi_o(x,y). \tag{6.3}$$

In analog holography, the hologram irradiance (Eq. 6.2) is transformed into amplitude or phase grating depending on the material. If the reference beam illuminates such grating, diffraction occurs, whatever the type of hologram. In any case, a wave containing the information of the object is diffracted $U_d(x,y)$. The latter propagates up to the image sensor plane, via an imaging lens. The diffracted wave can be written in the observation plane as $U_d(u,v) = A_d(u,v)\exp(i\varphi_d(u,v))$. It is generally assumed that the phase of the diffracted wave reproduces the phase of the object wave in the object plane $\varphi_d(u,v) = \varphi_o(u,v) \sim \varphi_o(X,Y)$. The latter is precisely the quantity, which is of interest in HI, as it is related to the object state at a given instant. In other hologram recording methods (speckle interferometry and DH), the phase can be obtained differently. However, it is the basic information that is made available for use in interferometric comparison for obtaining the optical path differences (OPDs). The latter can be observed from the reflection by solid objects (diffusive or specular) or from transmission in phase objects. Mostly we can distinguish the case of comparison of two waves and the integration of a very large number of waves (usually addressing the case of objects under vibration).

In this chapter, we will try to give an overview of almost 50 years in HI research and achievements. In Main Applications of HI section, we will start with the main applications of HI, which were developed through the use of analog holographic media. Most of these applications are still of interest today. However, we will see how the field evolved from these pioneering works to the modern approach using holographic recording by electronic devices. This evolution was mainly driven by pragmatic issues in view of industrial applications of HI, leading to the idea of an ideal system discussed in The Ideal Holographic System for HI section. In Analog HI section, we will review the attempts of industrialization of analog HI systems. Their limits led to consider electronic recording in HI in its first form known as ESPI, which is discussed in Electronic Speckle Pattern Interferometry section. The latter will also cover a particular configuration (speckle shearing interferometry), which is now recognized in the industry of NDT. Finally, in Digital Holographic Interferometry section, we will review the most recent achievements in digital HI.

MAIN APPLICATIONS OF HOLOGRAPHIC INTERFEROMETRY

Displacements and Deformations of Solids

Interferometry with two waves

The determination of displacements of solid objects with diffusive reflection was the first main application of HI. In two-wave interferometry, we consider an object, which is observed at two instants between which it has moved (rigid body motion) or has been deformed.

There are different ways to observe OPDs from the interference using wavefronts diffracted by a hologram: the real-time holographic interferometry (RTHI) and the double-exposure holographic interferometry (DEHI). Mainly this distinction is valid in the case of analog holography, whereas they are confused in the case of electronic holography. In the following, we will explain them in the case of analog holography.

RTHI consists in two steps, explained in Fig. 6.2. In the first step (Fig. 6.2A), a hologram $H(x,y)$ is recorded at instant t_0, when the object is in its first state (wavefield U_o). After processing of the hologram, the latter is readout. The diffracted wave U_d reproduces U_O. In the meantime, the object has changed, and the corresponding wavefield is transmitted through the hologram plate. Therefore, at the current instant t of the readout, two waves are present after the plate: the transmitted wave U_t, which represents the new object wave U'_O and the previous object wave U_O. Both waves interfere, giving rise to an irradiance pattern in the image plane:

$$I(u,v,t) = \left(U_o(u,v,t_0) + U_o^{'}(u,v,t)\right)\cdot\left(U_o^{*}(u,v,t_0) + U'^{*}_o(u,v,t)\right) \quad (6.4)$$

Assuming that there is a perfect conjugation between the image and object planes, respectively (u,v) and (X,Y), we can confuse these two planes. Furthermore, except in the cases we need it, we will omit dependency in the coordinate system. The irradiance above can be rewritten simply as

$$I = I_{av}[1 + m\cos\phi] \quad (6.5)$$

with $I_{av} = I_o + I'_o$, the average irradiance, $m = 2\frac{\sqrt{I_o \cdot I'_o}}{(I_o + I'_o)}$ the contrast, and the optical phase difference

$$\phi = \varphi'_o - \varphi_o \quad (6.6)$$

The latter can change in function of the time according to the state of the object, and the interference pattern shows moving fringes, sometimes called *live fringes*.

The second case is referred as the DEHI, which consists in three steps, as is shown in Fig. 6.3. Both first steps assume recording of two holograms in the same plate: H corresponding to U_o at time t_0 (Fig. 6.3A) and H' corresponding to U'_o at time t_1 (Fig. 6.3B). The third step (Fig. 6.3C) is the readout of both holograms with the reference beam: two diffracted waves are produced, U_d and U'_d. Like in the RTHI case, their superimposition leads to an interference pattern, which takes exactly the same form as Eq. (6.4), with the only difference that both diffracted waves are now frozen in time (*frozen fringes*). This method does not allow following continuously the object displacement but rather shows its changes between two defined instants.

A variant of DEHI is "sandwich holography" which consists in recording the two holograms on separate plates and recombine them at the readout [17]. The advantage of the latter is that, by tilting one of the plate with respect to the other one, it is possible to compensate spurious fringes appearing if rigid body motions appear between the two exposures.

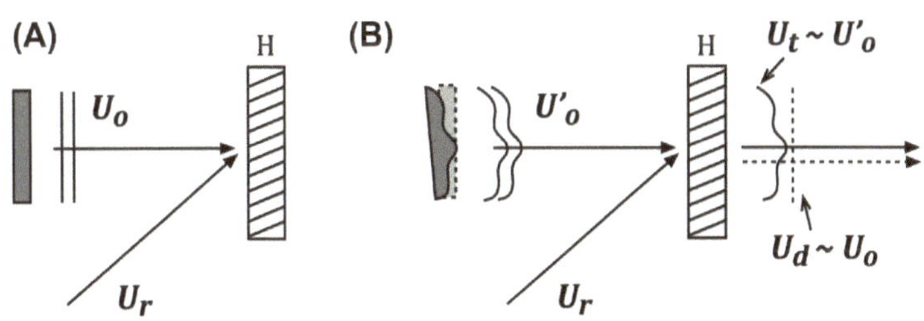

FIG. 6.2 Analog real-time holographic interferometry.

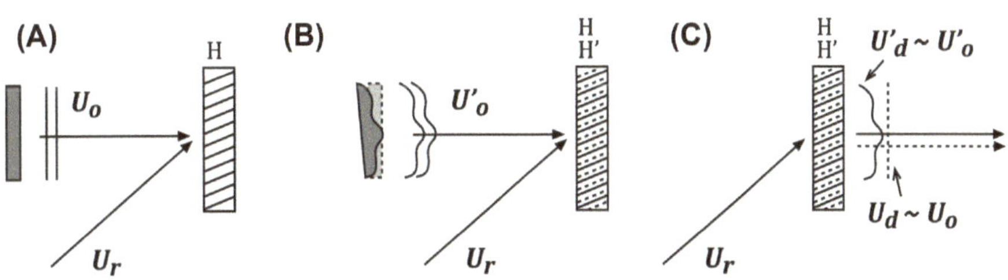

FIG. 6.3 Analog double-exposure holographic interferometry.

Remarks about the stability of the setup and the object

The most known application of HI is the measurement of displacement field of solid objects under external load. The way the hologram is captured influences strongly the applicability of HI in a given study case. Because the recording of holograms relies on the fluence of recording beams on the holographic material, the available laser power and the amount of light actually reaching the material will set the hologram recording time. The latter must be much smaller than the characteristic duration of the object change. It is generally assumed that, for properly recording a hologram, the stability of the whole experiment must be such that the hologram pattern $H(x,y)$ has residual movement smaller than a 10th of the laser wavelength. A direct consequence is that specific environment needs to be used for hologram recording, often limiting HI to laboratories. This influences the stability requirements of the environment but also the type and amplitude of object load that can be considered in the experiment. This is not an issue in the case of a controllable mechanical stimulus: a hologram is captured before loading and, when the load is applied, DEHI could be applied only if the object is stable while being deformed, allowing recording of a second hologram. In the case of thermal stimuli, the object generally undergoes changing deformation, and it is difficult to record a hologram under such circumstances. In this case, only RTHI can be applied, based on hologram recording prior to the stimulus. In the case of perturbed environments and/or rapidly evolving objects, the use of pulsed lasers (with a few nanosecond duration of light emission) is the only way to record holograms. Indeed, this type of laser system can bring the necessary hologram recording fluence while exposing the object to the light in such a short time (ns) that it literally freezes the object movement. After the pioneering experiments of HI, mostly achieved with continuous gas lasers, the beginning of the 1970s saw the rise of pulsed lasers in the first industrial HI attempts in NDT experiments (to be discussed later in this chapter) [18].

Interpretation of optical path differences in terms of displacement

It is important now to interpret the measured OPD in terms of the surface change undergone by the object under load. The general situation is depicted in Fig. 6.4. A laser beam is split in two parts by a beam splitter. One of them is used to illuminate the solid object, which is supposed scattering. Light reflected by the object partly travels toward the holographic plate where it interferes with the reference beam. Once a single hologram H or a double-exposed hologram is recorded, the readout is performed, giving rise to the interferogram (Eq. 6.5).

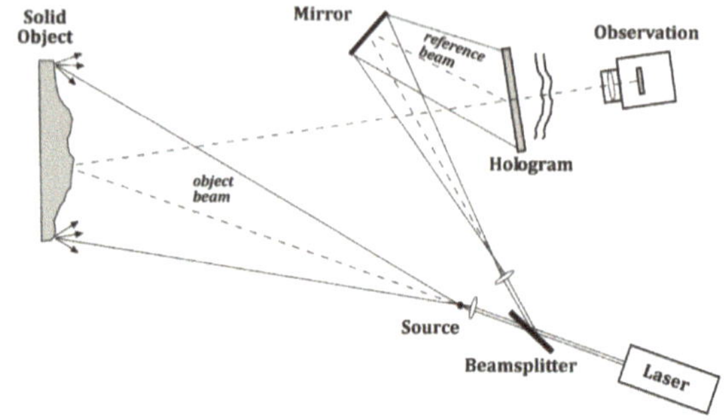

FIG. 6.4 Displacement of solid objects.

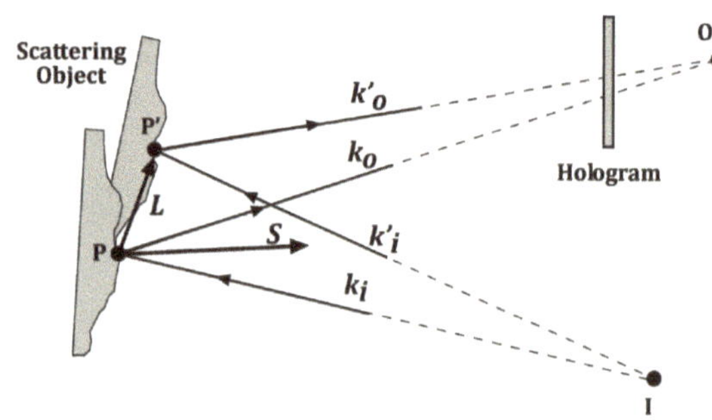

FIG. 6.5 Sensitivity vector.

We summarize here the interpretation model established by Sollid [19]. If the object has undergone some displacement, the two phases in Eq. (6.6) are different and the interferogram (Eq. 6.5) shows a series of fringes superimposed to the object image. The desired quantity is displacement of each point P of the object. This can be performed through the sensitivity vector S. The situation is explained in Fig. 6.5. We consider a scattering object with a point P(X,Y) in the first object state. It is illuminated by light ray with wave vector $\boldsymbol{k}_i$, and an observer receives the reflected light by a ray with wave vector $\boldsymbol{k}_o$. The phase of the object wave is defined by the travel between the source and the observation point via point P. When the object moves, the point is displaced to P$'(X,Y)$, and the vector $L(X,Y)$ is the displacement from P(X,Y) to P$'(X,Y)$. The illumination and observation vectors become, respectively, $\boldsymbol{k'}_i$ and $\boldsymbol{k'}_o$. The phase variation can be derived from vector calculation. In practice, the distance of both the observation O and illumination I locations to the object is much larger (a few tens centimeters) than the displacement (a fraction to a few micrometers). It comes that $\boldsymbol{k'}_o \approx \boldsymbol{k}_o$ and $\boldsymbol{k'}_i \approx \boldsymbol{k}_i$. Under such assumption, the phase difference observed can be simply related to the displacement by the following expression

$$\phi = S \cdot L = (k_o - k_i) \cdot L = \frac{2\pi}{\lambda}(\widehat{e}_o - \widehat{e}_i) \cdot L \tag{6.7}$$

The sensitivity vector S is defined in each point of the object surface as the bisector of illumination and observation wave vectors. The latter have a norm inversely proportional to the wavelength, as indicated in Eq. (6.7), where we have introduced the unit vectors along observation and illumination directions, respectively, $\widehat{e}_o$ and $\widehat{e}_i$. In general, if one wishes to determine the three components of displacements, we need three measurements of the phase difference ϕ for three different sensitivity vectors. This can be done, for instance, by considering one observation system with three illumination directions, which are not coplanar [10]. Eq. (6.7) gives rises to system of three equations with three unknowns, which can be inverted for finding the three components of $L(L_X,L_Y,L_Z)$ in every point of the object surface. It was noted by Stetson [20] that the above relationship allows measurement of displacements in every pixel relative to its neighbors. Nevertheless, we cannot access the absolute displacement in each pixel. For that we need to know a priori the displacement of one of the pixels. For introducing the capability of absolute measurement, we can rewrite the system (6.7) as $\phi = S \cdot L + N$, with N an unknown constant, which adds to the three unknown components of L. The new system may be solved only by adding a sensitivity vector, in our case four instead of three [20], e.g., by adding one illumination.

From Fig. 6.5 and Eq. (6.7), we see that only the component of L lying in the **IOP** plane and parallel to S can be measured. Moreover, components of L perpendicular to S will lead to $\phi = 0$. In most practical cases, a single illumination direction quasicollinear to the observation direction is used and the object is set perpendicular to the observation direction. Therefore the measurement is limited to the observation of out-of-plane displacements $L_\perp$ and Eq. (6.7) becomes simply

$$\phi = 4\pi L_\perp / \lambda \tag{6.8}$$

From the latter and the fact that the cosine in Eq. (6.5) is maximized for integer multiple of 2π, it comes that distance between two consecutive maxima in the fringe pattern corresponds to an out-of-plane displacement $L_\perp$ equal to $\lambda/2$.

Other simplified cases were presented in literature, which allow the determination of in-plane displacement as was first proposed by Ennos [21].

Once the surface displacement or deformation L has been determined, it can be the starting point of further experimental mechanics investigations, like determination of strains and stresses. This activity started in the early stage of HI, with works of Aleksandrov et al. [22] and Ennos [21]. Already then, HI showed its potentialities for materials evaluation, in particular for potential replacement of strain gauges, for it is contactless and over a very large number of measurement points (virtually, every discrete point (X,Y) in the interferogram is an optical strain gauge). Discussing the potentials of HI in experimental mechanics clearly overpasses the purpose of this chapter. Therefore, the reader should refer to textbooks already mentioned earlier [12–16].

Vibration Measurements

A specific technique can be applied to the case of vibrating objects: the TAHI. It was first put in evidence by Powell and Stetson [5,6]. In the case the object is stationary, the object wavefield in the hologram plane is $U_o(x,y) = A_o(x,y)\exp(i\varphi_o(x,y))$. If it is vibrating sinusoidally with angular frequency ω, an amplitude φ_A, and a phase retardation φ_v with respect to the excitation (also called the vibration phase), the object wavefield is time dependent as follows:

$$U_O(x,y,t) = A_O(x,y)exp\{i(\varphi_O(x,y) + \varphi_A(x,y)sin(\omega t + \varphi_v))\}. \tag{6.9}$$

In TAHI, the hologram is recorded during the vibration of the object and over a time T much longer than the vibration period $2\pi/\omega$. As illustrated in the example of Fig. 6.6A, an object vibrates with a mode shape consisting of three nodes n1, n2, and n3, which are the loci of points that do not move during vibrations. Between the nodes, the object moves toward two extreme positions (positive and negative) and the loci of points experiencing the largest displacements are antinodes. Therefore, during the recording step, the medium records an average hologram $H(x,y)$, in which the object wave is given by Eq. (6.9):

$$\begin{aligned} H(x,y) &= \frac{1}{T}\int_0^T U_o(x,y,t)dt \\ &= U_o(x,y)\frac{1}{T}\int_0^T exp\{i\varphi_A(x,y)sin(\omega t + \varphi_v)\}dt \end{aligned} \tag{6.10}$$

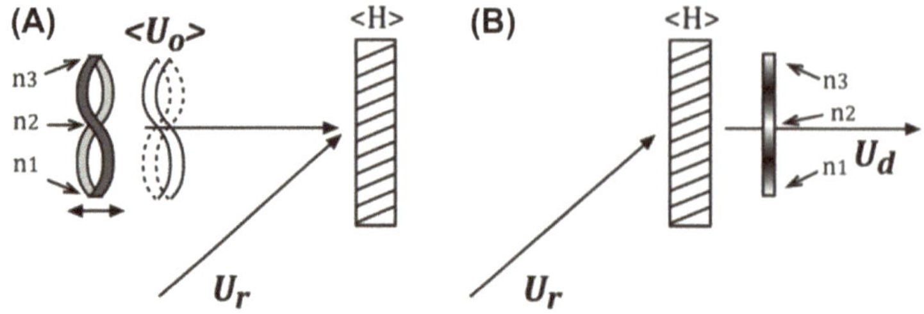

FIG. 6.6 Time-averaged holographic interferometry: **(A)** first recording step, **(B)** readout step.

It must be noted that, in the integration process, the vibration phase φ_v will be lost. It is useful to define the characteristic function as follows:

$$M(t) = \frac{1}{T}\int_0^T exp\{i\varphi_A(x,y)sin\omega t\}dt. \tag{6.11}$$

At the readout, we observe the irradiance of the diffracted image, hence the square of Eq. (6.10). For integration times much longer than the vibration period (taking the limit of Eq. (6.11) for $T\rightarrow\infty$), it comes that the characteristic function is proportional to the zero-order Bessel function with argument φ_A. Hence the irradiance of the diffracted beam in the image is

$$I_d(x,y) \div J_0^2(\varphi_A(x,y)) \tag{6.12}$$

whose maximum is found at the vibration nodes (n1, n2, n3) or where the object is not moving (i.e., attachment points). This results in a bright fringe. Out of the nodes, for increasing values of the phase, the intensity of fringes varies periodically and the fringe modulation decreases when φ_A increases.

Fig. 6.7A and C shows two modes of vibration of a cantilever beam clamped on its border. Different mode shapes appear at resonance frequencies. The bright fringes are clearly visible, as well as the decrease of intensity and visibility of fringes out of the nodes.

An alternative method to the TAHI for vibration analysis is the stroboscopic technique [23]. It consists in freezing the movement of the object at its maximum displacement position during the vibration. For that, an additional system like rotating chopper or deflection by acoustooptic devices is used to strobe the laser light in synchronization with the vibration, at a given angular frequency ω. Therefore, the interference pattern is stable and shows the vibration modes, but in this case, the fringe profile is sinusoidal and expressed as Eq. (6.5),

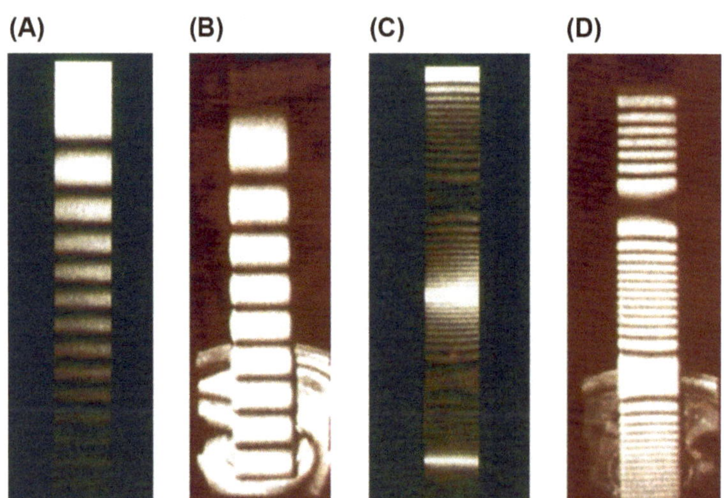

FIG. 6.7 Two resonance modes of a cantilever beam clamped on its border. **(A)** and **(C)** are obtained by TAHI with classical photoplates and **(B)** and **(D)** the corresponding modes observed by stroboscopic HI. (Courtesy P. Smigielski.)

as can be observed in Fig. 6.7B and D, corresponding to the same modes observed by TAHI (Fig. 6.7B and D, respectively). The contrast is the same everywhere, by opposition to the TAHI Bessel fringe profile.

Nondestructive Testing

One of the most impressive uses of HI is in NDT. NDT is generally associated to a group of methods dealing with localization of flaws (or defects) in materials or structures. These can be fractures in solids, debonds, delaminations in composite laminates, presence of foreign materials, etc. Because it is based on imaging, holographic NDT (HNDT) offers the potential of decreasing the inspection time compared with tedious pointwise inspection techniques, such as ultrasound testing.

To the best of our knowledge, the very first demonstrations of HNDT were performed almost simultaneously in 1969 by Friesem and Vest [24] and Grant and Brown [25]. The first ones used RTHI for finding microcracks around holes drilled in metal. The method involves giving some stress during the readout of the hologram. The authors tried thermal stresses (cooling and heating) and mechanical ones. In both cases, a fringe pattern can be observed, here in an out-of-plane configuration, and that shows a clear break in the pattern where the crack is present. They emphasize a very important feature of HNDT, which is the appropriateness of the applied stress to easily provide sufficient variations in the fringe pattern for flaw detection [26]. This is a major concern in HNDT application since then, for the local fringe pattern variation strongly depends on the nature of the flaw, its location and extent, the surrounding material and the geometry of the structure, as well as the loading method. The second group mainly used DEHI in the inspection of tires and rubber bonded on metal. They considered loading by pressure variation. This works fine for debonds where air is imprisoned into cavities in soft materials. The inspection of tires is still today one of the best successes of HNDT, which is marketed now in the form of shearography systems (which will be discussed later).

The early 1970s saw intensive researches in HNDT in the field of aeronautics. This can be found in the report by Erf et al. [27]. This vast study addressed different applications as well as the applicability of interferometric techniques in various cases of interest, including assessment of aspects like the environment of work, the surface properties, and the effect of rigid body motions on deformation measurements, to name a few. During all seventies, HNDT generated a lot of expectations in many engineering fields (aerospace, automotive, civil engineering) as reported in reviews of that period [10]. While pioneering activities of HNDT were done in the United States, European companies and institutes became very active soon after, especially in aerospace [28,29]. Additional investigation fields came afterward like art and cultural heritage [30].

After an intense decade of HNDT activities, Vest published in 1982 an excellent review paper [31] in which he analyzes all HNDT literature since its beginning. After documenting the different types of flaws observable in all reported HI applications, the author develops his views about the future of the activity. This visionary paper establishes what will be the bases of all subsequent researches that are still active today and that will be partly reviewed in this chapter. In particular, the use of new recording materials, the use of digital and electronic technologies for automated readout and analysis of fringe patterns, the development of analytical and computational methods for quantitative prediction and analysis of the results of HNDT and, finally, the development of optical components and more robust lasers for holographic systems are discussed.

Surface Contouring

In previous applications, the OPD resulted from a constant sensitivity vector **S**, whereas the positions of surface points change. In contouring, the approach is reversed: while the object remains unchanged, any of the parameters of the sensitivity vector can vary: the wavelength, the directions of illumination and observation, the refractive index.

Soon in the beginning of HI, holographic surface contouring was first discussed conceptually by Haines and Hildebrand in 1965 [32], with their first experimental demonstration in 1966 [33]. Holographic contouring of an object is performed by observing interference fringes on the object image arising from variations in the illumination of the object. Here, like in Displacements and Deformations of Solids section, the interference of two waves is required. Whether we work with a single hologram (RTHI) or two holograms (DEHI), the principle of contouring can be declined in three methods. The first one considers the translation of the illumination source between the two instants of observation of the object [34], the second one considers a change of wavelength [34], and the third one considers a change of the refractive index along the object beam [35,36].

The case of the source translation is shown in Fig. 6.8. In the case of RTHI, the hologram is recorded with a source **I**. At the readout, the object is illuminated with the source displaced to a new position **I'** (translation D). The OPD between the two resulting object

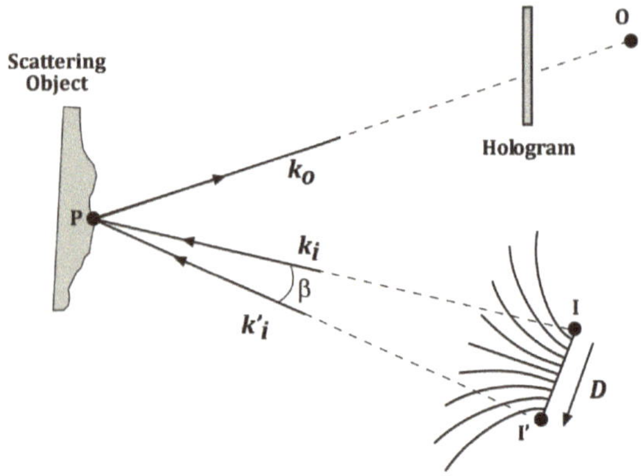

FIG. 6.8 Source translation for surface contouring.

wavefronts can be derived, similar to what is done in the case of displacement measurement. It is also assumed that D is very small compared with the distance between the source and the object. It comes that, for a given surface point **P**, the two illumination waves having traveled different distances with the wave vectors $\boldsymbol{k}_i$ and $\boldsymbol{k'}_i$, the OPD can be written as follows

$$\phi = \boldsymbol{D}\cdot(\boldsymbol{k}_i - \boldsymbol{k'}_i) = \frac{2\pi}{\lambda}\boldsymbol{D}\cdot(\widehat{\boldsymbol{e}}_i - \widehat{\boldsymbol{e}'}_i) \tag{6.13}$$

The phase difference (Eq. 6.13) tags the point **P** at the surface. Another point **P′** will be tagged with another value of phase because the geometry of rays will differ from that of point **P**. It is common to understand the situation as if the object was illuminated by the interference pattern of two close point sources **I** and **I′**. As is shown in Fig. 6.8, these two sources generate spherical waves, which interfere to form a set of hyperbolic equiphase lines, with a symmetry of rotation along the axis **I**–**I′**. In the simple case where the illumination beam is collimated, the resulting pattern is constituted by equidistant flat equiphase surfaces [14]. The pattern intersects the object and tags the latter with a set of phase values, which are related to a specific height. The interdistance between two fringes corresponds to variation of object height Δh, which is function of the wavelength and the angle between both beams:

$$\Delta h = \frac{\lambda}{2sin(\beta/2)} \tag{6.14}$$

This relationship also shows that the sensitivity of the method can be adapted by changing the angle β.

The case of the wavelength change is shown in Fig. 6.9. A hologram of the object is recorded wavelength λ_1. At the readout, the object is present but the laser source has another wavelength λ_2. This can be obtained by considering two lasers with different wavelengths or a single laser with tunable wavelength.

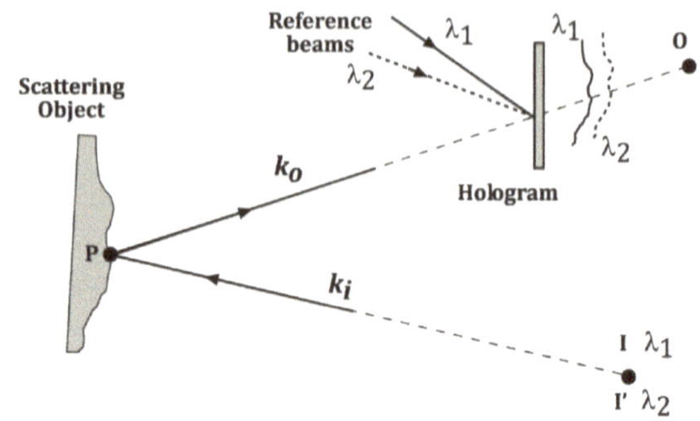

FIG. 6.9 Wavelength change for surface contouring.

At the readout, the hologram diffracts a wave reproducing the object as it was recorded with λ_1. However, as illustrated in Fig. 6.9, this requires to change the incidence angle of the reference beam to avoid lateral displacement that is usual to incorrect readout wavelength. Alternative schemes were developed to avoid that problem [14] . There is also an axial displacement along the line of sight (the z-direction) [16]. Usually, the illumination source and the observation locations are very close to one another, at a distance z from the object, the light travels a distance $l = 2z$ between **I** and **O**. Using Eq. (6.1), considering a refractive index $n = 1$, the OPD between the two waves can be expressed as:

$$\phi = \varphi_{\lambda_1} - \varphi_{\lambda_2} = 4\pi z\left[\frac{1}{\lambda_1} - \frac{1}{\lambda_2}\right] = \frac{4\pi z}{\lambda_{eq}}. \tag{6.15}$$

Their interference produces a fringe pattern projected to the object image and which has an equivalent wavelength λ_{eq} given by

$$\lambda_{eq} = \frac{\lambda_1\lambda_2}{|\lambda_1 - \lambda_2|} \tag{6.16}$$

The distance between two consecutive fringes corresponds to two values of ϕ that differ from one another by 2π; they correspond to a height variation given by

$$\Delta h = \frac{\lambda_{eq}}{2} \tag{6.17}$$

The choice of the couple of wavelengths sets the sensitivity of the method. The equivalent (sometimes called synthetic) wavelength is much larger than the two laser lines used. For instance, an argon ion laser with two lines $\lambda_1 = 514\ nm$ and $\lambda_2 = 488\ nm$ will lead to $\lambda_{eq} = 9.47\ \mu m$. For contouring objects with height variations of a few centimeters, one must find lasers with shorter wavelength separation to obtain a suitable λ_{eq}.

In deformation measurement by HI, methods using two wavelengths were used by extrapolation of the two-wavelength contouring [37,38]. The goal was to extend the displacement measurement range in the deformation application (Displacements and Deformations of Solids section).

The last contouring method is based on the variation of refractive index between the hologram and its readout. By using similar considerations as above, it comes that a difference of height between two fringes is obtained for some refractive index variation Δn along the line of sight:

$$\Delta h = \frac{\lambda}{2\Delta n} \tag{6.18}$$

It is much less used that the two other ones since it assumes that the object is placed in an immersion tank in which one can introduce liquid and change the latter (or relative concentration of compounds) for varying the refractive index.

Optical Path Differences in Transparent Media

Yet another important application of HI is the one of measurement of OPDs in beams going through transparent objects. To the best of our knowledge, the first case reported is the observation of refractive index variations in transparent media by Heflinger et al. [39].

A typical configuration is shown in Fig. 6.10. Without loss of generality, we consider a volume, represented here by a parallelepipedic cell, with a certain extent $[Z_1,Z_2]$ along the line of sight and which is illuminated by a collimated beam collinear to the line of sight. The volume can be a vessel with two windows allowing transmission of the object beam and which contains liquid or gas, where some phenomenon occurs that changes the refractive index. Before the recording of the hologram, the refractive index is supposed to be isotropic and equal to n. The readout takes place for a different situation where the refractive index has changed to $n'(X,Y,Z)$, because of some phenomenon occurring in the cell.

The OPD observed at the level of the object is related to the integral of the refractive index variation along the line of sight in a certain volume between Z_1 and Z_2.

$$\phi(X,Y) = \frac{2\pi}{\lambda}\int_{Z_1}^{Z_2} (n'(X,Y,Z) - n)dZ \tag{6.19}$$

Depending on the application, the above expression can be interpreted in terms of physical properties changing in the volume. In the case of aerodynamic flow visualization, the refractive index n of the gas inside the volume is related to the density by the Gladstone-Dale law:

$$n = K\rho + 1 \tag{6.20}$$

K is the Gladstone-Dale constant that depends on the gas and ρ is the density that depends on pressure P and temperature T, through the law of perfect gases. Therefore, HI allows observation of temperature or pressure fields in aerodynamics experiments, as was documented by many groups in aerospace and ballistic researches [40–42]. These works consider double pulse lasers because their extremely short pulse duration (a

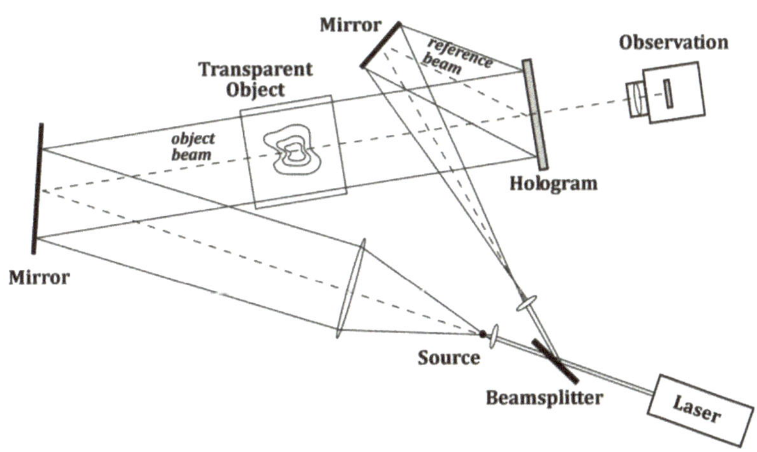

FIG. 6.10 Measurement of refractive index in transparent media.

few nanoseconds) allows freezing the holograms in time in a highly perturbed environment, characteristic of wind tunnels.

Another application was the study of heat transfer in gases and liquids [10]: variations of temperatures can be related to variations of n, through the Gladstone-Dale equation in gases or through empirical equations in liquids.

It is difficult to invert Eq. (6.19) for determining the refractive index changes along the three directions. In cases where the phase object has small dimension along Z, thus assuming no variation of n along that direction, $\Delta n(X,Y)$ can be determined. For phenomena occurring in the volume, only qualitative interpretation can be made, except for specific cases of variations with some symmetry or no variation along one direction (two-dimensional refractive index fields) [10,14]. In more general cases, a quantitative interpretation requires different lines of sight and tomographic reconstruction [10,14]. Alternative schemes of transparent object illuminations make use of a diffuser, placed either in front or after the cell in the object beam [14]. This increases the flexibility of the setup at the level of the observation system (which becomes similar to the one used for a diffusive solid object) and for the illumination alignment.

THE IDEAL HOLOGRAPHIC SYSTEM FOR HOLOGRAPHIC INTERFEROMETRY

The Ideal Holographic Measurement Device

Even though HI is unique and powerful technique for measuring deformation, mode shape visualization, or in NDT, it experienced difficulties to be accepted by industrial end users. A lot of developments have been achieved to make HI a reliable and attracting tool. In early stages of HI, most of the experiments remained at the laboratory level, requiring large cumbersome lasers and specialists to interpret the interferogram. Moreover, the holographic recording media generally necessitate complicated chemical processing (in the case of photoplates) or additional electrical charging and heating/cooling devices (in the case of photothermoplastics). These processes are time-consuming, and the hologram is not useable before some tens of seconds if not minutes. This was the main drawback of holography based on conventional analog recording media for its industrial transfer.

The recommendations, made by Vest early 1980s [31], and already discussed earlier, to make HI systems better accepted by potential users are still valid today. We will also add other features that can help make the holographic systems viable commercially. Hereafter we draw a list of such features for an ideal holographic measurement device, which we have established from our own experience in two decades of interaction with industries (end users, manufacturing of holographic components) as well as from other groups extremely active in the field.

1. *Compactness.* Some measurement cases need a small lightweight device that can be placed in any position with respect to the object under test.
2. *Low cost.* This condition is mainly influenced by the laser source but also by the quantity of consumables. The latter is generally important with photoplates and photothermoplastics. This also means that the measurement head has to incorporate the least complicated and expensive components possible.
3. *Versatility.* Because such a device is an investment for the potential user, it may be useful if it can be adapted or adaptable to different applications.
4. *User-friendliness.* The measurement procedure has to be as simple as possible in order to not require an optics specialist for handling a measurement. This means no adjustment or the fewest possible.
5. *Large area of measurement.* This can reduce inspection time, but intrinsic lateral resolution should be put in the balance.
6. *Real-time operation.* No significant processing time between taking the hologram capture and the display of results.
7. *Robustness.* More often than they should, optical devices are very sensitive to environment perturbation such as vibration, temperature fluctuation, humidity, and dust. The system should be designed so that it can be handled in industrial environment without the need for constant recalibration or worst operational failure.
8. *Allow quantified phase measurement.* This is probably one of the most important issues with the aim to determine the object surface displacement from the phase ϕ in the interferogram (Eq. 6.1). Because of its importance, we will review its different steps in the next subsection.

To summarize: an ideal system must be robust, compact, affordable, with fast operational time, and easy to use and allow direct interpretation of the results on a large variety of application cases.

Phase Quantification

The interpretation of the phase from the interferogram generally requires an associated processing, called phase quantification. This processing aims at computing automatically the value of phase ϕ, from the fringe pattern image, which in turn is used to calculate the displacement or refractive index variations. Evaluation of fringe pattern is an activity, which really started in the 1980s, benefiting from the rise of personal computers and ease of capture of digitized images from CCD cameras. The reader should refer to a very good textbook edited in 1993 for a broad view of the different fringe pattern analysis strategies [43]. Since then, the field of fringe evaluation became a strong topic of research, with many conferences devoted to the subject (for instance the "Fringe" conference series [44–46]). It is difficult to give an exhaustive list and of fringe processing methods, leading to phase quantification. Here we present what we think are the more applied ones.

The first methods used were based on the intensity of interferograms (Eq. 6.5). They consist in several steps: tracking the fringes to trace a skeleton of maxima pattern, interpolation, and fringe numbering to reconstruct phase values. A good review can be found in Ref. [47]. These simple methods are applied on single interferogram, with fringes of different natures (sinusoidal, Bessel-like in TAHI, etc.), and, contrarily to methods presented hereafter, they do not require adaptation of the interferometer. Their drawback is that they can suffer from noise and they do not provide the sign of the phase.

Modern automated phase quantification techniques are mostly based on heterodyning, i.e., on the inclusion of an additional phase term in the argument of the cosine of Eq. (6.5). Heterodyning can be implemented temporally or spatially. We present the two usual methods.

The first one is the phase-shifting or phase-stepping (PS). It was first proposed by Hariharan [48] and generated a lot of possible algorithms, which are deeply discussed in Refs. [14,49,50]. PS consists in acquiring several interferograms with known phase steps introduced between the acquisitions:

$$I_k = I_{av}[1 + m\cos(\phi + \beta_k)] \quad (6.21)$$

with $k = 1, \ldots, N$ (N is an integer greater than or equal to 3) and the phase step β_k. The computation of the phase ϕ is carried out following one or another algorithm depending on the number N and the value of the additional constant phase β_k at each step. Eq. (6.22) shows the case of the four-frame algorithm with four images phase-shifted by $\pi/2$:

$$\phi = \arctan\left(\frac{I_4 - I_2}{I_1 - I_3}\right) \quad (6.22)$$

The most usual PS is the temporal PS: several interferograms are acquired sequentially with a suitable phase increment β_k. In practice, it can be achieved by translating a folding mirror in the reference arm, but there are several alternative ways to shift the phase [49]. In the case of RTHI, the PS is applied during the readout step, shifting the phase of the diffracted object image. In DEHI, this is not applicable if one uses a single reference beam, as suggested in Fig. 6.3C. Indeed, shifting the phase of the reference beam will have the same effect on both diffracted images. Therefore, two separate reference beams need to be introduced in the setup, each of them recording a separate hologram. This configuration is the so-called two-reference DEHI, in which the PS is introduced on only one of the reference beam, thus on one of both diffracted wave, preserving the other one. This complexity caused that DEHI was much less interesting for industrial application of analog HI.

It is worth to note that PS generally requires that the phase step is the same between each acquisition and that both the average intensity I_{av} and the contrast m do not vary during the acquisition sequence; otherwise errors may arise in the phase computed by Eq. (6.22). Fig. 6.11 illustrates the result of applying PS in the case of a plate undergoing a tilt. Fig. 6.11A is one of the N interferograms described by Eq. (6.2), and Fig. 6.11B is a gray level image of the OPD (usually

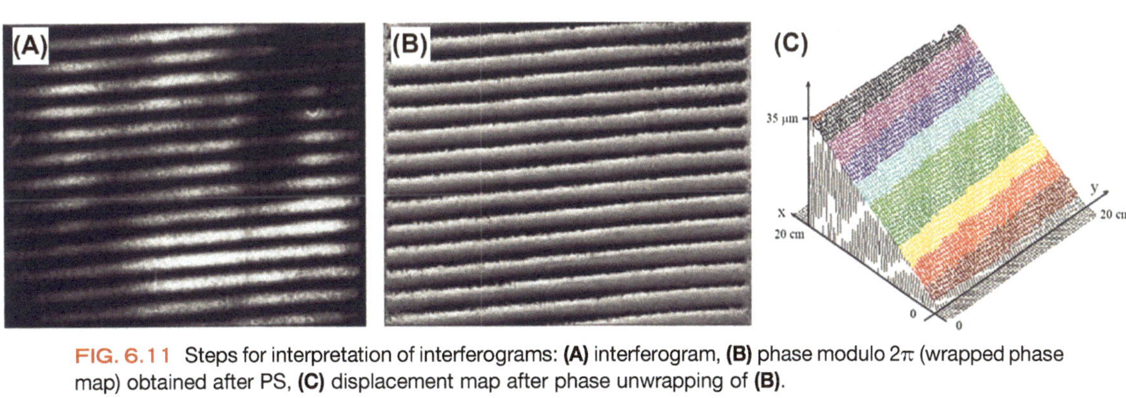

FIG. 6.11 Steps for interpretation of interferograms: **(A)** interferogram, **(B)** phase modulo 2π (wrapped phase map) obtained after PS, **(C)** displacement map after phase unwrapping of **(B)**.

called wrapped phase map) calculated through Eq. (6.3). The resulting phase values are ranged between 0 (black level) and 2π (white level) because of the cyclic nature of Eq. (6.3). In addition, and to illustrate the PS principle, Fig. 6.11 provides one of the two main types of displacement field measured by HI: a rigid body motion consisting of a rotation around an axis lying in the object surface and oriented parallel to the fringes.

In the case of dynamic objects, the application of PS is impossible because the phase is varying too fast compared with the acquisition time of the PS sequence. Therefore, PS is applicable for quasistatic or slowly varying objects, for which I_{av}, m, and ϕ do not vary significantly. For dynamic objects, other phase quantification techniques were proposed. One is PS applied spatially instead of temporally: various optical strategies (reviewed in Ref. [51]) produce a multichannel system in the interferometer in order to simultaneously obtain phase-shifted images captured by separate cameras.

Finally, other methods, based on Fourier transform (FT), use a single interferogram [51]. Fig. 6.12 illustrates the principle of the method. Fig. 6.12A shows the interferogram with the carrier frequency fringe pattern, which is written as follows:

$$I = I_{av}[1 + m\cos(\phi + 2\pi f_0 x)], \qquad (6.23)$$

with f_0 the carrier frequency. The Fourier spectrum of Eq. (6.23) shows a central peak with two symmetric side lobes (Fig. 6.12B). The latter contains information on the phase ϕ, which is extracted by suitable filtering in Fourier space and taking the inverse FT of the result, as is shown in Fig. 6.12C. The latter illustrates the second type of displacement field observable in HI: the deformation of a solid object, which generally provides fringe patterns that have no linear shape, contrarily to rigid body motions. Here the object is a plate attached on its four borders and mechanical finger pushes it in the middle of the rear side, causing a deformation looking like a hill.

The FT method requires adding a spatial carrier to the phase between the hologram recording and the readout in RTHI or between the two holograms recordings in DEHI. This can be achieved by translating the object illumination point source. Such translation creates a fringe pattern, which can be assimilated to projecting a pattern of hyperbolic fringes formed by interference between two point sources. At long distances, such pattern is similar to a pattern of rectilinear fringes.

The phase calculated by any of the above methods (PS or FT) is obtained modulo 2π (wrapped phase map), and a further step, called phase unwrapping, is used to eliminate the 2π phase jumps [52]. From this monotonic phase map, one can compute the variation of optical path in each point of the object scene. Fig. 6.11C shows the results of displacement map after phase unwrapping of Fig. 6.11B and conversion of phase to displacement through Eq. (6.8).

ANALOG HOLOGRAPHIC INTERFEROMETRY

Toward the Industrial Use of Analog Holographic Interferometry

The beginning of HI history is based on analog holography with silver halide photoplates [53]. The search for fast and user-friendly systems in HI yields to the development of liquid gates. They consisted in encapsulated plate holder with windows, in which different liquids for sensitization and chemical processing were acting

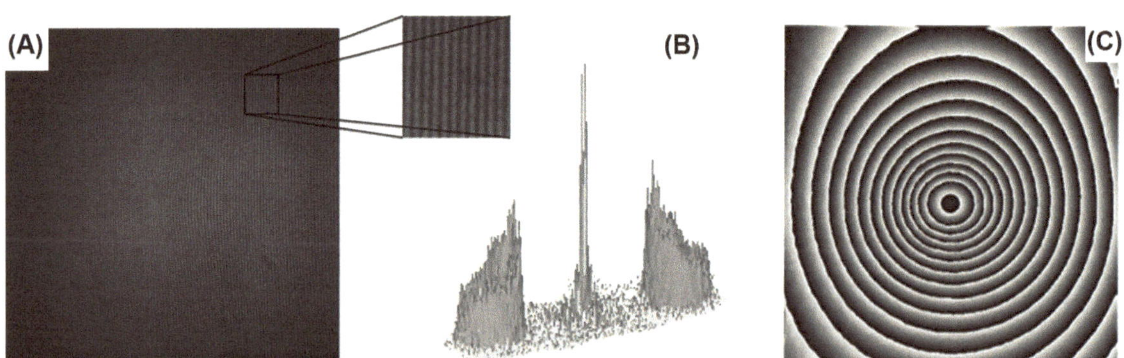

FIG. 6.12 Principle of FT application in single interferogram analysis: **(A)** interferogram with carrier frequency added, **(B)** FT of **(A)**, **(C)** phase modulo 2π obtained by inverse FT of one of the side lobes of **(B)**.

in situ [54–56]. Yet this remained unsatisfactory for two reasons. First, silver halides require consumables for processing; second, they are neither reversible nor refreshable/reusable. For that, dynamic materials were later developed and used in HI. Principally, photothermoplastics were introduced and led to industrial systems in the 1980s. Companies such as Rottenkolber (System HIK 1000-FE) [11] and Steinbichler (System TPC 200) worked with photothermoplastic rolls, whereas Newport Corporation proposed reusable plates (system HC-300). The photothermoplastic process involves electric charging and heating/cooling of the recording material, which requires cumbersome equipment. Despite this, the relatively good sensitivity of these materials was such that they could be used with low power He-Ne lasers in stable laboratory conditions. An argon laser of 700 mW allowed to make holograms of 1 m^2 object in the case of the Rottenkolber system [11]. Both silver halides on films and photothermoplastics were used in microgravity space experiments. The applications were mainly focused on transparent media, specifically monitoring crystal growth in liquid cells or various flow analysis experiments, where the absence of gravity is interesting to study. Let us cite the early experiments onboard the Soviet Salyut station [57], the American IML-1 [58,59], and different versions of the European HOLOP, based on Steinbichler TP 200 system [60,61], used in German spacelab and Columbus.

The middle of the 1980s saw also true implementation of HI in industrial applications. For instance, in aerospace, French company Aerospatiale (now part of Airbus Defense and Space) developed dedicated facilities for testing composite elements using pressure changes and used it routinely. These systems were mostly based on continuous lasers in confined environmental conditions, mainly for inspecting helicopter rotor blades and rocket launcher parts [62,63].

Pulsed laser-based systems were also developed to allow industrial applications. One major interest of such lasers is when pulse durations are in the range of nanoseconds, freezing the environmental disturbances during hologram recording. Another particularity of pulsed lasers is the ability to provide pulses at a certain repetition rate or to emit two pulses with variable delay from one another. Pulsed lasers are generally based on flash lamps, which pump the gain medium. The repetition rate depends on the laser technology. In any case, an optical Q-switch device is inserted in the laser cavity that blocks the laser emission until the Q-switch is open, delivering a brief pulse with a high energy. The first experiments were made with ruby lasers, which emit pulses of a few tens of nanoseconds, with relatively low repetition rates (0.5–1 Hz). Progressively, YAG lasers took the lead with pulses duration of a few nanoseconds at the useful 532 nm wavelength (after frequency doubling of the fundamental emission line), higher repetition rates (a few Hz to tens of Hz), and energies up to a few hundred millijoules per pulse.

Vibration and transient events were studied on the basis of double-pulse lasers, emitting sequences of two pulses [64]. Both pulses record a hologram (on thermoplastic or argentic rolls) during the object movement, and the readout is performed later with a continuous laser on each of the recorded holograms. In some cases, phase quantification can be applied using the two separate reference beams, as this setup is in a DEHI configuration. To cope with the dynamics of the phenomena studied, it is often necessary to vary the temporal separation between the two pulses and this is closely related to the laser technology itself.

The time interval depends on the laser configuration. Either a single laser can be used with the Q-switch device open twice during one flash of the lamp or used on two separate flashes. This can limit the operations, so another configuration using two separate lasers (one master, one slave, each delivering one of the pulses) can be used for a higher flexibility [64]. Such systems were demonstrated and used in aerospace and automotive industries [64,65], even some systems capable of 3D displacement vector measurements [66].

Despite these successful advances, other methods and systems using electronic hologram recording directly on CCD sensors (based on the speckle effect) took progressively the lead for most of the applications described in Main Applications of HI section. Nevertheless, this did not mean the end for analog materials in HI. In particular, one of the best alternative analog HI systems made use of inorganic photorefractive crystals (PRCs).

Dynamic Holographic Interferometry With Photorefractive Materials

PRCs have specific properties, which make them suitable for hologram recording [67,68]. When a PRC is illuminated by an interference pattern $H(x,y)$, electric charges are generated in the bright area of the crystal due to photoabsorption and migrate to the dark area, where they are trapped by lattice defects. A local space-charge field is then created, which in turn is transformed as refractive index grating by a Pockels effect. The process is dynamic and reversible: any change in the pattern $H(x,y)$ will result in a change in the hologram. There are various PRCs families, which can be split in two main categories: the ferroelectric (e.g.,

$LiNbO_3$, $BaTiO_3$) with high diffraction efficiencies but high recording fluence (i.e., they are considered as slow for hologram recording) and the sillenites (e.g., $Bi_{12}SiO_{20}$) with very low efficiency but lower recording fluence. Many HI experiments with PRCs were performed at the end of the 1970s with both types of crystals. The very first demonstration of PRC-HI was made by Huignard in the case of transparent media refractive index changes [69] and later with TAHI on transparent membrane [70]. Contrarily to most other holographic materials, PRCs offer a wide range of properties depending on crystal species, crystal-cut orientation, and process for charge transfers (assisted or not with an external electric field) [64,68]. Many possibilities were exploited and demonstrated in one or another HI application: surface contouring with two-wavelength [71], deformation of solids [71], vibration patterns observation with TAHI [72,73], flow analysis with multiple hologram recording and selective readouts [74]. The reader should refer to reviews of dynamic HI with PRCs in Refs. [75,76]. One particular configuration was extensively studied in the beginning of the 1990s, which was the most promising for HI and tends to cope with the ideal HI system discussed in The Ideal Holographic System for HI section. It is based on diffraction anisotropy arising under a particular crystal configuration [77]: the diffracted wave U_d has a polarization at 90 degrees from the transmitted wave U_t. This is the particular case for $Bi_{12}SiO_{20}$ (BSO) crystal used in RTHI. BSO crystals are good candidates for HI because they have the lowest recording fluence among other PRC species; unfortunately, they exhibit a poor efficiency. However, placing a polarizer after the crystal allows equalizing the intensities of the diffracted and transmitted waves in RTHI or even cancels the transmitted beam in DEHI, to obtain a very good contrast in the interferogram, as was demonstrated in Ref. [78]. Based on that configuration, an industrial system for HI was developed and used in many applications. It is described in the next section.

An Industrial Holographic Interferometry System Based on Dynamic Holography

The system works with a continuous frequency–doubled YAG laser, emitting a few Watts at 532 nm (Fig. 6.13). The crystal is a $Bi_{12}SiO_{20}$ (BSO). Details of the development are provided elsewhere [79–81]. The principle is that both object and reference beams are always present in the crystal, which records the hologram within a certain time. When the object changes, the hologram is erased, while a new one is recording due to the new object beam. Consequently, the

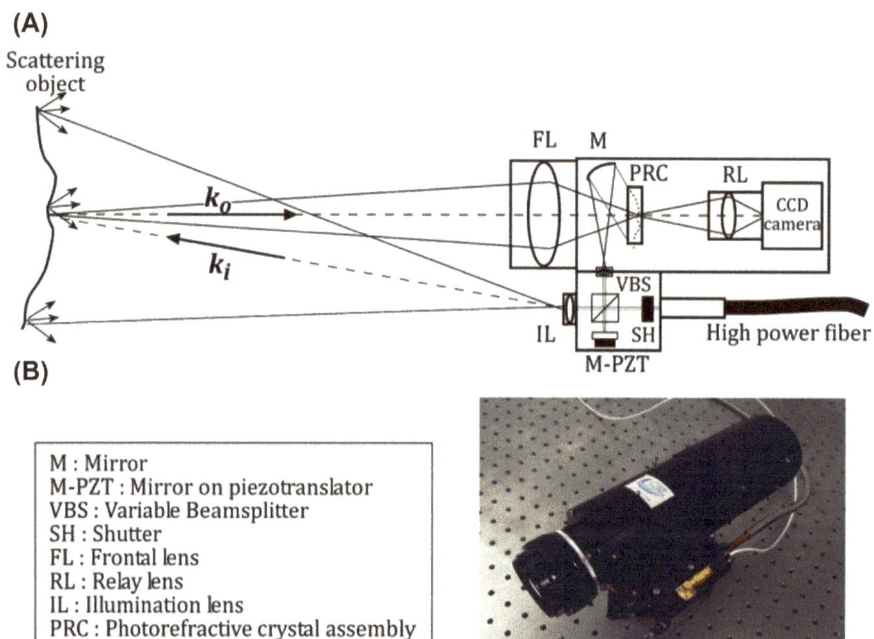

FIG. 6.13 **(A)** Scheme of the photorefractive holographic camera for observation of large scattering objects, **(B)** picture of the instrument.

interferogram showing the OPD appears just when the object has changed. Then it disappears within the erasure time, which is equal to the recording time. The latter depends on the intensity of the incoming beams, mainly driven by the reference beam, which is more intense than the object beam in the general case of scattering objects.

It was demonstrated that, due to specific property of PRCs, the diffracted intensity is proportional to the object wave intensity and not the ratio of reference and object beams, which record the hologram. Therefore, the visibility of the interferogram depends on the power of light used to illuminate the object, whereas the response time depends on the reference beam intensity. It was demonstrated that a power of 500 mW was enough to illuminate a 0.25 m^2 object covered with white scattering powder [79,82]. A very small part of the light is taken from the laser source to form the reference beam. The recording time is set in such a way that the application of temporal PS is feasible [83]. Alternatively, phase quantification with carrier frequency and FT was also used for dynamic objects [79]. The stroboscopic technique was incorporated in the device for the case of vibrating objects [84].

The system presented above was used in a wide variety of applications, which have been reported in the literature [75,76,85]. They include NDT of composite structures with internal damages [79,83], vibration mode shapes observation of aeronautic turbine blades [75,84], metrology of deformation of microstructures (MEMS) with additional optical system [75], monitoring of painting conditions for cultural heritage preservation [86], observation of refractive index changes in heated air [76].

A noteworthy application is the measurement of thermomechanical deformation of various space structures for refining finite element analysis by convergence of both approaches. Some results have been reported already [75]. Fig. 6.14 shows an example of satellite representative part built as a sandwich structure, with aluminum honeycomb core and carbon fiber–reinforced polymer skins. Various solicitations were performed on these structures: thermal, mechanical, and combination of both. Fig. 6.14A shows the setup for measuring the deformation of a representative sandwich structure made of honeycomb and aluminum skins, undergoing thermal load (heater on the backside). Fig. 6.14B shows the phase map of the deformation, Fig. 6.14C shows the deformation simulated by FEM, and Fig. 6.14D shows the corresponding experimental deformation map after phase unwrapping and computation of displacement (Eq. 6.8 in the case of a pure out-of-plane case).

The system was further expanded for measuring in-plane displacement fields, with two symmetric illumination beams [76] and further to full displacement vector L [87]. For the latter, four illuminations beams are used, as shown in Fig. 6.15A, where an appendix has been developed for this application. The four illuminations are useful for measuring the absolute displacement, as already explained in Interpretation of OPDs in term of displacement section. When the object is at rest, all illumination beams are used simultaneously and four holograms are recorded with the same reference beam. When the object is loaded, four sequential readouts are performed, with each of the illumination beams. For each one, a series of interferograms are acquired with temporal PS, allowing

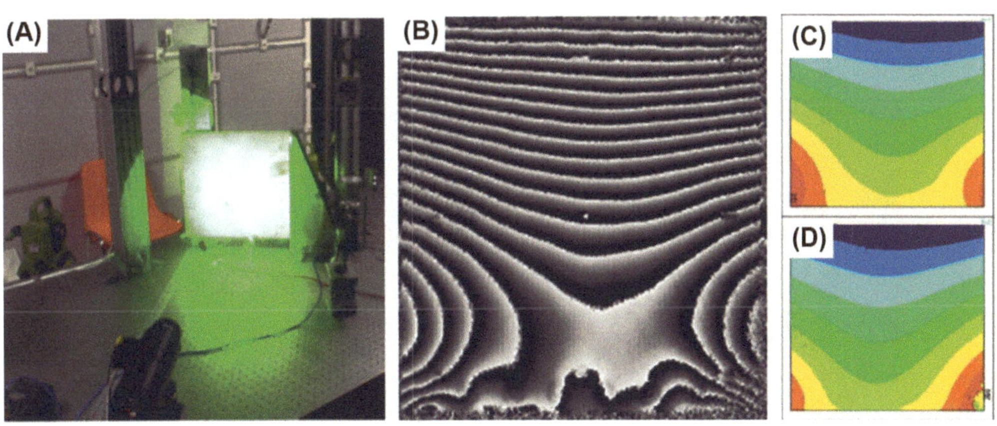

FIG. 6.14 **(A)** Test bed with composite space structure and holographic camera in front, **(B)** phase map due to heating, **(C)** simulation of the deformation of the structure under thermal load, **(D)** experimental deformation map obtained in the same condition as the one simulated.

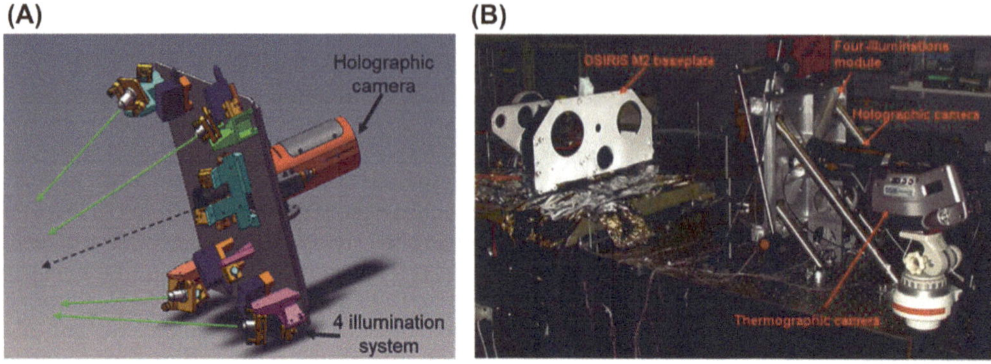

FIG. 6.15 **(A)** Sketch of the four illumination system and the holographic camera, **(B)** experimental setup with object (left) and holographic camera.

obtaining four OPDs, ϕ_1, ϕ_2, ϕ_3, and ϕ_4. The three components of displacements L_x, L_y, and L_z are calculated by solving a matrix system. Details can be found in Refs. [20,87]. Fig. 6.15B shows the setup used for measuring the full-field vector displacement of the structure of OSIRIS space hardware made of SiC material and provided by EADS company.

Fig. 6.16 shows the four wrapped phase maps obtained with each of the illumination beams, after a local thermal stress has been applied to the object. From these OPDs, the experimental displacements $L_{x,y,z}(\text{exp})$ were derived and can be compared with the ones simulated by finite element modeling $L_{x,y,z}(\text{sim})$.

Most of the applications proposed were performed in laboratory conditions. In Ref. [75], we have presented some examples of NDT where the holographic device was used in field conditions, although in relatively quiet area. To cope with industrial environments, the use of injection-seeded pulsed lasers led to some additional developments and applications for dynamic events and vibrations [76,85]. However, since then, the niche market of such specific lasers declined to almost disappear nowadays. For coping with difficult environments using the basic configuration of the photorefractive holographic camera, phase stabilization techniques were studied and proved to be efficient up to some point [88].

As we have seen, analog HI based on dynamic holography with PRC allows a lot of industrial applications under moderately quiet environment. To our opinion, it still remains a good alternative to the more convenient methods based on electronic hologram recording (presented in the next sections), mainly because of the better hologram resolution (which allows observation of more complex objects) and better interferogram quality. Organic self-processable materials were used, but none reached so far the level of industrial maturity of the system described above (see Ref. [75] for a review

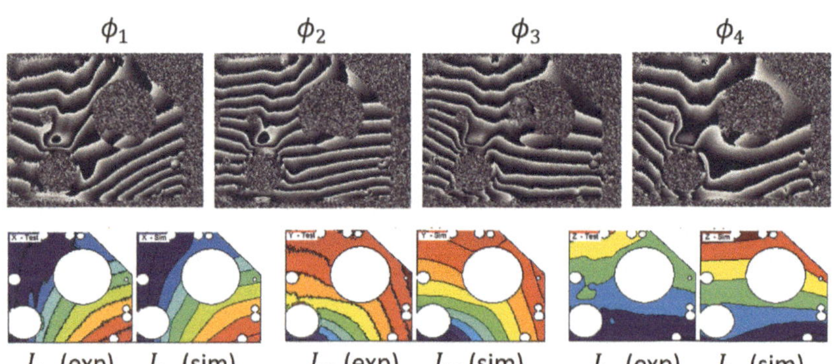

FIG. 6.16 Top row: four wrapped phase map of the OSIRIS hardware after thermal stress and obtained for each of the four illumination directions. Bottom row: components of the vector displacement: experimental ones (exp) obtained from the four measurements and those obtained by simulation (sim).

of these). In the two next sections, we will present HI based on electronic holographic recording, which has imposed itself as the alternative to analog HI. We will start with ESPI.

ELECTRONIC SPECKLE PATTERN INTERFEROMETRY

Basics

ESPI, sometimes called digital speckle pattern interferometry, electronic holography, or TV holography, is based on the correlation of speckle patterns captured for two states of an object at two different instants. Such correlation was first put in evidence by Leendertz [89] who recorded speckle patterns on a photographic plate and the readout produced correlation. The correlation patterns look like interference fringes and are related to the object changes. The very first demonstration of ESPI, where the speckle patterns are recorded on a camera, was achieved in parallel by works of Butters and Leendertz [90] and Macovski [91].

As a reminder, the speckle effect is due to the microscopic relief variations (roughness) of a surface illuminated by a coherent source, the relief inducing variations of optical paths larger than the wavelength. The waves scattered by the different surface points have a random phase distribution at a given location away from the surface. Their interference produces the so-called speckle grains, which are zones of constructive interference with a very limited spatial extent. The size of the speckle is related to the object size, the distance at which it is observed (e.g., on a screen), the wavelength, and the angle with respect to the normal of the object [92]. It is found that the speckle grains have rice grain shapes with their longest size along the longitudinal direction. This is the so-called objective speckle. In the case of the ESPI, an image holography scheme is used, as is shown in Fig. 6.17. A lens is used to image the object on a camera sensor, where interference occurs with the reference beam. Therefore, the hologram plane (x,y) and the sensor plane (u,v) are confused. In this case, speckle is also observed superimposed to the object; one speaks about subjective speckle. In this case, the speckle size is limited by the diffraction of the imaging lens; hence it is related to the airy function. For an effective diameter D of the imaging system (lens diameter or any aperture stop), the distance d between the lens and the sensor, and wavelength λ, the transverse speckle size in the (u,v) sensor plane is given by

$$\sigma_u = \sigma_v = 2.44\frac{\lambda d}{D} \tag{6.24}$$

In the case of ESPI, the camera records the interference $H(u,v)$ that is similar to the hologram in HI but is often referred as specklegram. Fig. 6.18A shows the speckle pattern and (b) the specklegram. The main difference with the hologram is that the specklegram is formed with a reference wave, which does not depart too much from the object beam curvature; otherwise the interference pattern is not resolved by the camera. For that purpose, as is shown in Fig. 6.17, the reference beam is introduced in an in-line configuration via a

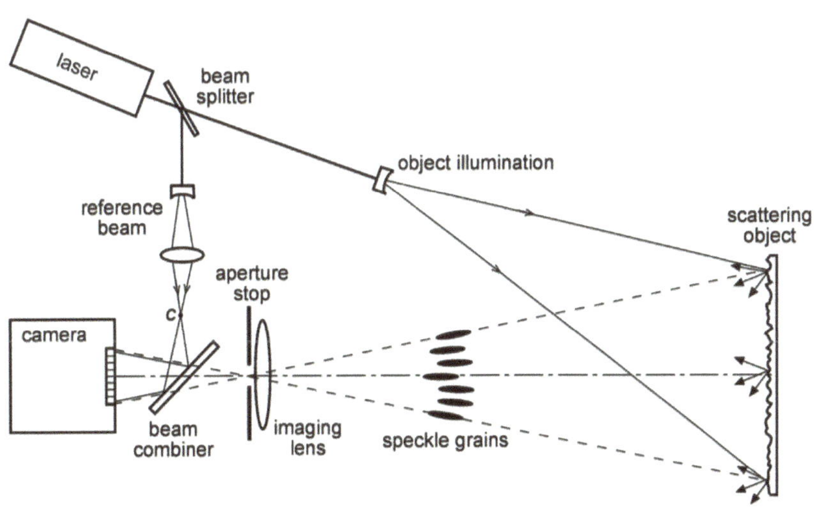

FIG. 6.17 Principle of ESPI setup.

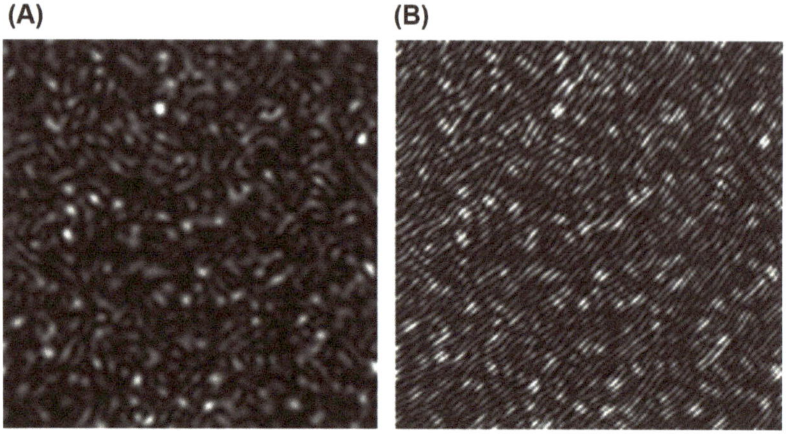

FIG. 6.18 **(A)** Speckle pattern, **(B)** specklegram

beam combiner (BC). Its wavefront is spherical, issued from a point **C**, which is conjugated to the center of the aperture stop of the imaging lens. The figure shows the reference beam forming by two lenses, but alternative systems use an optical fiber whose output is placed in **C**. The size of the speckle is generally larger than the pixel size so it can be resolved.

The working principle of ESPI is as follows. Two specklegrams $H(x,y)$ and $H'(x,y)$ are recorded at two separate instants between which the object phase has changed, from, respectively, φ_o to φ'_o. Once recorded, these two patterns can be either added or subtracted numerically [93], the latter being more commonly used, to form a correlation pattern. Using Eqs. (6.2) and (6.6), and considering that the reference and object waves intensities remain constant, the correlation pattern can be written as

$$|H' - H| = 4\sqrt{I_r I_o} sin\left[\Delta\varphi + \frac{\phi}{2}\right] sin\left(\frac{\phi}{2}\right) \tag{6.25}$$

where we already defined $\Delta\varphi = \varphi_r - \varphi_o$. Because the pattern (Eq. 6.25) is produced numerically, it can only be visible after calculation by computer. The first sine function has a phase argument with a high spatial frequency, which contains the random speckle noise. It is modulated by the second sine function at lower spatial frequency and which depends on the OPD. Only the latter is of interest. The correlation pattern looks like an interference pattern with added noise (Fig. 6.19A).

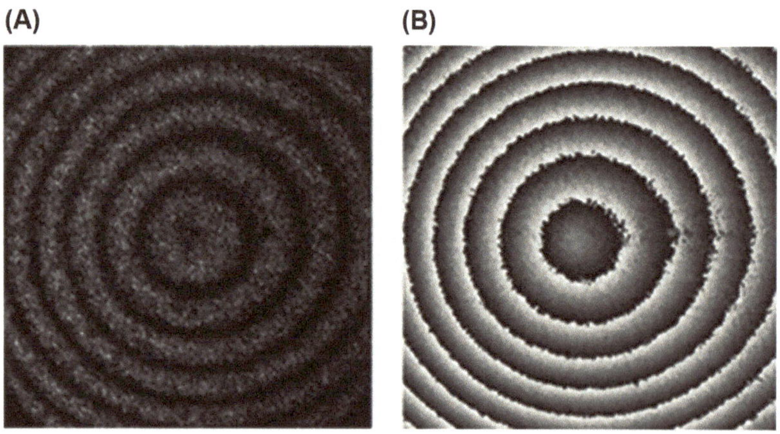

FIG. 6.19 **(A)** Correlation fringe pattern, **(B)** corresponding phase difference. (Courtesy R. Tatam, reprinted with permission from D. Francis, D. Masiyano, J. Hodgkinson, R. Tatam, A mechanically stable laser diode speckle interferometer for surface contouring and displacement measurement, Measurement Science and Technology 26, 055402 (2015), IOP.)

In the case of ESPI, one does not speak anymore of real time or double exposure. Usually, once the first specklegram H is recorded, it is subtracted to all subsequent specklegrams, showing fringes in real-time or more precisely at frame rate depending on the camera and computer calculation time. By extension, ESPI allows recording a large number of specklegrams (as much as the computer memory allows), and the correlation pattern (Eq. 6.25) can be computed between any specklegram of the series. Generally, the first one is the "reference state" and is subtracted from the next ones. However, when the displacement becomes large, it is possible that the fringe pattern cannot be resolved anymore (higher fringe density than pixel pitch, which does not satisfy the Nyquist sampling condition). Therefore, it is necessary to define another specklegram in the series as new reference state for the subsequent part of the sequence. This is a flexibility that is not allowed in analog holography, where a new hologram must be recorded. This was so far the case with dynamic PRC HI (Dynamic HI With Photorefractive Materials section), but in ESPI, this is a truly straightforward and fast procedure.

Modern applications of ESPI make use of phase quantification techniques, which can be applied to retrieve the phase of each specklegram and compute the phase ϕ. For instance, considering the PS technique, a series of specklegrams are recorded in the first state of the object (e.g., at rest), and the phase φ_o is deduced. The process is repeated during the object stress and φ_o is subtracted from the current one and the phase modulo 2π is directly displayed (Fig. 6.19B).

Interesting Applications of Electronic Speckle Pattern Interferometry

All applications of HI were demonstrated using ESPI. We will not review all of them; an abundant literature can be found on this in scientific journals and conference proceedings. In particular, and as was the case for HI a decade earlier, ESPI generated series of specific conferences devoted to that topic, under the sponsorship of the International Society for Optics and Photonics (SPIE) since 2003 [94–100].

Besides usual applications, ESPI allowed unique configurations, which were simply not possible with holography. In other cases, ESPI allowed breakthroughs that were not achievable with analog recording materials. Mainly we will describe the possibility of 3D strain measurements, the use of double-pulse lasers and ESPI at nonvisible wavelengths. Another specific application based on speckle shearing interferometry will be described at the end of Electronic Speckle Pattern Interferometry section.

Electronic Speckle Pattern Interferometry for 3D strain measurements

A popular engineering application of holographic methods is the measurement of solid object deformation [13–15]. The strain is defined as the change of length with respect to a reference length. There are strains normal to the surface and other ones parallel to it (shear strains). On a general point of view, it is necessary to measure the vector displacement in every point of the surface, as well as the distance between two points serving as basis for the strain calculation. For flat surface, this is relatively obvious; however, it is not the case for curved shapes. In the latter, one must associate shape and deformation measurements. Although HI allowed these two types of measurements (Interpretation of OPDs in term of displacement and Surface Contouring sections), their practical implementation was largely improved with ESPI.

For the simple case of flat objects, more basic displacement measurement configurations exist and the shape could be easily determined without holographic contouring. The out-of-plane configuration has been presented already in Fig. 6.17. The interpretation of fringes as surface displacement is totally similar to what was discussed in HI (Interpretation of OPDs in term of displacement section). Soon in the beginning of ESPI, a particular configuration allowing in-plane displacement measurement was demonstrated [90]. It does not make use of reference and object beams, but instead of two object illumination beams incident at symmetric angles with respect to the object normal (Fig. 6.20A). Both form two object beams, which interfere at the level of the camera sensor because they have exactly the same curvature. This configuration is said to be self-referenced. Specklegrams are recorded before and after stress and correlation fringes are produced. In the case of phase quantification is applied, the phases at each instant can be retrieved and subtracted to provide the OPD.

It can be shown that, in the case of the setup of Fig. 6.20A, the OPD is written as

$$\phi = (4\pi/\lambda)L_Y sin\theta \qquad (6.26)$$

with L_Y the displacement along the axis Y, which lies in the plane of incidence defined by vectors $\boldsymbol{k}_{i+}$ and $\boldsymbol{k}_{i-}$. This equation is valid for the practical case of objects with dimensions smaller than the distance to the camera, and if the angles θ and $-\theta$ are constant on the entire surface (collimated illuminations). In the case of a rotation of a rectangular plate around an axis perpendicular to the surface, a phase map such as the one of Fig. 6.20B is observed [101].

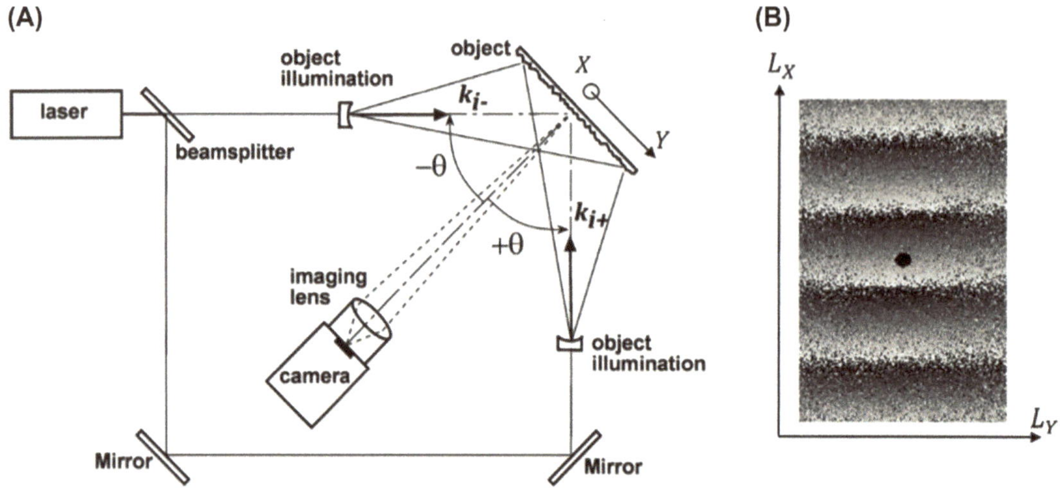

FIG. 6.20 **(A)** In-plane ESPI setup, **(B)** OPD due to rotation around the normal. (Adapted with permission from J.-F. Vandenrijt, M. Georges, Electronic speckle pattern interferometry and digital holographic interferometry with microbolometer arrays at 10.6 μm, Applied Optics 49, 5067-5075 (2010), OSA.)

For more complex objects, systems combining vector displacement and contouring were developed [102,103]. Applying the contouring in a 3D displacement setup is quite easy. Indeed, several object illuminations are required (at least 3, see Interpretation of OPDs in term of displacement section) to form a basis of sensitivity vectors and capture several OPD maps ϕ to retrieve the $\boldsymbol{L}$ vector components. The contouring is applied on one of the illuminations by using the source translation method explained in Surface contouring section (Fig. 6.8). Based on this principle, commercial systems were proposed by the company Ettemeyer [103], and some are still marketed today (under Dantec Dynamics brand).

Pulsed and double-pulsed Electronic Speckle Pattern Interferometry

Despite its generally lower interferogram quality (subject to speckle noise), it was soon recognized that ESPI would be superior to HI for analysis of fast phenomena (like vibrations or transient events), making use of pulsed lasers [104]. As discussed in Towards the Industrial use of Analog HI section, pulsed lasers were already used in industrial applications of HI. However, ESPI dramatically improved the readout process. The first attempts with ESPI were made with television cameras, which are based on vidicon tube and not the matrix sensor technologies such as CCDs [104,105]. Here we will discuss the advances made early in the 1990s with the introduction of CCD cameras.

As was already mentioned in Basics section, ESPI allows storing a long series of specklegrams, which can be used two by two to determine fast displacements. Several methods were applied. The first one makes use of single-pulse lasers. The specklegrams are recorded on separate CCD frames and subsequently subtracted on a pixel-to-pixel basis [106]. However, this is of limited interest because any environmental disturbances at low frequency (i.e., at the laser pulses repetition rate, a few Hz or even less) are seen in the correlation pattern. For that reason, double-pulse mode was introduced in ESPI [106].

The double pulsing can be brought by double opening of the laser cavity during a single flash lamp operation or by considering two lasers in parallel, as already evoked in Towards the Industrial use of Analog HI section. A first possibility was demonstrated by Spooren [106]: two pulses are launched during object movement and are integrated on a unique camera frame. In this case, both specklegrams are added to one another. A correlation pattern can be produced in addition mode, as discussed in Refs. [93,106], but the contrast is lower than in the subtraction mode, like in Eq. (6.25). In order to use subtraction, both specklegrams must be recorded separately. For that, interline-transfer CCD provided a nice solution for recording specklegrams with a few microseconds delay. Once the first specklegram is recorded with pulse 1, it is transferred to a gate and the second specklegram is now recorded with pulse 2. Therefore, both specklegrams coexist but are not mixed together [106]. They can be readout by the electronic circuit and then the CCD is ready for a new pair of pulses. Pedrini et al. went a step further by adding the possibility of acquiring phase-shifted images by a spatial carrier method [107]. They apply it on the study of vibrations arising after a shock. The latter generates a few vibration modes on the surface. By recording sequences of double-exposed specklegrams, with different temporal pulse separations, one can retrieve the different modes present in the deformation [108]. They also applied the technique to a rotating object, making use of a derotation device. Fernandez et al. made similar works but using the FT method [109]. They apply it to the study of crack detection by observing the evolving surface deformation with an increasing interpulse separation [110].

Electronic speckle pattern interferometry at nonvisible wavelengths

One major interest of electronic hologram recording is to allow holography applications at wavelengths for which analog materials do not exist or have poor performances. An electronic holographic device just requires a laser, a camera, and some optical elements to form a setup like in Fig. 6.17.

On the shorter wavelengths side, the use of ultraviolet was proposed to study micrometric devices [111,112]. The interest of shorter wavelengths is to increase the imaging resolution, which otherwise is reduced by diffraction if one uses visible wavelengths. Also, as the scattering properties of surfaces are directly related to wavelength, diffusive reflection appears more easily on polished or low-roughness surfaces at short wavelengths [113]. In particular, reflection by micromirror structures induces speckle at 266 nm wavelengths, rendering possible the development of ESPI on such objects [111,112].

On the longer wavelength side, near-infrared (NIR) diodes have been used as a compact replacement of continuous laser such as He-Ne [114]. The interest of laser diodes is that one can change the wavelength, allowing the contouring by the two-wavelength method [115], as explained in Surface Contouring section. At such wavelengths (around 800 nm), the optical elements usually are the same than in visible. Also CCD cameras (or more recent CMOS) have sufficient sensitivities to allow ESPI. At the level of measurement range, in terms of displacement, these wavelengths do not bring much difference with respect to their visible counterpart.

More recently, ESPI has been envisaged at long infrared wavelengths (usually referred as LWIR), combining CO_2 lasers (emitting around 10 μm) and microbolometer array cameras, sensitive between 8 and 14 μm [101,116]. The interest of wavelengths about 20 times larger than in visible is twofold. First, the stability requirements are relaxed, allowing holographic experiments out of laboratory conditions. A portable ESPI system has been demonstrated in different industrial conditions, with a continuous CO_2 lasers, much affordable than pulsed visible lasers [117]. The second interest is that large displacements can be measured in a single shot [118], because they are proportional to the wavelength, as can be seen in Eq. (6.8). Large displacements are usually observed in aerospace applications, where often small stresses are not relevant. On the opposite to what happens in ultraviolet, objects appear more specular in infrared than in visible, for a given roughness. Nevertheless, we have found that many aerospace composites of interest had roughness, which allows to produce speckle [119].

Another interesting outcome of LWIR ESPI is the possibility of simultaneous observation of surface temperature change and deformation. This is because the technique uses a thermographic camera for specklegram recording. Therefore, the interference between reference and object beams (Eq. 6.2) contains an additional term $I_{th}(x,y)$, which is the incoherent thermal image of the object. This can be exploited in various NDT experiments as was demonstrated in Refs. [120,121]. Indeed, many NDT applications aim at correlating deformation and temperature changes. LWIR ESPI offers this at once, correlating every pixel in both OPD and temperature variations. As an example, Fig. 6.21A shows an NDT experiment consisting in observing the deformation of a repaired helicopter panel. Heating with lamp is used. A series of phase-shifted specklegrams is recorded prior to the heating, and a second one after. An adaptation of PS algorithm [120] allows to obtain temperature variations (Fig. 6.21B) and OPD modulo 2π (Fig. 6.21C) from the same image set. After phase unwrapping, the out-of-plane displacement can be deduced (Fig. 6.21D) and data fusion between deformation and temperature variation can be processed easily pixel-to-pixel (Fig. 6.21E). That way, it is possible to easily understand the relationship between temperature variations and deformation of the structure. Other applications such as detection of defects were proposed [121].

Speckle Pattern Shearing Interferometry (Shearography)

Shearography is a variant of ESPI that has many applications nowadays. It was proposed first by Leendertz and Butters [122] and is referred as speckle pattern shearing interferometry, or more simply shearography. Shearography does not make use of the interference of

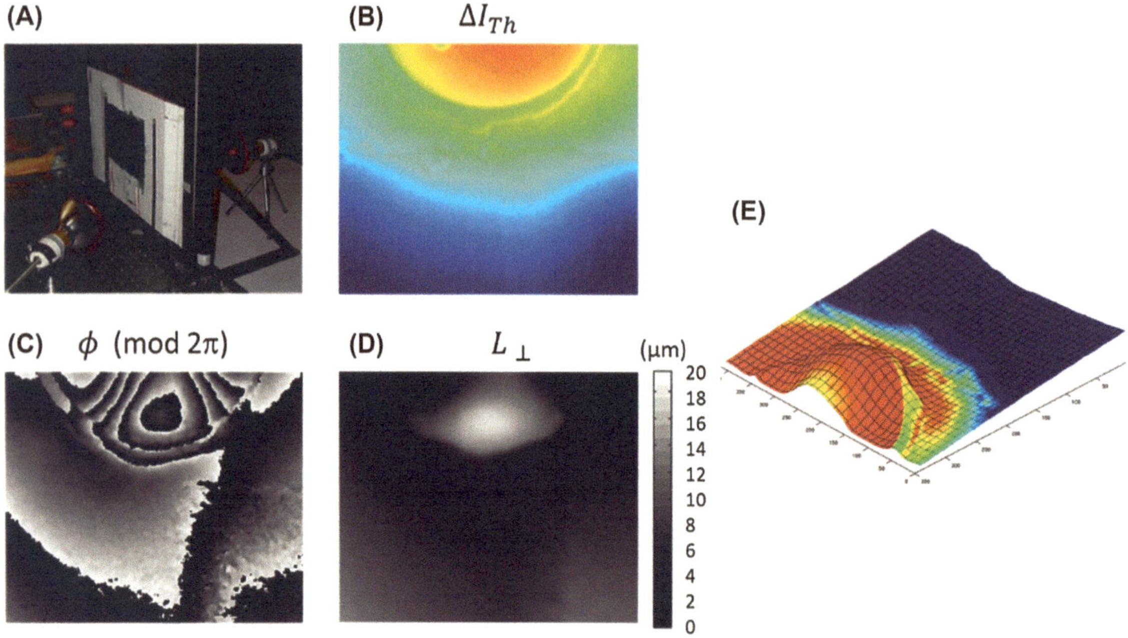

FIG. 6.21 **(A)** Helicopter panel with repair, **(B)** temperature increase due to heating, **(C)** OPD due to heating, **(D)** out-of-plane deformation corresponding to **(C)**, **(E)** fused deformation and temperature data.

reference and object waves contrarily to HI and ESPI. Instead a single beam coming from the laser is used to illuminate the object. The latter scatters light toward the observation system, which is preceded by a shearing device. In the pioneering works, interference was recorded on holographic photoplates [123] and electronically [122], but rapidly the latter became more prominent. In literature, one can find various configurations of shearing devices. They can be based on (1) insertion of a wedged glass plate in half of the field-of-view [123], (2) birefringent elements introducing polarization separation in images followed by polarizer [124], or (3) the mostly used modified Michelson interferometer [122]. Fig. 6.22A shows principle of a shearography setup based on a Michelson interferometer.

A beam splitter cube/BC splits the image in two similar equal intensity images: the transmitted one (represented by solid lines) and the reflected one (represented by dot lines). Both are reflected by mirrors, respectively, M1 and M2, to be further recombined by the same BC at the level of the camera. One of the two mirrors (M2) is tilted by an angle α, which produces a lateral shear between the two images in the (u,v) sensor plane. Fig. 6.22B shows the situation where the shear is oriented along direction u. However shear direction can be adjusted more generally in the image plane (u,v). The two electric fields of the object waves reflected by M1 and M2 can be written, respectively, $U_{o,M1}(u,v)$ and $U_{o,M2}(u+\Delta u,v)$. At the overlap of both waves on the sensor (the gray zone in Fig. 6.22B), an interference pattern $H(u,v)$ is formed, here called shearogram (in reference to specklegram and hologram). The pattern is of the usual form of Eq. (6.2), with the phase difference (Eq. 6.3) now given by

$$\Delta\varphi_{sh}(u,v) = \varphi_{o,M1}(u,v) - \varphi_{o,M2}(u+\Delta u,v) \tag{6.27}$$

Like in ESPI, shearograms can be recorded at different instants during different object states. A correlation pattern similar to Eq. (6.25) can be computed, showing fringes related to the phase difference due to the object change. Here also, if a phase quantification technique (like phase-shifting) is applied, we obtain for each instant the phase difference $\Delta\varphi_{sh}$.

In the usual case of illumination and observation normal to the object (the out-of-plane configuration), the phase change ϕ_{sh} between two object states is given by Ref. [123].

$$\phi_{sh} = \frac{4\pi}{\lambda}\frac{\partial L_\perp}{\partial u}\Delta u, \tag{6.28}$$

where u is the optical shear direction and Δu is the shear amount. By comparing Eq (6.28) and Eq. (6.8), we see

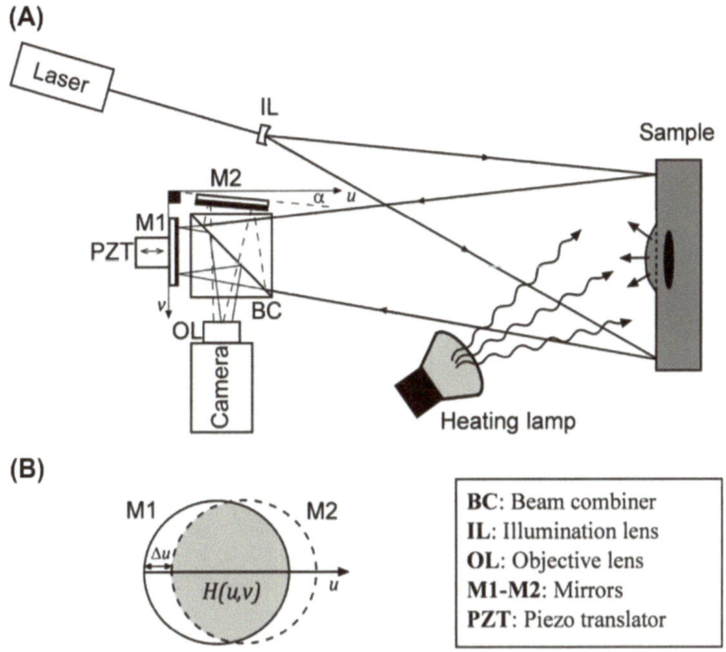

FIG. 6.22 **(A)** Principle of shearography setup, **(B)** shearogram ***H*(*u*,*v*)** recorded at overlap of object beams.

that, unlike in ESPI and HI, the OPD observed in shearography is not related to the object displacement field, but rather to its derivative along the optical shear direction u. A more general formulation can be provided if we consider multiple illumination schemes (like those discussed in Interpretation of OPDs in term of displacement and ESPI for 3D-strain measurements section), giving access to derivatives of the different components of the vector displacement L [127]. Such approach allows to determine the surface strains [125,126]. It is more straightforward than ESPI, which requires an a posteriori derivation of the displacements and knowledge of surface point positions (see ESPI for 3D-strain measurements section). Further from strain is the determination of residual stress with specific advanced systems, such as radial shear interferometry, in the works of Albertazzi et al. [127].

The particularity of shearography setup is that it is a self-referenced configuration (like in-plane ESPI). Any environmental perturbation on the object will impact both interfering waves in a similar way. Therefore, their interference will not be affected by this perturbation, by opposition to other configurations making use of an external reference beam (HI, ESPI). For that reason, shearography is much more immune to environment perturbations found in industries, compared with both others. Shearography is traditionally used for defect detection, industrial environment [128], like explained in Nondestructive Testing section. A defect in the observed material will provoke a local surface deformation due to the effect of a stress. Generally, the latter induces also a global deformation. Shearography allows a better view of the defect response compared with HI or ESPI as is explained hereafter.

Because of the derivation along the shear direction, the interference patterns observed in shearography look different than those in HI and ESPI. Fig. 6.23A shows the simulation of a surface deformation with a Gaussian profile. Fig. 6.23B shows the simulated OPD observed by ESPI, which consists in a circular fringe pattern. Fig. 6.23C is the equivalent obtained by shearography with a diagonal shear (black arrow). Two lobes are observed (one negative, one positive). When phase unwrapping is applied for removing the artificial 2π phase jumps of figures (b) and (c), a linear profile passing through the maximum of both images will look like Fig. 6.23D. In the case of a global deformation pattern also present, its effect can be reduced by the derivation process. In Fig. 6.23E, an additional tilt was simulated and appears as parallel fringes in ESPI, whereas the corresponding shearographic interferograms in Fig. 6.23F show only the typical two-lobe pattern.

A true example of NDT of a composite sample is shown in Fig. 6.24. The sample is a sandwich with a

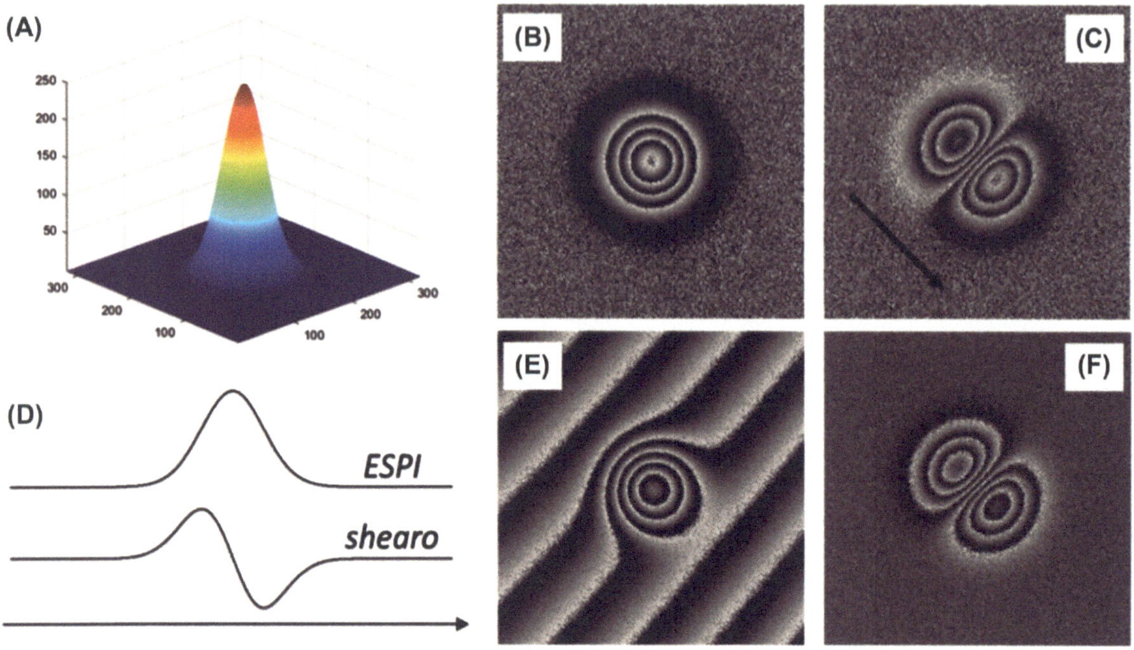

FIG. 6.23 Simulation of ESPI versus shearography. **(A)** Actual deformation, **(B)** ESPI phase map, **(C)** shearography phase map corresponding to **(B)** with diagonal shear (arrow), **(E)** and **(F)** are similar to **(B)** and **(C)** with additional diagonal tilt.

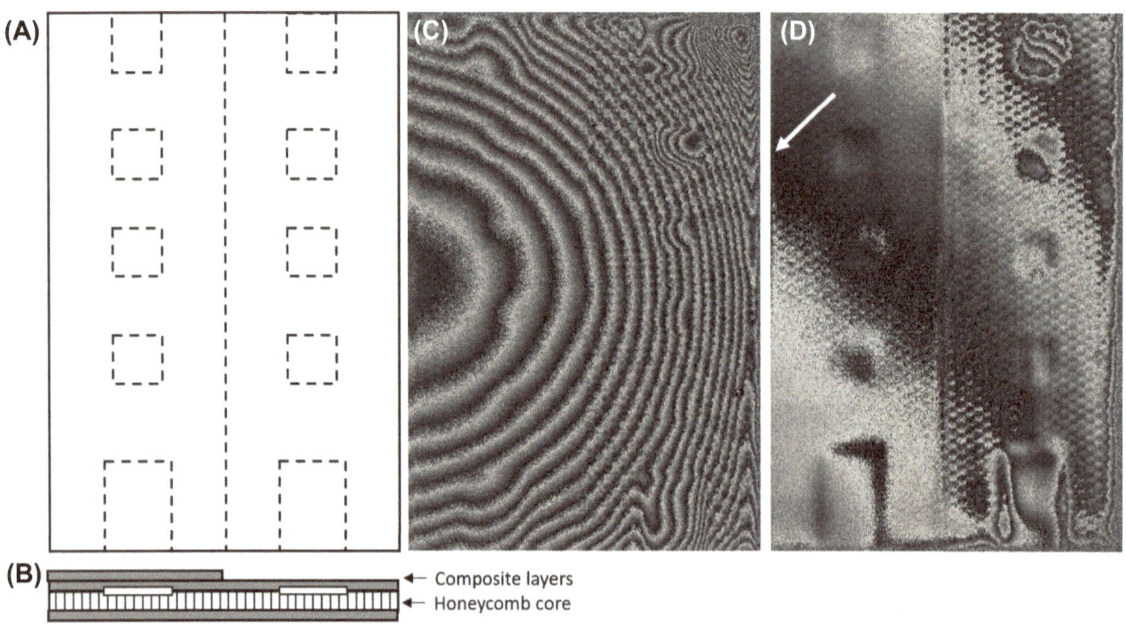

FIG. 6.24 Composite sandwich panel with artificial defects. **(A)** Front view with rectangular defects, **(B)** cross section of the sample, **(C)** OPD due to thermal loading observed by ESPI, **(D)** corresponding OPD obtained by shearography with diagonal shear (white arrow).

metallic honeycomb core with carbon fiber skins as detailed in the cross section (Fig. 6.24B), with zones of different thickness of the upper skin. Defects (lack of glue) are present between the upper skin and the core, in the form of rectangular pattern arranged following the sketch shown in Fig. 6.24A. Fig. 6.24C and D shows, respectively, the fringe patterns obtained by ESPI and shearography in the same conditions of thermal stress (heating by a lamp). The white arrow represents the diagonal direction of shearing used for shearography. Local defect signatures are more clearly visible in shearography.

This typical example shows why shearography has taken the lead compared with other holographic methods (HI, ESPI) for the particular application of defect detection. The fringe patterns are much easier to interpret for identifying the location of defects, in addition to the highest immunity against environmental perturbations.

Industrial Electronic Speckle Pattern Interferometry and Shearography Systems

ESPI has benefited from the rapid evolution of CCD cameras and the ever-increasing calculation power of personal computers. For that reasons, it took the lead on HI in industries. Systems with automated phase quantification were made available by some companies since mid of the 1990s. Steinbichler (now Zeiss), Ettemeyer (now Dantec Dynamics), or LTI were mainly the market drivers. Many versions of speckle interferometry devices were proposed by all these companies: small compact systems for out-of-plane displacements or for full 3D strains measurements (including contouring), big systems with double-pulse lasers for shock and vibrations measurements, etc. Most of the commercial ESPI equipment disappeared from the market, being left aside by some of the manufacturers. Besides ESPI, all of them proposed speckle shearing systems (shearography) for defect detections. Nowadays, the market is as follows: LTI with ESPI systems on an isolated table, as well as shearography for NDT [129]; Optonor with ESPI and shearography [130]; Laser Optical Elements with portable shearography and vacuum hood [131]; ISI-SYS with shearography and vibration excitation [132]; Dantec Dynamics with 3D strain measurement system [133] and portable shearography with vacuum hood [134]; V2i/Optrion that recently introduced a combined shearography and thermography system; and finally Zeiss that is a world leader in tire testing machines working pressure vessels and shearography [135]. The last application is even now a standardized method for tire testing in aeronautical industry.

DIGITAL HOLOGRAPHIC INTERFEROMETRY

We will close this chapter by discussing the more recent holographic method used in interferometry: digital holographic interferometry (DHI) based on DH. Like for ESPI, in DH the hologram is recorded directly on the image sensor. Therefore the planes (u,v) and (x,y) overlap. Different mathematical processes allow to reconstruct the diffracted order of interest, which contains the information about the object, recorded in the hologram $H(x,y)$ (refer to Chapter 5 on Digital Holography of this book for the different processing methods). The result is a complex object field $U_0(X,Y)$ from which the phase $\varphi_0(X,Y)$ can be deduced directly.

Like for ESPI, there is no distinction between RTHI and DEHI. The power of modern computer is such that the reconstruction of digital holograms is very fast. Once the first hologram irradiance $H(x,y)$ is recorded, the object phase $\varphi_0(X,Y)$ is computed. When the object has changed, a new hologram $H'(x,y)$ is recorded and the corresponding phase $\varphi'_0(X,Y)$ is computed. In DHI, one just computes the phase difference between the current phase and the one recorded instants before, allowing following the object movements in real time. The situation is exactly the same as for ESPI. However, the notable differences are the following. Although ESPI requires an imaging system with in-line injection of the reference beam (see Fig. 6.17), DHI can use any of the configuration of DH: lensless, off-axis, in-line, and so on, with the appropriate numerical reconstruction method. Also the reconstructed object image can be refocused numerically by selecting the correct distance of reconstruction in the computation. Different object parts can then lie in different planes; they can be refocused independently from one another. We will not review the various reconstructions schemes as they were described largely in Chapter 5 in this book. We will present here some notable applications.

Dynamic Deformation of Solids

The most usual application of DHI is measuring the deformation of solid objects, under various stress conditions. The simplest cases of out-of-plane displacement were demonstrated in the early papers of Schnars et al. [136,137] and also for NDT [138]. Pedrini et al. later demonstrated an important advantage of DH in the off-axis configuration: since the phase can be extracted from a single hologram, dynamic deformation (e.g., during vibration or transient events) can be quantitatively evaluated by recording sequences of digital holograms [139]. They used an image plane DH scheme for out-of-plane sensitive measurement. A similar arrangement and hologram processing was used by Mendoza's group in connection with high-speed CMOS cameras, with 4000 frames per second for observing the shape change of a balloon [140]. Later they showed the flapping of a butterfly with a 500 frames per second temporal sampling (Fig. 6.25) [141].

Pedrini et al. extended DHI to measurement with sensitivities in two directions (out-of-plane and in-plane) [142]. The basic idea is to split laser beam in two, each beam j ($j = 1,2$) giving rise to a pair of object and reference waves, respectively, $U_{o,j}$and $U_{r,j}$, obtained by a second beam splitter. The object beams are arranged so as to illuminate the object from two separate directions. The reference beams are also arranged to illuminate the camera with tilts in separate directions [142]. Two holograms $H_1(u,v)$ and $H_2(u,v)$ are then recorded simultaneously. The tilt angles of the references are such that the spectra of both holograms are in two orthogonal directions (e.g., along u and v). Both holograms can be separated by the two-directional spatial carrier technique. To avoid unwanted interference between both pairs of beams, they can be made mutually incoherent by using a delay line on one of the pairs, with the condition that the delay exceeds the laser coherence length. Suitable order filtering in the Fourier plane allows reconstructing of two separate images of the object, each with a specific sensitivity direction. After computation of the phase, the OPD can be calculated from the phases at different instants. The optical arrangement is such that the sum of the OPD provides the out-of-plane displacement component, whereas their difference gives the in-plane component. Pedrini et al. have demonstrated the technique with both continuous and pulsed lasers [143].

With the same purpose of two-dimensional measurements, Picart et al. have used a lensless DH configuration where the incoherent multiplexing is provided by separate polarizations for beams 1 and 2 [144], instead of the delay lines used in Pedrini's scheme. The setup is shown in Fig. 6.26. Again, the two-dimensional spatial carrier technique is used and the object observed with two orthogonal sensitivity directions appears in the Fourier plane at two separate places. Fig. 6.27A shows the amplitude reconstructed by the Fresnel method. Fig. 6.27B and C shows the final OPD obtained from calculating phase difference between two instants, respectively, with out-of-plane and in-plane sensitivities.

The object is a metallic washer, which was deformed after the recording of the first hologram. The authors have applied this technique in fracture mechanics [145] and in industrial vibration measurements [146]. Later they proposed multiplexing through the use of two mutually incoherent lasers with different wavelengths for 2D sensitivity measurements [147] and with three lasers (red (R), green (G), and blue (B)) for real-time 3D sensitivity [148]. It is worth to note that, in the latter, a single reference beam direction is used (no more spatial multiplexing) in association with a stacked RGB sensor, as is shown in Fig. 6.28. All three color channels are mixed in each pixel. A similar optical arrangement was also used on the basis of a 3-CCD camera, with separate color channels [149]. Fig. 6.29 shows the results obtained in deformation of a metallic object. The wrapped phase map obtained in each of the R, G, and B channels simultaneously is shown in Fig. 6.29A–C, respectively. Fig. 6.29D–F shows, respectively, the three object displacements computed using the three measurements (a), (b), and (c).

Transparent Objects

Mainly, DHI for transparent objects distinguished itself in the particular case of flow analysis with the three-color method, already discussed in the previous section.

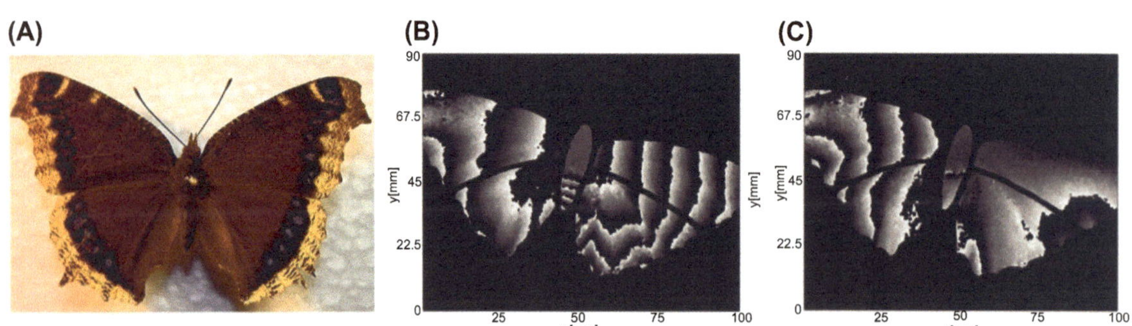

FIG. 6.25 DHI for observing fast movement of butterfly wings. **(A)** Picture of the butterfly, **(B)** and **(C)** wrapped OPD at different instants of the flapping. (Courtesy Mendoza Santoyo.)

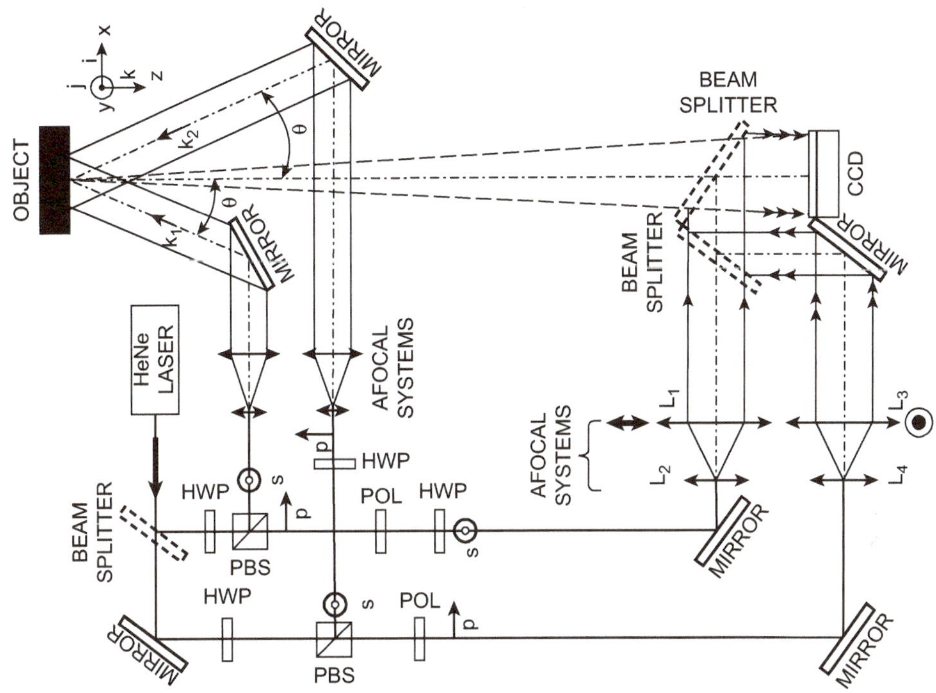

FIG. 6.26 DHI setup for simultaneous out-of-plane and in-plane measurements, based on polarization separation. (Reprinted with permission from P. Picart, E. Moisson, D. Mounier, Twin sensitivity measurement by spatial multiplexing of digitally recorded holograms, Applied Optics 42 (2003), 1947–1957, OSA.)

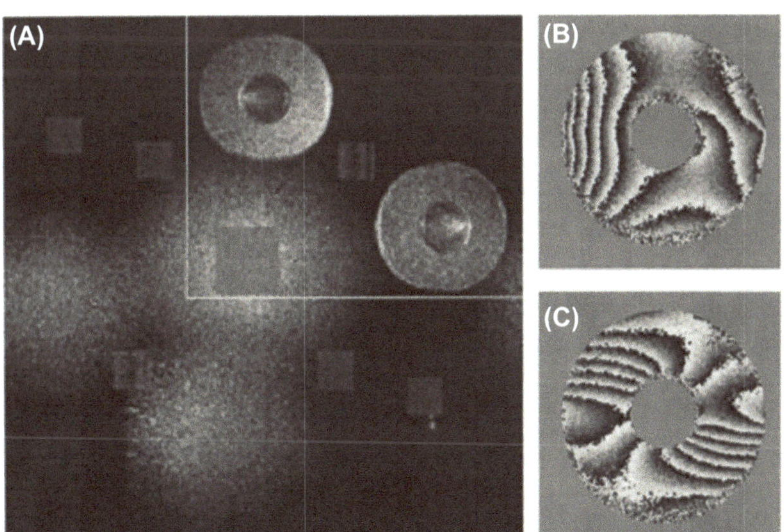

FIG. 6.27 **(A)** Amplitude of spatially multiplexed holograms, **(B)** out-of-plane deformation, **(C)** in-plane deformation. (Courtesy of Picart.)

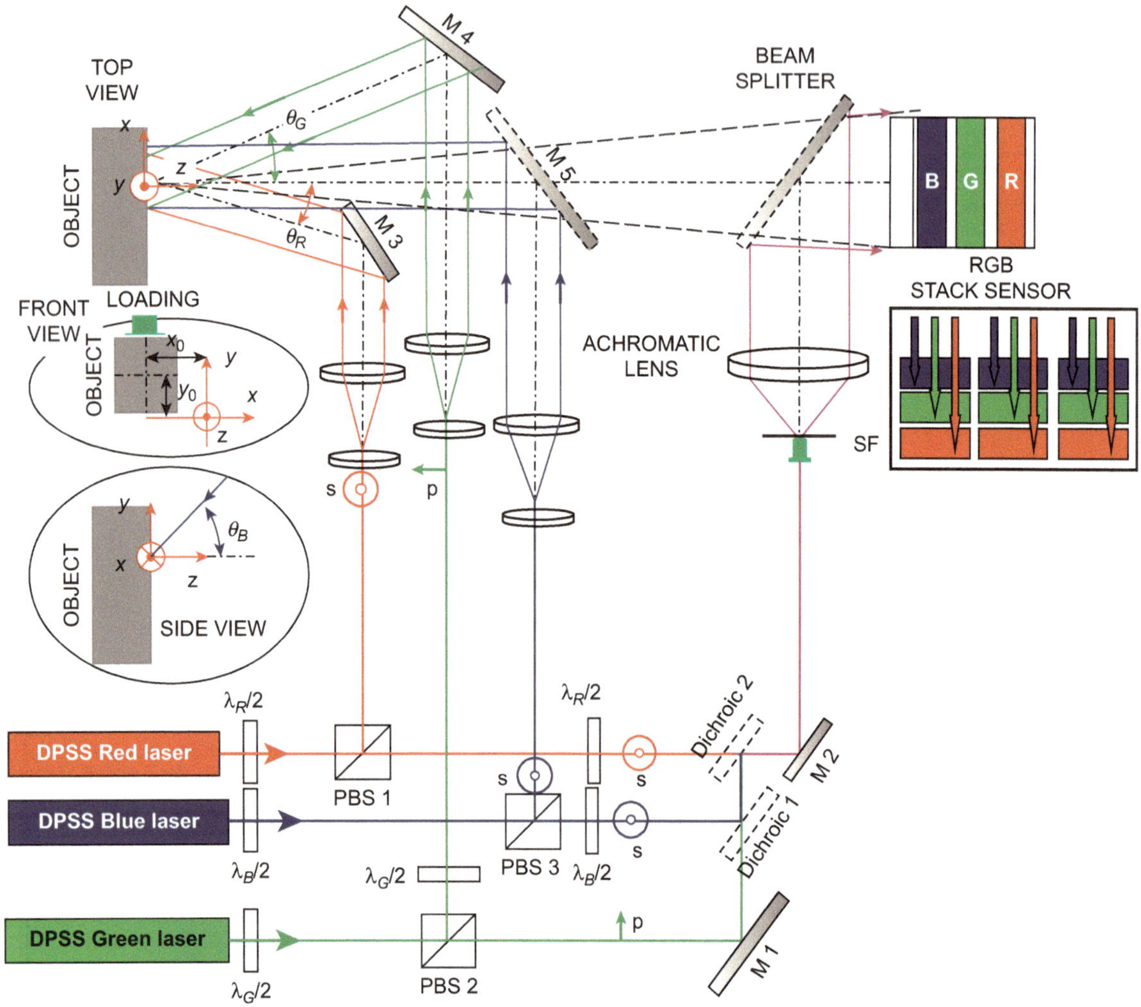

FIG. 6.28 Three-color DHI setup for simultaneous 3D sensitivity deformation measurements. (Reprinted with permission from P. Tankam, Q. Song, M. Karray, J. Li, J-M. Desse P. Picart, Real-time three-sensitivity measurements based on three-color digital Fresnel holographic interferometry, Optics Letters 35 (12) (2010), 2055–2057, OSA.)

Indeed, HI at a single wavelength λ applied in flow analysis gives only relative values for the OPD, yielding density variations (see OPDs in transparent media section). Extending DH to three-color scheme, similar to the one discussed in the previous section, allows absolute measurements by combining the OPD measured in the three channels R, G, and B [149]. Fig. 6.30A shows a typical arrangement for three-color DHI for flow analysis. Detailed explanation of this setup can be found elsewhere [149–151]. The object cell (on the right) is the test section of a wind tunnel in which the flow circulates. Two windows allow optical access. The three-laser beams are coaligned, and a beam splitter cube splits the diverging beam in two parts. In the transmitted arm, the object beam is collimated before entering the windows and, being retroreflected by a flat mirror, back to the camera. On the latter, it interferes with the reference beam, which travels through a concave mirror. Before reaching the camera, both the reference and object beams are recollimated.

In DH mode, a reference hologram is captured without the flow phenomena occurring, with an artificial fringe carrier induced by tilting a mirror in the path. This hologram contains also phase information that is present in the whole setup, for example, related to the windows, the flat mirror, etc. The resulting

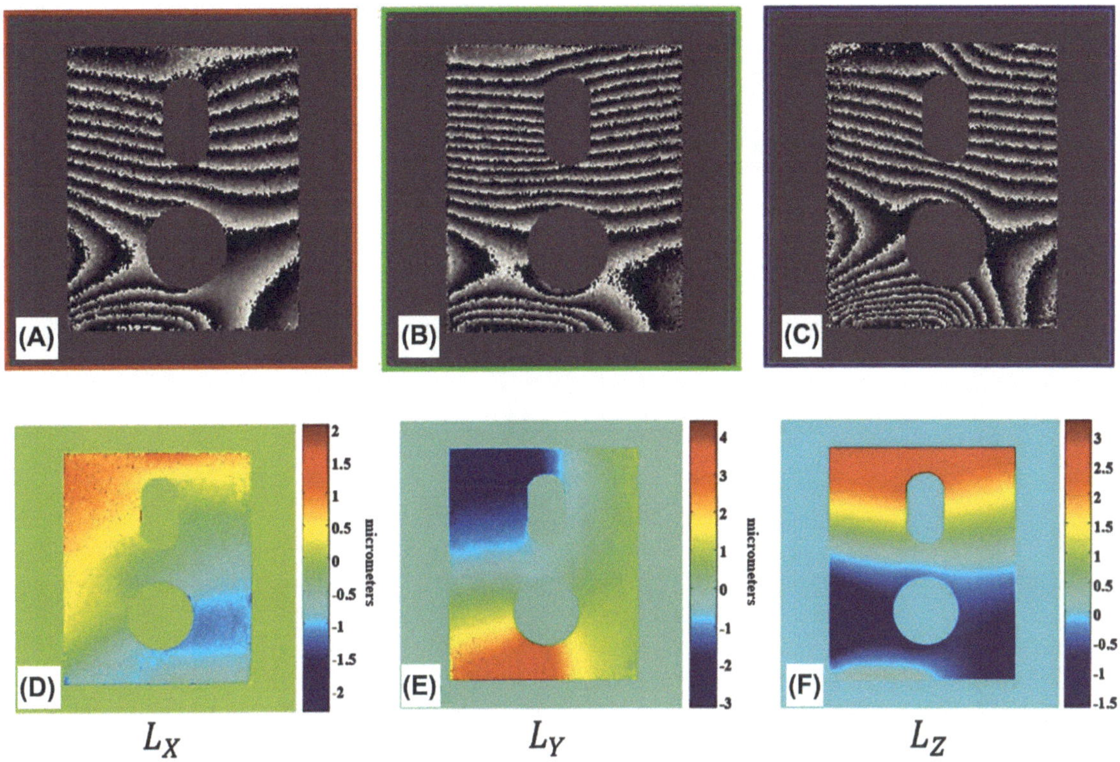

FIG. 6.29 Three-dimensional deformation measured simultaneously by DHI. **(A–C)** OPD obtained by red, green, and blue channels, respectively; **(D–F)** object deformation along *X*, *Y*, and *Z*, respectively. (Courtesy of Picart.)

spectrum of the reference hologram is shown in Fig. 6.30B. After some refractive index variation occurs in the test section, the phase of the object beam is altered and the resulting "probe" (or "object") spectrum is shown in Fig. 6.30C. Isolating the side lobes allows computation of phase, and the phase difference can be deduced. Fig. 6.30D shows the phase difference for one of the three wavelengths in the case of air flow around a cylinder perpendicular to the flow direction (as shown in the test section in Fig. 6.30A). The carrier fringes present in both the reference and the probe holograms naturally cancel each other during the phase subtraction.

All three lobes are processed separately and the phase differences ϕ_λ are calculated at each wavelength λ. A composite interferogram I_{RGB} is then formed, based on the three interferograms colored in R, G, and B.

$$I_{RGB} = \sum_{\lambda=R,G,B} I_\lambda, \quad \text{with } I_\lambda = I_{av,\lambda}[1 + m_\lambda cos\phi_\lambda] \tag{6.29}$$

A typical result is shown in Fig. 6.31A, together with an interferogram taken in similar experiments performed earlier with analog holography with three lasers and panchromatic holographic plates [152].

The interest of the three-color interferogram is the white fringe, which is interpreted as the locus where no refractive index change occurred between the reference and the probe capture (similarly to the white fringe in TAHI, which represents the locus of fringes where no movement is present during vibration, i.e., at the nodes). Analysis of this white fringe, and its evolution, is of particular importance in fluid experiments [149].

Vibrations Analysis

Observation of vibrations modes is one important engineering application of holography. The simplest method for vibration analysis, based on time averaging, was first demonstrated by Picart et al. in 2003 [153]. The configuration is a lensless off-axis one with Fresnel reconstruction. Unlike its analog counterpart, time-averaged DH offers the possibility to obtain directly the phase in the image, which improves interpretation [154]. Picart et al. expanded their method to the multiple sensitivity case already discussed

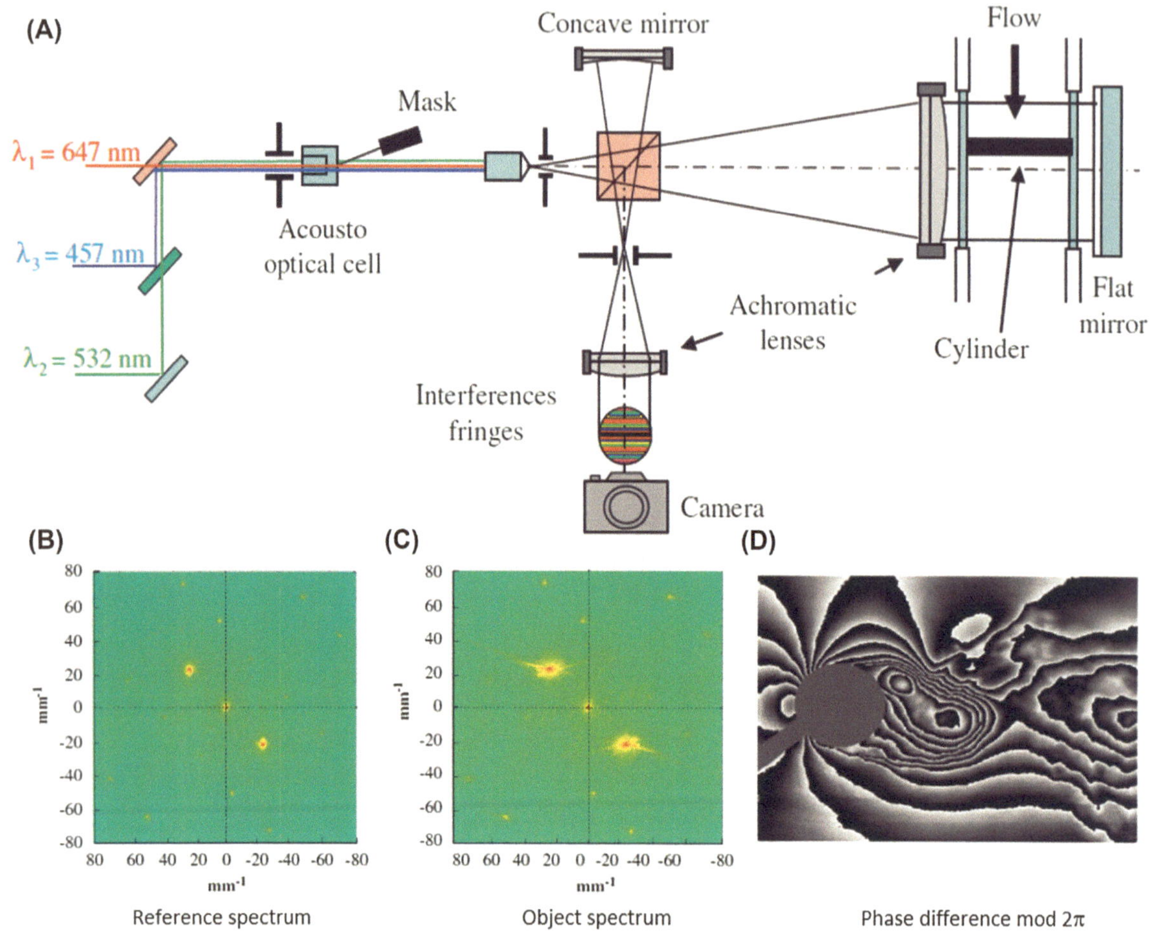

FIG. 6.30 Three-color DHI flow analysis. (Courtesy of Desse.)

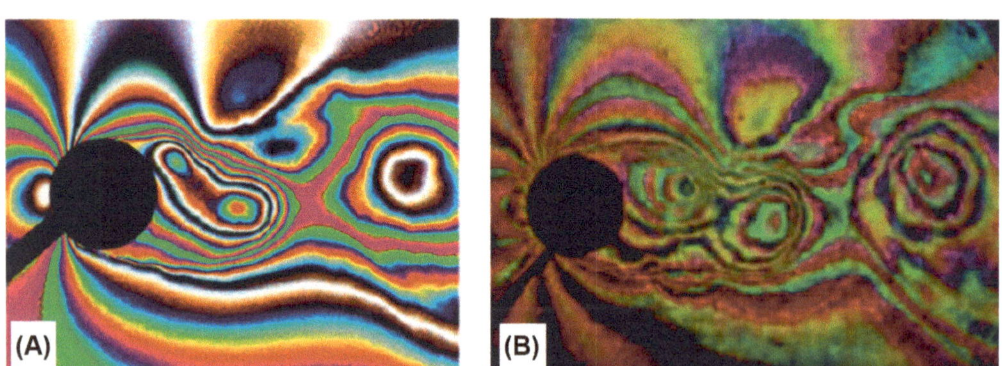

FIG. 6.31 **(A)** Three-color interferogram obtained by DHI in flow analysis; **(B)** similar interferogram with analog holography. (Courtesy of Desse.)

previously [146]. Later the same group demonstrated the potentiality of DH for a full-field vibrometry measurement.

In vibration analysis, let us recall that when an object is under sinusoidal excitation with frequency $\omega/2\pi$, its displacement takes the general form $\varphi_A sin(\omega t + \varphi_v)$, where φ_A is the amplitude of vibration and φ_v is the vibration phase (retardation between the object response and the excitation). Vibrometry, namely laser Doppler vibrometry (LDV), is a widely spread tool for measuring the frequency response of an object under vibration. It allows easy determination of both the amplitude and the phase of vibration in every object point [155]. Its drawback is that it requires scanning the object, point by point. Picart's group developed a method based on stroboscopic three-step phase-shifting of digital holograms [156]. The method allows to obtain both the amplitude and phase, for all pixels at a time, which LDV method cannot. Fig. 6.32A shows the amplitude of vibration of a membrane at 2.44 kHz, and Fig. 6.32B is the corresponding phase. A frequency scanning can be performed to display the resonance peaks, by computing the mean quadratic velocity in function of the frequency (Fig. 6.32C). Picart at al. later used their technique for studying a wide variety of acoustic phenomena [157,158].

Time-averaged DH associated by two-exposure DHI was demonstrated by Demoli et al. [159]. This allows the observation of vibration mode shapes, together with hidden static deformation. Demoli et al. used an off-axis approach, like in works by Picart [153]. On their side, Asundi et al. proposed an in-line DH configuration, which allows larger field-of-view than the off-axis configuration [160]. In this case too, a mix of mean static deformation was observed superposed to the vibration mode itself. The group of Asundi pursued a considerable effort in DHI for vibration analysis of microsystems (MEMS) [161,162].

Vibration analysis by DH was also considered in the biomedical field, in particular for studying the behavior of tympanic membranes, mostly through stroboscopic illumination [163–166], sometimes with a three-dimensional measurement sensitivity. Fig. 6.33 shows in-plane and out-of-plane sounds–induced deformations of a human tympanic membrane at several excitation frequencies, induced by a loudspeaker. The left and central columns represent the two in-plane deformation measurements, with sensitivities perpendicular to one another. The right column represents the out-of-plane deformation. The lines (a) to (c) are related to three excitation frequencies. It can be observed that the deformation is mainly out-of-plane.

Long-Wave Infrared Digital Holographic Interferometry

The electronic hologram recording allows application of DH at any wavelength, where lasers, cameras, and suitable optics can be found. We already pointed out this fact in ESPI at non visible wavelengths section devoted to ESPI at nonvisible wavelengths. Long-wave infrared DH was first introduced by Allaria et al. [167] and followed by intense works by the group of Ferraro, mostly for imaging purposes [168–170]. There are many advantages to extend DH at long wavelengths [171] as already discussed in ESPI at non visible wavelengths section. In interferometry, a larger wavelength allows measurement of larger deformations. Also, in the case of lensless DH, the size of reconstructed objects is 7–10 times larger than in the visible. For that reasons, LWIR DHI seems well suited to measure deformations of large space structures, which undergo important temperature variations when on orbit. As the parabolic reflectors have a specular reflectivity, they require a specific and expensive illumination optical element (null lens) for wavefront error measurement by classical interferometry. For this application, the aim is to

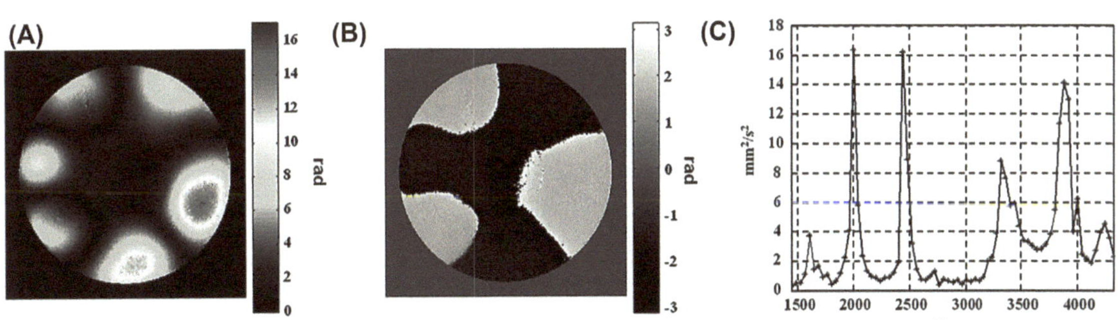

FIG. 6.32 Vibration mode of membrane: **(A)** amplitude at 2.44 kHz, **(B)** phase at 2.44 kHz, **(C)** mean quadratic velocity. (Courtesy Picart.)

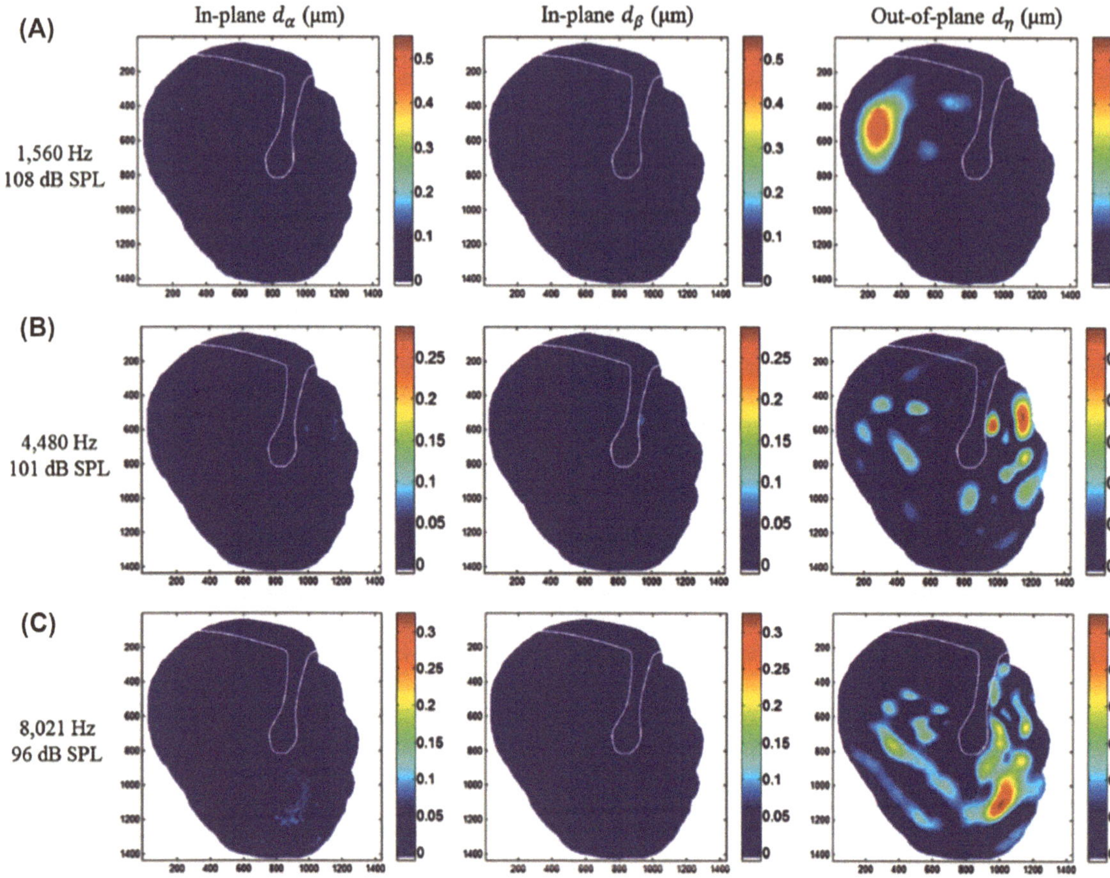

FIG. 6.33 In-plane and out-of-plane sound–induced motions of a human tempanic membrane excited with frequencies 1560 Hz **(A)**, 4480 Hz **(B)**, and 8021 Hz **(C)**; left and central columns: two in-plane deformation, right column: out-of-plane deformation. (Reprinted with permission from M. Khaleghi, C. Furlong, M. Ravicz, J. Tao Cheng, J. Rosowski, Three-dimensional vibrometry of the human eardrum with stroboscopic lensless digital holography, Journal of Biomedical Optics 20 (5) (2015), 051028, SPIE.)

measure the deformation of the reflector due to temperature variations. For that, a relative measurement method such as HI is well suited and does not necessarily require a costly reference element. It is sufficient to illuminate the reflector and to bring back the object wave to a camera sensor where it interferes with a reference wave. Illumination through a diffuser was found suitable for a flexible object illumination system [172,173], resulting in the DH interferometer shown in Fig. 6.34, in which the object behaves as a scattering one, in a classical HI arrangement. The space optics testing requires a vacuum chamber for thermal radiative transfer, enabling cryogenic temperature variations on the large reflector. However, the CO_2 laser and the thermal infrared camera have to stay outside the chamber because they are not vacuum compatible. Therefore, relay optics are used to form collimated beams entering the main part of the interferometer, which was placed inside the vacuum chamber.

Fig. 6.35A shows a picture of the thermal vacuum chamber of the Centre Spatial de Liège used for the test. Fig. 6.35B is a schematic of the whole test bed (not a scale), Fig. 6.35C shows the parabolic reflector (1.1 m diameter) inside a thermal shroud enabling the radiative transfer, Fig. 6.35D is the OPD obtained between 107 and 295 K, and Fig. 6.35E is the corresponding 3D plot.

Together with the large deformation capability, LWIR DHI showed an excellent immunity against vibrations that can occur in such large-scale facility.

Later, the same instrument was used for the cryogenic test of a mosaic of 4 × 4 detectors (Fig. 6.36A)

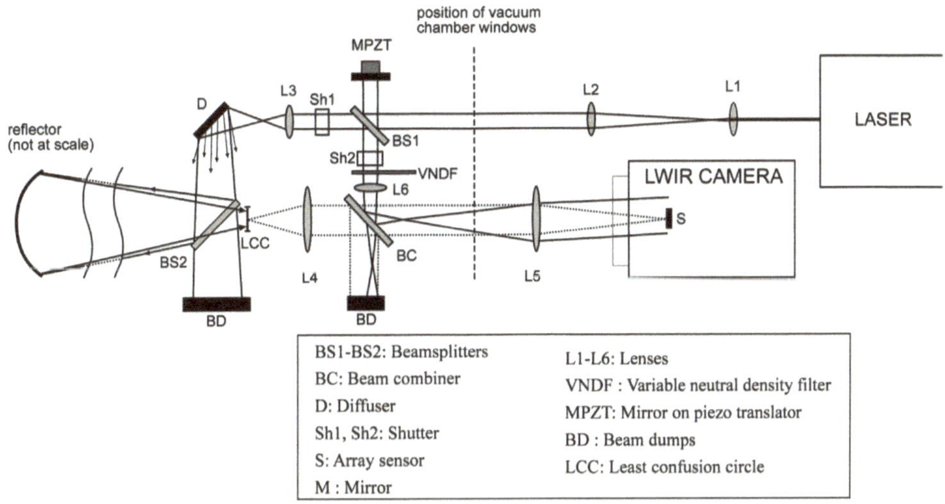

FIG. 6.34 Schematic of LWIR DHI for large space reflector testing with diffuser illumination. (Reprinted with permission from M. Georges, J.F. Vandenrijt, C. Thizy, Y. Stockman, P. Queeckers, F. Dubois, D. Doyle, Digital holographic interferometry with CO2 lasers and diffuse illumination applied to large space reflector metrology, Applied Optics 52 (2013), A102, OSA.)

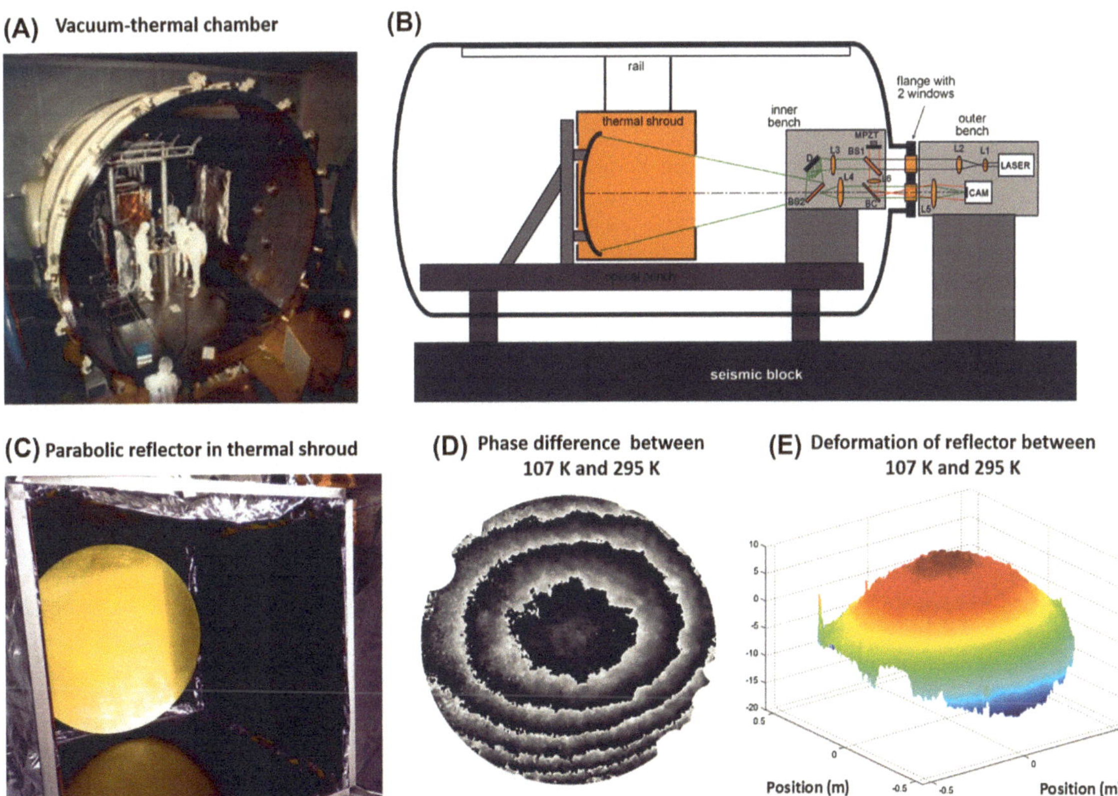

FIG. 6.35 Testing of large space reflectors under vacuum chamber. **(A)** Vacuum testing facility at Centre Spatial de Liege, **(B)** schematic of the test bed, **(C)** the reflector inside thermal shroud, **(D)** OPD measured by LWIR DHI of deformation between 107 and 295 K, **(E)** 3D plot of deformation.

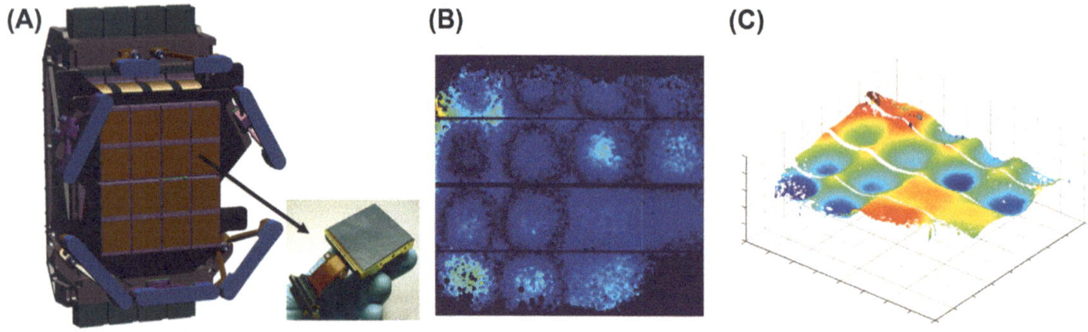

FIG. 6.36 **(A)** Mosaic of 4 × 4 detectors for EUCLID space mission, **(B)** OPD residual deformation between 293 and 90 K measured by LWIR DHI, **(C)** 3D plot of **(B)**.

used for the EUCLID space mission [174]. The different detectors can experience separate deformations and rigid body movements from one another. Measuring disconnected object parts is feasible considering the phase measurement in every pixel independently of its neighbors. For this purpose, the so-called temporal phase unwrapping [175] was applied to a sequence of digital holograms during a measurement campaign, which lasted for a few tens of hours, with acquisition of holograms every 30 s. Postprocessing of the phase allows retrieving the rigid body motion (rotation and piston effects) of each detector, as well as deformation. Fig. 6.36B shows the OPD resulting from rigid body motion removal, keeping the residual deformation; Fig. 6.36C is the corresponding 3D plot.

Two-Wavelength Contouring

The two-wavelength holographic interferometry principle, already presented in Surface Contouring section, is easily applied in DH and constitutes probably the most impressive industrial application of DH, which apply in a wide variety of domains. Unlike two-wavelength analog holography (Surface Contouring section), in DH there is no need for a readout angle correction when changing wavelength. The phases of two digital holograms can be computed directly and subtracted to one another to obtain the phase at the equivalent wavelength, like in Eqs. (6.15) and (6.16).

The applicability of the technique is only limited by the way digital holograms are recorded, whether sequentially (one wavelength per camera frame) or simultaneously (two wavelengths on a single camera frame), the latter being the most interesting for analysis of dynamic events. Pedrini et al. have first demonstrated the use of a pulsed ruby laser (694 nm) where different longitudinal modes of the laser cavity are rapidly selected sequentially and captured separately, taking benefit of the interline transfer of a CCD camera [176]. Two frames at separate wavelengths are recorded, one after another, as already considered in double-pulse ESPI for analysis of fast events [107,108] (Pulsed and double-pulsed ESPI section). An example of a vase was contoured with height steps of 7 mm and precision of 0.5 mm. After that measurement, the shape of the vase was refined by generating a UV line obtained by the second harmonic generation in a BBO crystal [177]. This method allows to obtain more details in some zones of the object. To discriminate both recording, Pedrini et al. use an image plane holography scheme with two separate reference beams (each at a given wavelength) reaching the sensor at different angles. Spatial multiplexing is achieved by separating the images in the reconstruction plane. The authors pursued their works by considering two separate lasers (HeNe at 633 nm and DPSS at 532 nm) for vibration measurements. The equivalent wavelength is obtained being around three times larger than the initial wavelengths; the object velocities measured are accordingly higher [178].

In parallel, the group of Depeursinge of the EPFL (Lausanne, Switzerland) carried out works in digital holographic microscopy (DHM) under various configurations. One of them is the application of two-wavelength contouring inside DHM (TW-DHM) for microobjects characterization [179]. They also consider spatial multiplexing using two reference beams at different angles but in a lensless DH configuration and Fresnel reconstruction. The technique generated a lot of applications in contouring, as well as biomedical application, for its single-shot ability, in both static and dynamic analyses [180]. Fig. 6.37 shows different applications: (1) measurement of certified step heights, (2) observation of rough surfaces, and (3) topography of microlenses. All of them concern static objects, whereas Fig. 6.37D

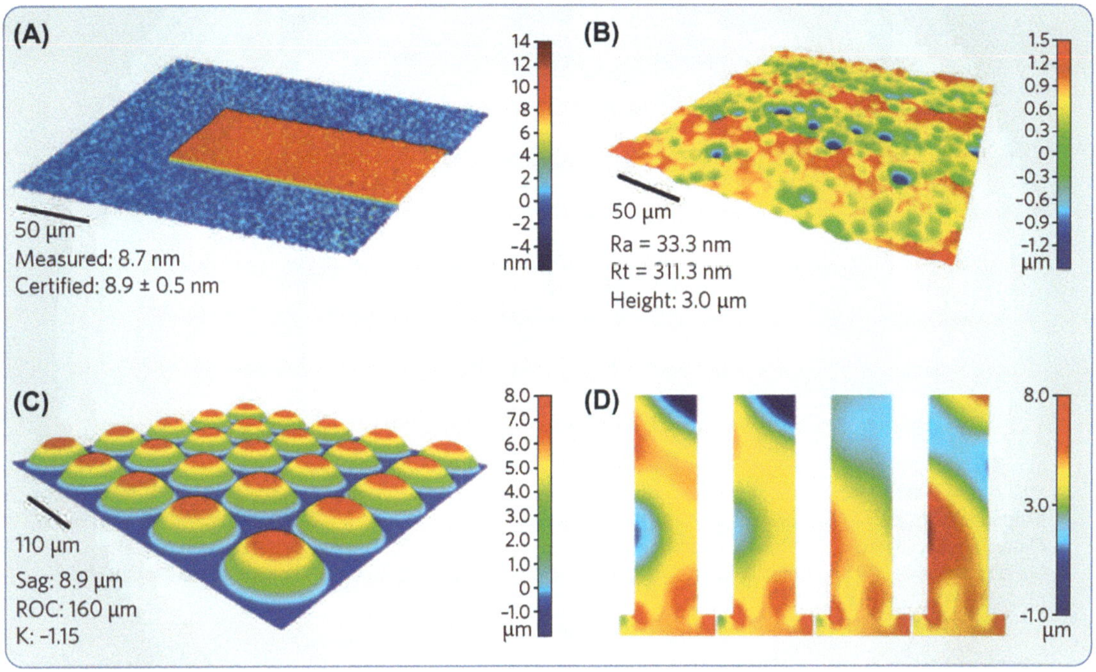

FIG. 6.37 Typical applications of two-wavelength contouring in DHM: **(A)** step height measurements, **(B)** roughness measurements, **(C)** microlenses topography, **(D)** cantilever MEMS vibrations characterization. (Reprinted with permission from E. Cuche, Y. Emery, F. Montfort, Microscopy: one-shot analysis, Nature Photonics 3(11) (2009), 633–635.)

shows vibration modes of cantilever microsystems (MEMS). In this case, a stroboscopic device is used in synchronization with a sinusoidal excitation of the object.

Coming back to our definition of an ideal holographic measurement system described in The Ideal Holographic Measurement Device section, the TW-DHM presented in Ref. [180] certainly fulfills most of it. Instruments based on this principle are now commercially available at Lyncee Tec [181].

Other groups proposed interesting achievements in two-wavelength contouring. Kühn et al. recently showed a TW-DHM approach that includes integrated optics and applies their systems in small manufactured parts analysis [182]. Industrial applications of two-wavelength contouring on production lines were also intensively developed by Carl's group at the Fraunhofer IPM [183–185]. In their case, the configuration is not DHM but still based on the spatial multiplexing, with different angles of reference beams at different wavelengths. Carl's application is the contour measurement of objects moving at a few meters per second in production line or under processing (e.g., milling machine) [183,184]. They recently showed the interest of numerical autofocusing (a widely known feature that is specific to DH) to extend the depth of view in contouring complex-shaped objects [185]. This "shape-from-focus" capability associated with DH contouring and the computational implementation directly in a graphic processing unit (GPU) makes these works one of the most impressive industrial achievements of DH metrology to date.

At last we will finish this chapter with a very promising application of DH contouring under development by Pedrini's group. This focuses on the measurement of the internal walls erosion of the future International Thermonuclear Experimental Reactor (ITER) tokamak, which is under construction in Cadarache (France). Erosion is expected to occur during the operation of ITER and needs to be assessed. The challenge is to perform continuous measurement of small height variations in difficult to access zones inside the tokamak and at long remote distance. Different strategies are undertaken and described in Refs. [186,187]. The first one uses a free space optical illumination scheme for the object, where the latter is located at 23 m from the DH head. The fast sequential recording takes benefit of the interline transfer of CCD, as was already exploited in

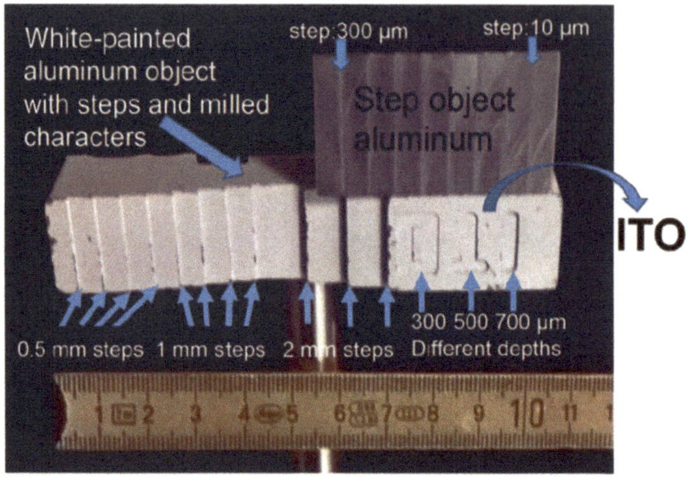

FIG. 6.38 Test object for demonstration of the endoscopic DH contouring prototype for the ITER tokamak. (Reprinted with permission from G. Pedrini, I. Alekseenko, G. Jagannathan, M. Kempenaars, G. Vayakis, W. Osten, Feasibility study of digital holography for erosion measurements under extreme environmental conditions inside the International Thermonuclear Experimental Reactor tokamak [invited], Applied Optics 58 (5) (2019) A147–A155, OSA.)

earlier works by Pedrini [107,108,176]. The source is a tunable Ti:Sapphire laser from Sirah (Matisse model) emitting between 700 and 820 nm, allowing a large choice of wavelengths and thus of measurement resolutions by suitable line selection. A second strategy was studied: an endoscopic system, which is more adapted to the tiny aperture (2–3 mm diameter) allowing optical access to the inner part of the tokamak from outside. For assessing the performance of the technique, a calibration object shown in Fig. 6.38 was developed. It consists of steps of various heights and milled characters of Pedrini's institute (ITO).

Fig. 6.39 shows the DH contouring results obtained. A first couple of wavelengths were used and correspond to height variations of 6.47 mm. The measured OPD (Fig. 6.39A) was corrected from convergence and tilt effect, resulting in Fig. 6.39B. A second measurement was performed for another couple of wavelengths (c) and (d), corresponding to height variations of 0.52 mm. The actual resolution is better than this value because of the subfringe resolution allowed by phase quantification. Combining OPDs obtained with different equivalent wavelengths allows to retrieve the shape with adapted resolution in different object parts, from the lower one in Fig. 6.39E to higher ones in Fig. 6.39F and G.

An important parameter that was also studied by the authors is the speed of measurements, which must be short enough to not be impacted by the phase changes related to the facility operation. For that, they considered both the sequential and the simultaneous recording of digital holograms at the different wavelengths. The simultaneous recording uses the spatial multiplexing approach with two reference beams, already discussed earlier. They reached typically 600 μs of recording time for both approaches. The sequential approach provides better results because the use of two references can lead to decorrelation between holograms [186]. Pedrini et al. further decreased the recording time by activation of acoustooptic modulators placed at the output of the laser beams and studied the quality of their measurements. They found that acceptable results are obtained for 100 μs recording time. This time needs to be shortened again, as well as other optical aspects for the methods to be fully integrated in ITER.

CONCLUSION

HI is one of the most impressive applications of holography since its discovery. We have seen that a wide range of utilizations were soon found, from the analysis of the deformation of solid objects to the observation of vibration modes by time averaging. In the two first decades of HI, experiments were based on photosensitive materials, requiring physicochemical processing to reveal the holograms (analog holography). The pioneering years were devoted to experimenting a lot of configurations, such as contouring, NDT, refractive index variations measurement in transparent objects, to name of few.

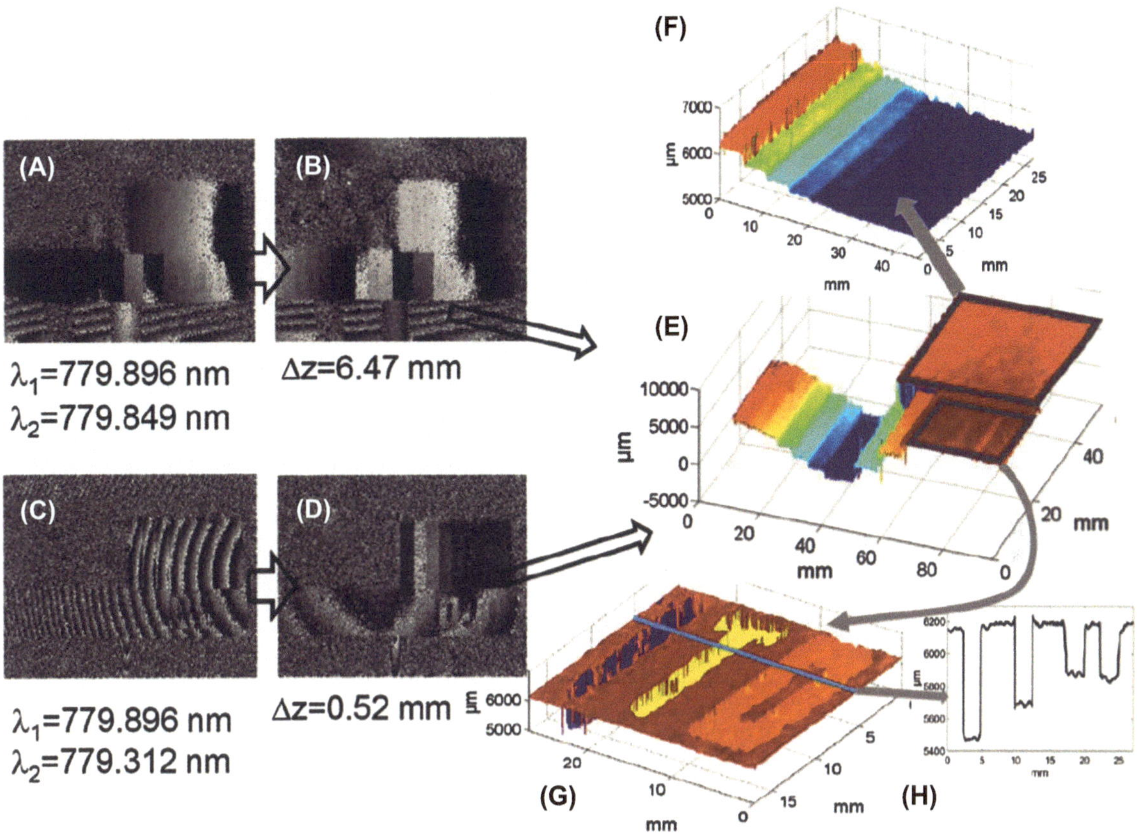

FIG. 6.39 DH contouring. **(A)** OPD due to wavelength change, **(B)** same of **(A)** with tilt and convergence compensation; **(C)** and **(D)** are similar to **(A)** and **(B)** for another wavelength combination; **(E)** shape of object, with details in **(F)** and **(G)**; line profile across milled characters **(H)**. (Reprinted with permission from G. Pedrini, I. Alekseenko, G. Jagannathan, M. Kempenaars, G. Vayakis, W. Osten, Feasibility study of digital holography for erosion measurements under extreme environmental conditions inside the International Thermonuclear Experimental Reactor tokamak [invited], Applied Optics 58 (5) (2019) A147–A155, OSA.)

The potentials of HI for industrial applications were well understood, and some visionary pioneers have clearly paved the way of a tremendous current of researches and developments that have taken place until now. Search for new analog holographic materials and improved processing was one way, but the interpretation of the fringe patterns was another. The tedious processes of holographic materials chemical processing did not cope with a wish for routine industrial application of HI. Therefore, new dynamic and refreshable materials surfaced, such as PRCs, which led to high-resolution holographic interferometer. However, recording holograms directly on the CCD sensors made its way and imposed itself, first in the form of ESPI. Despite the lower interferogram quality than analog HI, ESPI was extremely successful and still is marketed by some companies, for working under laboratory conditions, principally for 3D strain measurements. We have seen that electronic recording allows the use of exotic wavelengths such as UV and LWIR with lot of advantages. Also, we have seen that the particular configuration of speckle shearing interferometry (shearography) is highly successful in NDT for the detection of defects, eventually becoming a standardized technique in routine tire testing.

Finally, since the mid-1990s, intensive researches were carried out with DHI. The possibility of directly extracting the phase with single-shot holograms makes it very attractive for a large number of applications. We have seen that DH offered interesting possibilities for the observation of very fast events, the multisensitivities measurements in solid mechanics, for flow analyses with color DH, or the observation of large deformations with long wavelengths. Last but not least, we have seen that two-wavelength contouring has led to scientific and commercial success in the case of digital holographic microscope for analyzing the behavior of small structures, but also contouring highly demanding industrial applications like measurement of shape of manufactured parts with high velocities, or inside the ITER tokamak.

REFERENCES

[1] M. Horman, An application of wavefront reconstruction to interferometry, Applied Optics 4 (3) (1965) 333–336.

[2] R. Brooks, L. Heflinger, R.F. Wuerker, Interferometry, with a holographically reconstructed comparison beam, Applied Physics Letters 7 (1965) 248–249.

[3] J. Burch, The application of lasers in production engineering (The Viscount Nuffield Memorial Paper), Production Engineering 44 (1965) 431–442.

[4] R. Collier, E. Doherty, K. Pennington, Application of Moiré techniques to holography, Applied Physics Letters 7 (1965) 223–225.

[5] R. Powell, K. Stetson, Interferometric vibration analysis of three-dimensional objects by wavefront reconstruction, in: Paper FA16 Presented at the OSA Spring Meeting, March 31 – April 2, Dallas, 1965.

[6] R. Powell, K. Stetson, Interferometric vibration analysis by wave front reconstruction, Journal of the Optical Society of America 55 (12) (1965) 1593–1598.

[7] K. Stetson, R. Powell, Hologram interferometry, in: Paper WF13 Presented at the OSA Spring Meeting, October 5–8, Philadelphia, 1965.

[8] K. Stetson, Hologram interferometry: then and now, in: L. Lamberti, et al. (Eds.), Advancement of Optical Methods & Digital Image Correlation in Experimental Mechanics, Conference Proceedings of the Society for Experimental Mechanics Series, vol. 3, 2019, pp. 257–264. https://doi.org/10.1007/978-3-319-97481-1_35.

[9] K. Stetson, 50 years of holographic interferometry", invited paper, Proceedings of SPIE 9442 (2015) 944202.

[10] C.M. Vest, Holographic Interferometry, John Wiley & Sons, New York, 1979.

[11] Y. Ostrovsky, M. Butusov, G. Ostrovskaya, Interferometry by Holography, Springer Series in Optical Sciences, vol. **20**, Springer-Verlag, Berlin, 1980.

[12] R. Jones, C. Wykes, Holographic and Speckle Interferometry, second ed., Cambridge University Press, Cambridge, 1989.

[13] P. Rastogi (Ed.), Holographic Interferometry, Principle and Methods, Springer Series in Optical Sciences, vol. **68**, Springer-Verlag, Berlin, 1994.

[14] T. Kreis, Holographic Interferometry. Principle and Methods, Akademie Verlag, Berlin, 1996.

[15] T. Kreis, Handbook of Holographic Interferometry – Optical and Digital Methods, Wiley-VCH Verlag GmbH & Co. KGaA, Weinheim, 2005.

[16] U. Schnars, C. Falldorf, J. Watson, W. Jüptner, Digital Holography and Wavefront Sensing. Principles, Techniques and Applications, second ed., Springer, Heidelberg, 2015.

[17] N. Abramson, Sandwich hologram interferometry: a new dimension in holographic comparison, Applied Optics 13 (1974) 2019–2025.

[18] R. Pitlak, R. Page, Pulsed lasers for holographic interferometry, Optical Engineering 24 (1985) 639–644.

[19] J. Sollid, Holographic interferometry applied to measurements of small static displacements of diffusively reflecting surfaces, Applied Optics 8 (1969) 1587–1595.

[20] K. Stetson, Use of sensitivity vector variations to determine absolute displacements in double exposure hologram interferometry, Applied Optics 29 (1990) 502–504.

[21] A. Ennos, Measurements of in-plane surface strain by hologram interferometry, Journal of Physics E: Scientific Instruments 2 (1) (1968) 731–746.

[22] E. Aleksandrov, A. Bonch-Bruevich, Investigation of surface strains by the hologram technique, Soviet Physics – Technical Physics 12 (1967) 258–265.

[23] E. Archbold, A. Ennos, Observation of surface vibration modes by stroboscopic hologram interferometry, Nature 217 (1968) 942–943.

[24] A. Friesem, C. Vest, Detection of microfractures by holographic interferometry, Applied Optics 8 (1969) 1253–1254.

[25] R. Grant, G. Brown, Holographic nondestructive testing (HNDT), Materials Evaluation 27 (1969) 79–84.

[26] C. Vest, E. McKague, A. Friesem, Holographic detection of microcracks, Transactions of the ASME. Series D, Journal of Basic Engineering 93 (1971) 237–246.

[27] R. Erf, J. Waters, R. Gagosz, K. Stetson, F. Michael, G. Whitney, Nondestructive Holographic Techniques for Structures Inspection, Technical Report AFML-TR-72-204, United Aircraft Research Laboratories, East Hartford, CT, October 1972, pp. 27–54.

[28] P. Smigielski, General review on the investigations conducted at Institute Saint-Louis (ISL) in the field of holographic nondestructive testing, Proceedings of SPIE 349 (1982) 72–90.

[29] C. Le Floc'h, B. Gagnage, Industrial application of holography in Aerospatiale, Proceedings of SPIE 349 (1982) 118–129.

[30] S. Amadesi, F. Gori, R. Grella, G. Guattari, Holographic methods for painting diagnostics, Applied Optics 13 (1974) 2009–2013.

[31] C. Vest, Status and future of holographic nondestructive evaluation, Proceedings of SPIE 349 (1982) 186–198.

[32] K. Haines, B. Hildebrand, Contour generation by wavefront reconstruction, Physics Letters 19 (1) (1965) 10–11.

[33] B. Hildebrand, K. Haines, The generation of three-dimensional contour maps by wavefront reconstruction, Physics Letters 21 (4) (1966) 422–423.

[34] B. Hildebrand, K. Haines, Multiple-wavelength and multiple-source holography applied to contour generation, Journal of the Optical Society of America 57 (2) (1967) 155–162.
[35] T. Tsuruta, N. Shiotake, J. Tsujiuchi, K. Matsuda, Holographic generation of contour map of diffusely reflecting surface by using immersion method, Japanese Journal of Applied Physics 6 (5) (1967) 661–662.
[36] J. Zelenka, J. Varner, Multiple-index holographic contouring, Applied Optics 8 (7) (1969) 1431–1434.
[37] P. Lam, J. Gaskill, J. Wyant, Two-wavelength holographic interferometer, Applied Optics 23 (1984) 3079–3081.
[38] N. Ninane, M. Georges, Holographic interferometry using two-wavelength holography for the measurement of large deformations, Applied Optics 34 (1995) 1923–1928.
[39] L. Heflinger, R. Wuerker, R. Brooks, Holographic interferometry, Journal of Applied Physics 37 (2) (1966) 642–649.
[40] C. Reinheimer, C. Wiswall, R. Schmiege, R. Harris, J. Dueker, Holographic subsonic flow visualization, Applied Optics 9 (9) (1970) 2059–2065.
[41] A. Hirth, P. Smigielski, A. Stimpfling, Use of holography for visualization of the wake of projectiles in hypersonic flight at Mach 6, Optics & Laser Technology 3 (4) (1971) 195–199.
[42] J. Trolinger, Flow visualization holography, Optical Engineering 14 (5) (1975) 470–481.
[43] D.W. Robinson, G.T. Reid, in: D.W. Robinson, G.T. Reid (Eds.), Interferogram Analysis: Digital Fringe Pattern Measurement Techniques, Inst. Of Physics, London, 1993.
[44] W. Osten (Ed.), Fringe '89. Automatic Processing of Fringe Patterns, Akademie-Verlag, Berlin, 1989.
[45] W. Jüptner, W. Osten (Eds.), Fringe '93. Automatic Processing of Fringe Patterns, Akademie-Verlag, Berlin, 1993.
[46] W. Jüptner, W. Osten (Eds.), Fringe '97. Automatic Processing of Fringe Patterns, Akademie-Verlag, Berlin, 1997.
[47] T. Yatagai, Intensity bases analysis methods, in: D.W. Robinson, G.T. Reid (Eds.), Interferogram Analysis: Digital Fringe Pattern Measurement Techniques, Inst. Of Physics, London, 1993, pp. 72–93.
[48] P. Hariharan, B. Oreb, N. Brown, A digital phase-measurement system for real-time holographic interferometry, Optics Communications 41 (6) (1982) 393–396.
[49] K. Creath, Temporal phase measurement methods, in: D.W. Robinson, G.T. Reid (Eds.), Interferogram Analysis: Digital Fringe Pattern Measurement Techniques, Inst. Of Physics, London, 1993, pp. 94–140.
[50] K. Creath, Phase-shifting holographic interferometry, in: P. Rastogi (Ed.), Holographic Interferometry. Principle and Methods, Springer Series in Optical Sciences, vol. 68, Springer-Verlag, Berlin, 1994, pp. 109–150.
[51] M. Kujawinska, Spatial phase measurement methods, in: D.W. Robinson, G.T. Reid (Eds.), Interferogram Analysis: Digital Fringe Pattern Measurement Techniques, Inst. Of Physics, London, 1993, pp. 141–193.
[52] D. Ghiglia, M. Pritt, Two-Dimensional Phase Unwrapping:Theory, Algorithms, and Software, Wiley, New York, 1998.
[53] H. Bjelkhagen, Silver-halide Recording Materials for Holography and Their Processing, Springer Series in Optical Sciences, 1993.
[54] K. Biedermann, N.-E. Molin, Combining hypersensitization and rapid in situ processing for time-average observation in realtime hologram interferometry, Journal of Physics E: Scientific Instruments 3 (1970) 669–680.
[55] B. Achia, A simple plateholder for on site wet processing of holograms in real-time holographic interferometry, Journal of Physics E: Scientific Instruments 5 (1972) 128–129.
[56] P. Hariharan, B. Ramprasad, Rapid in situ processing for real-time holographic interferometry, Journal of Physics E: Scientific Instruments 6 (1973) 699–701.
[57] S. Gurevich, V. Konstantinov, D. Chernikh, Interferometry and holography studies in space, Proceedings of SPIE 1183 (1989) 479–485.
[58] R. Wuerker, L. Heflinger, J. Flannery, A. Kassel, A. Rollauer, Holography on the NASA space shuttle, AIP Conference Proceedings 65 (1980) 450–459.
[59] J. Trolinger, R. Lal, A. Batra, Holographic instrumentation for monitoring crystal growth in space, Optical Engineering 30 (10) (1991) 1608–1614.
[60] K.H. Sill, K. Heimann, M. Roth, HOLOP D-2, a holographic interferometer for microgravity research in space, in: Proceedings of the European Space Agency, ESA SP-295, VIIth European Symposium on Materials and Fluid Sciences in Microgravity, 1989, pp. 575–580.
[61] E. Bennett, W. Geist, K. Heimann, D. Heyland, G. Hibsch, K. Schmidt, Telescience with MARCO/HOLOP on board the spacelab D2-mission as a preparation for Columbus, Acta Astronautica 26 (8–10) (1992) 551–561.
[62] P. Smigielski, Holographie Industrielle (in French), Teknea, Toulouse, 1994.
[63] Patent EP0499559B1, "Procedure and Apparatus for the Non-destructive Control of Composite Material Containers by Holographic Interferometry".
[64] P. Smigielski, Cineholography in nondestructive testing, in: Proc. SPIE 10268, Optical Inspection and Testing: Critical Review, 102680E, June 1, 1993.
[65] H. Steinbichler, Holographic non destructive testing with automatic evaluation, Proceedings of SPIE 654 (1986) 242–248.
[66] H. Steinbichler, G. Gehring, TV-holography and holographic interferometry: industrial applications, Optics and Lasers in Engineering 24 (1996) 111–127.
[67] J. Frejlich, Photorefractive Materials. Fundamental Concepts, Holographic Recording and Materials Characterization, Wiley, Hoboken (NJ), USA, 2007.

[68] M. Petrov, S. Stepanov, A. Khomenko, Photorefractive Crystals in Coherent Optical Systems, Springer Series in Optical Sciences, vol. 59, Springer Verlag, Berlin, 1991.

[69] J.P. Huignard, J.P. Herriau, Real-time double exposure interferometry with $Bi_{12}SiO_{20}$ crystals in transverse electrooptic configuration, Applied Optics 16 (7) (1977) 1807–1809.

[70] J.P. Huignard, J.P. Herriau, T. Valentin, Time average holographic interferometry with photoconductive electrooptic $Bi_{12}SiO_{20}$ crystals, Applied Optics 16 (11) (1977) 2796–2798.

[71] H. Tiziani, Real-time metrology with BSO crystals, Optica Acta: International Journal of Optics 29 (4) (1982) 463–470.

[72] A. Marrakchi, J.P. Huignard, J.P. Herriau, Application of phase conjugation in $Bi_{12}SiO_{20}$ crystals to mode pattern visualization of diffuse vibrating structures, Optics Communications 34 (1980) 15–18.

[73] J.P. Huignard, A. Marrakchi, Two-wave mixing and energy transfer in $Bi_{12}SiO_{20}$ crystals: application to image amplification and vibration analysis, Optics Letters 6 (1981) 622–624.

[74] X. Wang, R. Magnusson, A. Haji-Sheikh, Real-time interferometry with photorefractive reference holograms, Applied Optics 32 (11) (1993) 1983–1986.

[75] M. Georges, Photorefractives for holographic interferometry and nondestructive testing, in: P.A. Blanche (Ed.), Photorefractive Organic Materials and Applications, Springer Series in Materials Science, vol. 240, Springer, Cham, 2016, pp. 283–312.

[76] P. Lemaire, M. Georges, Dynamic holographic interferometry: devices and applications, in: P. Günther, J.-P. Huignard (Eds.), Photorefractive Materials and Their Applications 3. Applications, Springer, New York, USA, 2007, pp. 223–251.

[77] A. Kamshilin, M. Petrov, Continuous reconstruction of holographic interferograms through anisotropic diffraction in photorefractive crystals, Optics Communications 53 (1985) 23–26.

[78] R. Troth, J.C. Dainty, Holographic interferometry using anisotropic self-diffraction in $Bi_{12}SiO_{20}$, Optics Letters 16 (1991) 53–55.

[79] M. Georges, P. Lemaire, Real-time holographic interferometry using sillenite photorefractive crystals. Study and optimization of a transportable set-up for quantified phase measurements on large objects, Applied Physics B 68 (1999) 1073–1083.

[80] M. Georges, V. Scauflaire, P. Lemaire, Compact and portable holographic camera using photorefractive crystals. Applications in various metrological problems, Applied Physics B 72 (2001) 761–765.

[81] M. Georges, V. Scauflaire, P. Lemaire, Compact and portable holographic camera based on photorefractive crystals and application in interferometry, Optical Materials 18 (2001) 49–52.

[82] M. Georges, P. Lemaire, Holographic interferometry using photorefractive crystals for quantitative phase measurement on large objects, Proceedings of SPIE 2652 (1996) 248–257.

[83] M. Georges, P. Lemaire, Phase-shifting real time interferometry that uses bismuth silicon oxide crystals, Applied Optics 34 (32) (1995) 7497–7506.

[84] M. Georges, P. Lemaire, Real-time stroboscopic holographic interferometry using sillenite crystals for the quantitative analysis of vibrations, Optics Communications 145 (1998) 249–257.

[85] M. Georges, C. Thizy, V. Scauflaire, S. Ryhon, G. Pauliat, P. Lemaire, G. Roosen, Holographic interferometry with photorefractive crystals: review of applications and advances techniques, Proceedings of SPIE 4933 (2003) 250–255.

[86] C. Thizy, M. Georges, M. Doulgeridis, E. Kouloumpi, T. Green, S. Hackney, V. Tornari, Role of dynamic holography with photorefractive crystals in a multifunctional sensor for the detection of signature features in movable cultural heritage, Proceedings of SPIE 6618 (2007) 661828.

[87] C. Thizy, F. Eliot, D. Ballhause, K.R. Olympio, R. Kluge, A. Shannon, G. Laduree, D. Logut, M. Georges, Holographic interferometry based on photorefractive crystal to measure 3D thermo-elastic distortion of composite structures and comparison with finite element models, Proceedings of SPIE 8788 (2013) 878807.

[88] C. Thizy, M. Georges, P. Lemaire, Y. Stockman, D. Doyle, Phase control strategies for stabilization of photorefractive holographic interferometer, Proceedings of SPIE 6341 (2006) 63411O.

[89] J. Leendertz, Interferometric displacement measurement on scattering surfaces utilizing speckle effect, Journal of Measurement and Control 3 (1970) 214–218.

[90] J. Butters, J. Leendertz, Holographic and video techniques applied to engineering measurement, Journal of Measurement and Control 4 (1971) 349–354.

[91] A. Macovski, S. Ramsey, L. Schaefer, Time-lapse interferometry and contouring using television systems, Applied Optics 10 (12) (1971) 2722–2727.

[92] Q. Li, F. Chiang, Three-dimensional dimension of laser speckle, Applied Optics 31 (1992) 6287–6291.

[93] R. Jones, C. Wykes, Holographic and Speckle Interferometry, in: Cambridge Studies in Modern Optics, Second ed., vol. 6, Cambridge University Press, Cambridge, 1989.

[94] P. Jacquot, J.-M. Fournier (Eds.), Interferometry in Speckle Light. Theory and Applications, Springer-Verlag Berlin Heidelberg, 2000.

[95] K. Gastinger, O. Lokberg, S. Winther (Eds.), Speckle Metrology 2003, Proc. SPIE 4933, SPIE, Bellingham, 2003.

[96] P. Slangen, C. Cerruti (Eds.), Speckle, from Grains to Flowers, Proc. SPIE 6341, SPIE, Bellingham, 2006.

[97] A. Albertazzi, G. Kaufmann (Eds.), Speckle 2010: Optical Metrology, Proc. SPIE 7387, SPIE, Bellingham, 2010.

[98] A. Doval, C. Trillo, J. Carlos Lopez-Vazquez (Eds.), Speckle 2012: V International Conference on Speckle Metrology, Proc. SPIE 8413, SPIE, Bellingham, 2012.

[99] F. Mendoza, E. Mendez (Eds.), Speckle 2015: VI International Conference on Speckle Metrology, Proc. SPIE 9660, SPIE, Bellingham, 2015.
[100] M. Kujawinska, L. Jaroszewicz (Eds.), Speckle 2018: VII International Conference on Speckle Metrology, Proc. SPIE 10834, SPIE, Bellingham, 2018.
[101] J.-F. Vandenrijt, M. Georges, Electronic speckle pattern interferometry and digital holographic interferometry with microbolometer arrays at 10.6 μm, Applied Optics 49 (2010) 5067–5075.
[102] S. Winther, 3D strain measurements using ESPI, Optics and Lasers in Engineering 8 (1988) 45–57.
[103] L. Yang, A. Ettemeyer, Strain measurement by three-dimensional electronic speckle pattern interferometry: potentials, limitations and applications, Optical Engineering 42 (5) (2003) 1257–1266.
[104] T. Cookson, J. Butters, H. Pollard, Pulsed lasers in electronic speckle pattern interferometry, Optics & Laser Technology 10 (1978) 119–124.
[105] J. Tyrer, Application of pulsed holography and double pulsed electronic speckle pattern interferometry to large vibrating engineering structures, Proceedings of SPIE 599 (1985) 181–188.
[106] R. Spooren, Double-pulse subtraction TV holography, Optical Engineering 31 (5) (1992) 1000–1007.
[107] G. Pedrini, B. Pfister, H. Tiziani, Double pulse-electronic speckle interferometry, Journal of Modern Optics 40 (1) (1993) 89–96.
[108] G. Pedrini, H. Tiziani, Double-pulse electronic speckle interferometry for vibration analysis, Applied Optics 33 (34) (1994) 7857–7863.
[109] A. Fernández, J. Blanco-García, A. Doval, J. Bugarín, B. Dorrío, C. López, J. Alén, M. Pérez-Amor, J. Fernández, Transient deformation measurement by double-pulse-subtraction TV holography and the Fourier transform method, Applied Optics 37 (16) (1998) 3440–3446.
[110] A. Fernández, A. Moore, C. Pérez-López, A. Doval, J. Blanco-García, Study of transient deformations with pulsed TV holography: application to crack detection, Applied Optics 36 (10) (1997) 2058–2065.
[111] P. Aswendt, Micromotion analysis by deep-UV speckle interferometry, Proceedings of SPIE 5145 (2003) 17–22.
[112] P. Aswendt, Promise and challenge of DUV speckle interferometry, Proceedings of SPIE 4933 (2003) 117–122.
[113] I. Yamaguchi, Fundamentals and applications of speckle", Proceedings of SPIE 4933 (2003) 1–8.
[114] S. Guntaka, V. Sainov, V. Toal, S. Martin, T. Petrova, J. Harizanova, Compact electronic speckle pattern interferometer using a near infrared diode laser and a reflection holographic optical element, Journal of Optics A: Pure and Applied Optics 8 (2006) 182–188.
[115] D. Francis, D. Masiyano, J. Hodgkinson, R. Tatam, A mechanically stable laser diode speckle interferometer for surface contouring and displacement measurement, Measurement Science and Technology 26 (2015) 055402.
[116] J.-F. Vandenrijt, M. Georges, Infrared electronic speckle pattern interferometry at 10 μm, Proceedings of SPIE 6616 (2007) 661672.
[117] J.F. Vandenrijt, C. Thizy, I. Alexeenko, G. Pedrini, J. Rochet, B. Vollheim, I. Jorge, P. Venegas, I. Lopez, W. Osten, M.P. Georges, Mobile speckle interferometer in the long-wave infrared for aeronautical nondestructive testing in field conditions, Optical Engineering 52 (2013) 101903.
[118] I. Alexeenko, J.-F. Vandenrijt, G. Pedrini, C. Thizy, B. Vollheim, W. Osten, M.P. Georges, Nondestructive testing by using long-wave infrared interferometric techniques with CO_2 lasers and microbolometer arrays, Applied Optics 52 (1) (2013) A56–A67.
[119] J.-F. Vandenrijt, C. Thizy, I. Alexeenko, I. Jorge, I. López, I.S. de Ocáriz, G. Pedrini, W. Osten, M. Georges, Electronic speckle pattern interferometry at long infrared wavelengths. Scattering requirements, in: W. Osten, M. Kujawinska (Eds.), Fringe 2009 – 6th International Workshop on Advanced Optical Metrology, Springer, 2009, pp. 596–599.
[120] M.P. Georges, J.F. Vandenrijt, C. Thizy, I. Alexeenko, G. Pedrini, J. Rochet, B. Vollheim, I. Jorge, P. Venegas, I. Lopez, W. Osten, Combined holography and thermography in a single sensor through image-plane holography at thermal infrared wavelengths, Optics Express 22 (2014) 25517.
[121] M.P. Georges, Speckle interferometry in the long-wave infrared for combining holography and thermography in a single sensor: applications to nondestructive testing: the FANTOM Project, Proceedings of SPIE 9525 (2015) 952557.
[122] J. Leendertz, J. Butters, An image-shearing speckle-pattern interferometer for measuring bending moments, Journal of Physics E: Scientific Instruments 6 (1973) 1107–1110.
[123] Y. Hung, A speckle-shearing interferometer: a tool for measuring derivatives of surface displacements, Optics Communications 11 (1974) 132–135.
[124] S. Debrus, Speckle shearing interferometer using a Savart plate, Optics Communications 20 (1977) 257–267.
[125] W. Steinchen, L. Yang, Digital Shearography, SPIE Press, Bellingham, 2003.
[126] D. Francis, R. Tatam, R. Groves, Shearography technology and applications: a review, Measurement Science and Technology 21 (2010) 102001.
[127] M. Viotti, A. Albertazzi, Robust Speckle Metrology: Techniques for Stress Analysis and NDT, SPIE Press, Bellingham, 2014.
[128] M. Kalms, W. Osten, Mobile shearography system for the inspection of aircraft and automotive components, Optical Engineering 42 (5) (2003) 1188–1196.

[129] http://www.laserndt.com.
[130] https://optonor.com.
[131] http://laseroptical.co.uk/shearography/.
[132] http://www.isi-sys.com/isi-sys-homepage/.
[133] https://www.dantecdynamics.com/electronic-speckle-pattern-interferometry.
[134] https://www.dantecdynamics.com/shearography-non-destructive-testing.
[135] https://optotechnik.zeiss.com/en/products/tire-testing-intact.
[136] U. Schnars, Direct phase determination in hologram interferometry with use of digitally recorded holograms, Journal of the Optical Society of America 11 (7) (1994) 2011–2015.
[137] U. Schnars, W. Jüptner, Digital recording and reconstruction of holograms in hologram interferometry and shearography, Applied Optics 33 (20) (1994) 4373–4377.
[138] U. Schnars, H.-J. Hartmann, W. Jüptner, Digital recording and numerical reconstruction of holograms for nondestructive testing, Proceedings of SPIE 2545 (1995) 250–253.
[139] G. Pedrini, H. Tiziani, Digital double-pulse holographic interferometry using Fresnel and image plane holograms, Measurement 15 (1995) 251–260.
[140] C. Perez-Lopez, M. De La Torre-Ibarra, F. Mendoza Santoyo, Very high speed cw digital holographic interferometry, Optics Express 14 (21) (2006) 9709–9715.
[141] D. Aguayo, F. Mendonza Santoyo, M. De La Torre-Ibarra, M. Salas-Araiza, C. Caloca-Mendez, D. Asael Gutierrez Hernandez, Insect wing deformation measurements using high speed digital holographic interferometry, Optics Express 18 (6) (2010) 5661–5667.
[142] G. Pedrini, Y.-L. Zhou, H. Tiziani, Simultaneous quatitative evaluation of in-plane and out-of-plane deformations by use of a multidirectional spatial carrier, Applied Optics 36 (4) (1997) 786–792.
[143] G. Pedrini, H. Tiziani, Quantitative evaluation of two-dimensional dynamic deformations using digital holography, Optics & Laser Technology 29 (5) (1997) 249–256.
[144] P. Picart, E. Moisson, D. Mounier, Twin sensitivity measurement by spatial multiplexing of digitally recorded holograms, Applied Optics 42 (2003) 1947–1957.
[145] P. Picart, B. Diouf, J.M. Berthelot, Investigation of fracture mechanisms in resin concrete using spatially multiplexed digital Fresnel holograms, Optical Engineering 43 (2004) 1169–1176.
[146] P. Picart, J. Leval, J.-C. Pascal, J.-P. Boileau, M. Grill, J.-M. Breteau, B. Gautier, S. Gillet, 2D full field vibration analysis with multiplexed digital holograms, Optics Express 13 (22) (2005) 8882.
[147] P. Tankam, P. Picart, D. Mounier, J.-M. Desse, J. Li, Method of digital holographic recording and reconstruction using a stacked color image sensor, Applied Optics 49 (3) (2010) 320–328.
[148] P. Tankam, Q. Song, M. Karray, J. Li, J.-M. Desse, P. Picart, Real-time three-sensitivity measurements based on three-color digital Fresnel holographic interferometry, Optics Letters 35 (12) (2010) 2055–2057.
[149] J.-M. Desse, P. Picart, P. Tankam, Digital color holography applied to fluid and structural mechanics, Optics and Lasers in Engineering 50 (2012) 18–28.
[150] J.-M. Desse, P. Picart, Digital holography for analyzing unsteady wake flows, in: P. Picart (Ed.), New Techniques in Digital Holography, Wiley-ISTE, London, 2015, pp. 107–136.
[151] J.-M. Desse, P. Picart, P. Tankam, Digital three-color holographic interferometry for flow analysis, Optics Express 16 (8) (2008) 5471–5480.
[152] J.-M. Desse, J.-L. Tribillon, Real-time three-color reflection holographic interferometter, Applied Optics 48 (36) (2009) 6870–6877.
[153] P. Picart, J. Leval, D. Mounier, S. Gougeon, Time-averaged digital holography, Optics Letters 28 (20) (2003) 1900–1902.
[154] P. Picart, J. Leval, D. Mounier, S. Gougeon, Some opportunities for vibration analysis with time-averaging in digital Fresnel holography, Applied Optics 44 (3) (2005) 337–343.
[155] P. Castellini, G.M. Revel, E.P. Tomasini, Laser Doppler vibrometry: a review of advances and applications, The Shock and Vibration Digest 30 (1998) 443–456.
[156] J. Leval, P. Picart, J.-P. Boileau, J.-C. Pascal, Full-field vibrometry with digital Fresnel holography, Applied Optics 44 (27) (2005) 5763–5772.
[157] J. Poittevin, P. Picart, C. Faure, F. Gautier, C. Pezerat, Multi-point vibrometer based on high-speed digital in-line holography, Applied Optics 54 (11) (2015) 3185–3196.
[158] J. Poittevin, F. Gautier, C. Pezerat, P. Picart, High-speed holographic metrology: principle, limitations, and application to vibroacoustics of structures, Optical Engineering 55 (12) (2016) 121717.
[159] N. Demoli, D. Vukicevic, Detection of hidden stationary deformations of vibrating surfaces by use of time-averaged digital holographic interferometry, Optics Letters 29 (20) (2004) 2423–2425.
[160] A. Asundi, V. Singh, Time-averaged in-line digital holographic interferometry for vibration analysis, Applied Optics 45 (11) (2006) 2391–2395.
[161] V. Singh, J. Miao, Z. Wang, G. Hegde, A. Asundi, Dynamic characterization of MEMS diaphragm using time averaged in-line digital holography, Optics Communications 280 (2) (2007) 285–290.
[162] A. Asundi (Ed.), Digital Holography for MEMS and Microsystem Metrology, John Wiley & Sons, Chichester, UK, 2011.
[163] M. del Socorro Hernandez-Montes, C. Furlong, J. Rosowski, N. Hulli, E. Harrington, J. Cheng, M. Ravicz, F. Mendoza Santoyo, Optoelectronic holographic otoscope for measurement of nano-

displacements in tympanic membranes, Journal of Biomedical Optics 14 (3) (2009) 034023.
[164] M. del Socorro Hernandez-Montes, F. Mendoza Santoyo, C. Perez Lopez, S. Munoz Solis, J. Esquivel, Digital holographic interferometry applied to the study of tympanic membrane displacements, Optics and Lasers in Engineering 49 (2011) 698–702.
[165] D. De Greef, J. Soons, J. Dirckx, Digital stroboscopic holography setup for deformation measurement at both quasit-static and acoustic frequencies, International Journal of Optomechatronics 8 (2014) 275–291.
[166] M. Khaleghi, C. Furlong, M. Ravicz, J. Tao Cheng, J. Rosowski, Three-dimensional vibrometry of the human eardrum with stroboscopic lensless digital holography, Journal of Biomedical Optics 20 (5) (2015) 051028.
[167] E. Allaria, S. Brugioni, S. De Nicola, P. Ferraro, S. Grilli, R. Meucci, Digital holography at 10.6 μm, Optics Communications 215 (2003) 257–262.
[168] M. Paturzo, A. Pelagotti, A. Finizio, L. Miccio, M. Locatelli, A. Geltrude, P. Poggi, R. Meucci, P. Ferraro, Optical reconstruction of digital holograms recorded at 10.6 μm: route for 3D imaging at long infrared wavelengths, Optics Letters 35 (12) (2010) 2112–2114.
[169] A. Pelagotti, M. Paturzo, A. Geltrude, M. Locatelli, R. Meucci, P. Poggi, P. Ferraro, Digital holography for 3D imaging and display in the IR range: challenges and opportunities, 3D Research 1 (4) (2010) 1–10. Springer Publishing.
[170] M. Locatelli, E. Pugliese, M. Paturzo, V. Bianco, A. Finizio, A. Pelagotti, P. Poggi, L. Miccio, R. Meucci, P. Ferraro, Imaging live humans through smoke and flames using far-infrared digital holography, Optics Express 21 (5) (2013) 5379–5390.
[171] M. Georges, Long-wave infrared digital holography, in: P. Picart (Ed.), New Techniques in Digital Holography, Wiley-ISTE, London, 2015, pp. 219–254.
[172] M. Georges, J.F. Vandenrijt, C. Thizy, Y. Stockman, P. Queeckers, F. Dubois, D. Doyle, Digital holographic interferometry with CO_2 lasers and diffuse illumination applied to large space reflector metrology, Applied Optics 52 (2013) A102.
[173] J.F. Vandenrijt, C. Thizy, P. Queeckers, F. Dubois, D. Doyle, M. Georges, Long-wave infrared digital holographic interferometry with diffuser or point source illuminations for measuring deformations of aspheric mirrors, Optical Engineering 53 (2014) 112309.
[174] J.F. Vandenrijt, C. Thizy, L. Martin, F. Beaumont, J. Garcia, C. Fabron, E. Prieto, T. Maciaszek, T, M. Georges, Digital holographic interferometry in the long-wave infrared and temporal phase unwrapping for measuring large deformations and rigid body motions of segmented space detector in cryogenic test, Optical Engineering 55 (2016) 121723.
[175] G. Pedrini, I. Alekseenko, W. Osten, H. Tiziani, Temporal phase unwrapping of digital hologram sequences, Applied Optics 42 (29) (2003) 5846–5854.
[176] G. Pedrini, P. Fröning, H. Tiziani, M. Gusev, Pulsed digital holography for high-speed contouring that uses a two wavelength method, Applied Optics 38 (1999) 3460–3467.
[177] G. Pedrini, H. Tiziani, M. Gusev, Pulsed digital holographic interferometry with 694- and 347-nm wavelengths, Applied Optics 39 (2) (2000) 246–249.
[178] Y. Fu, G. Pedrini, B. Hennelly, R. Groves, W. Osten, Dual-wavelength image-plane digital holography for dynamic measurement, Optics and Lasers in Engineering 47 (2009) 552–557.
[179] J. Kühn, T. Colomb, F. Montfort, F. Charrière, Y. Emery, E. Cuche, P. Marquet, C. Depeursinge, Real-time dual-wavelength digital holographic microscopy with a single hologram acquisition, Optics Express 15 (12) (2007) 7231–7242.
[180] E. Cuche, Y. Emery, F. Montfort, Microscopy: one-shot analysis, Nature Photonics 3 (11) (2009) 633–635.
[181] https://www.lynceetec.com/category/4d_profilometry/.
[182] M. Agour, C. Falldorf, R. Bergmann, Spatial multiplexing and autofocus in holographic contouring for inspection of micro-parts, Optics Express 26 (22) (2018) 28576.
[183] A. Schiller, T. Beckmann, M. Fratz, D. Belzer, A. Bertz, D. Carl, K. Buse, Digital holography on moving objects: multiwavelength height measurements on inclined surfaces, Proceedings of SPIE 10329 (2017) 103290D.
[184] T. Seyler, M. Fratz, T. Beckmann, A. Schiller, A. Bertz, D. Carl, Miniaturized multiwavelength digital holography sensor for extensive in-machine tool measurement, Proceedings of SPIE 10329 (2017) 103290F.
[185] T. Seyler, M. Fratz, T. Beckmann, A. Schiller, A. Bertz, D. Carl, Extending the depth of field beyond geometrical imaging limitations using phase noise as a focus measure in multiwavelength digital holography, Applied Sciences 8 (2018) 1042.
[186] G. Pedrini, I. Alekseenko, G. Jagannathan, M. Kempenaars, G. Vayakis, W. Osten, Feasibility study of digital holography for erosion measurements under extreme environmental conditions inside the International Thermonuclear Experimental Reactor tokamak [invited], Applied Optics 58 (5) (2019) A147–A155.
[187] G. Pedrini, I. Alekseenko, G. Jagannathan, M. Kempenaars, G. Vayakis, W. Osten, Digital holography for erosion monitoring inside the ITER Tokamak,"(invited paper), Proceedings of SPIE 10677 (2018) 1067722.

CHAPTER 7

Holographic Sensors

IZABELA NAYDENOVA, PHD, MSC

INTRODUCTION

Holography and Optical Sensing

Holography is a technique for recording information about an optical wavefront by interfering it with a reference coherent wave and was invented by Dennis Gabor [1]. The pattern produced by the interference of the two coherent waves can be stored as an analog record in a photosensitive material by modifying accordingly its optical properties or the pattern can be digitized and stored in a computer [2–4]. The recorded hologram, analog or digital, can be used to reconstruct the original wavefront by illumination of the hologram with the reference wave. For an exact replica of the original wavefront to be reproduced, both the hologram and the reference beam must be the same as that used at the recording stage. Any deviation of the reference beam from its original condition, or change in the hologram, will cause a change in the properties of the reconstructed wavefront [1–6]. This basic principle has been exploited in optical sensors utilizing holography to detect small dimensional changes or changes in the refractive index of the layer containing the hologram. In Table 7.1, we present a simplified description of optical sensors utilizing the principle of holography. Generally, one could divide these sensors in two main groups. Wavefront sensors are used to register a wavefront change by comparing the reconstructed wavefront to the originally recorded one by utilizing analogue [7] or digital holography [8]. Hologram sensors are capable of detecting changes in their environment, by detecting changes in the properties of the hologram [9,10].

This chapter will focus on one particular type of optical sensor utilizing holography, which in the literature is often referred to as holographic sensor. This type of sensor utilizes a hologram, which is capable of responding to changes in its environment, by changing its properties. Thus a change in the concentration of a chemical substance, the presence of a specific target analyte, a change in stimuli such as pressure, temperature, humidity in the environment will trigger a change in the hologram and thus in its ability to reconstruct the original wavefront [9]. The change can be either a dimensional change such as swelling/shrinking of the layer containing the hologram or a change of the effective refractive index or refractive index modulation in the layer (Fig. 7.1A) (see also Chapter 1: Media Thickness Variation section). The change introduced by the external stimulus is typically demonstrated in a change in the

TABLE 7.1
Holographic Sensors Utilizing Holography.

Type of Sensor	Changing Element	Observed Change	Detected Property
Holographic wavefront sensor	The aberrated wavefront entering the holographic wavefront sensor	Deviation of the aberrated wavefront from the originally recorded (reference) wavefront	Quantitative characterization of the changes in optical wavefronts
Hologram sensor (often referred in the literature to as "holographic sensor")	Hologram	Diffraction properties of the hologram: - Diffraction efficiency - Angular and wavelength response	Concentration of target analytes; quantitative characterization of external stimuli, such as magnetic, electric field, pressure, temperature variation

Optical Holography-Materials, Theory and Applications. https://doi.org/10.1016/B978-0-12-815467-0.00007-4

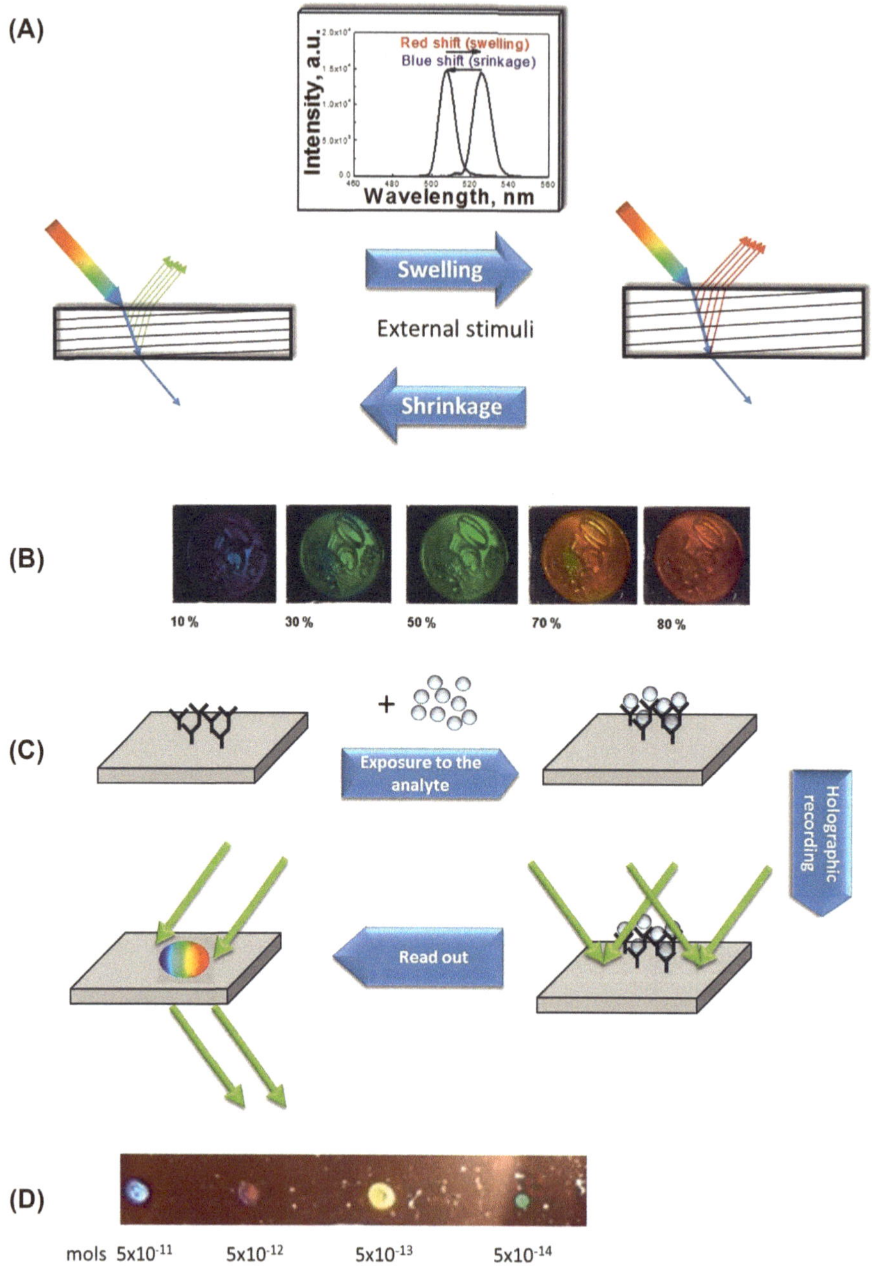

FIG. 7.1 **(A)** Simplified presentation of the principle of detection in holographic sensors based on swelling/ shrinkage; **(B)** visual detection of the response of the hologram to humidity. **(C)** detection by creation of a hologram; **(D)** visual detection of analyte by appearance of diffraction pattern (**(B)** Reprinted from I. Naydenova, R. Jallapuram, S. Martin, V. Toal Humidity sensors: types, in: C.T. Okada (Ed.), Nanomaterials and Environmental Monitoring, Nova Science Publishers, Hauppauge, 2011, p. 117. **(D)** Reprinted with permission from I. Naydenova, S. Martin, V. Toal, J. Eur. Opt. Soc., Rapid Publ. 4 (2009) 09042. Copyright 2009 European Optical Society.)

diffraction efficiency of the hologram and/or its spectral characteristics (Fig. 7.1A). The latter feature is particularly useful when a visual detection of the change is sought, as the response of the hologram could be simply a change in the color of the reconstructed image (Fig. 7.1B).

In addition to using an existing hologram, the actual creation of a hologram can be the mechanism for detection of an analyte. In photopolymers, this was first described in Ref. [12] and the principle of operation is shown in Fig. 7.1C. An unsensitized photopolymer layer contains a receptor molecule, which is capable of capturing the dye-labeled analyte molecule. When placed in contact with the unsensitized layer, the dye label is capable of triggering photopolymerization when illuminated with light. This principle was applied to detect dye-labeled DNA molecules. The outcome of a test to detect dye-labeled DNA strand with four different concentrations is shown in Fig. 7.1D.

History of the Development of Holographic Sensors

A detailed account of the historical development of holographic sensors is provided in a recently published review on holographic sensors [9]. Here, we discuss some of the most significant developments in this research area and cover the most recent advances.

The sensitivity of holograms to their environment has been known for a long time before being utilized for development of sensor devices. In some cases, this was an advantage and the sensitivity of the recording layer to moisture has been used to tune the wavelength of the reconstructed beam [13,14]. In other cases, measures to avoid sensitivity to humidity had to be taken to prevent damage to the holograms [14]. In Ref. [15], Spooncer et al. reported their attempt to utilize the sensitivity of gelatin to humidity to fabricate a humidity sensor. Their device consisted of a holographic filter and an optical fiber. The authors reported a strong hysteresis in humidity response, temperature dependence, and variable sensitivity, which made this particular arrangement unsuitable for use as a sensor.

The first successful attempts to develop holographic sensors were reported by the group of Christopher R. Lowe at the University of Cambridge [16–18]. For the following two decades, these authors extensively studied reflection holograms that change their properties due to exposure to the target analyte or other external stimuli such as pressure or moisture. The holographic photonic structures were recorded in different materials, and this allowed the group to create a variety of different sensors. The first examples were for detection of proteases [17], water in solvents [18], trypsin [19], and alcohol [20]. At about the same time, two other groups reported the first sensors based on transmission holographic gratings for detection of moisture [21] and salinity of water [22].

Over the years, a large number of different target analytes have been investigated by different groups following these first steps. A detailed summary of the different target analytes for which holographic sensors have been developed over the years is given in Ref. [9]. This information, modified and updated by including the newest publications, is presented in Table 7.2 from which it can be seen that holographic sensors are very flexible and can be modified to detect a variety of analytes.

The different sensor platforms are covered in Sensor Platforms section, which also outlines the main factors that must be considered when a sensor based on a hologram is to be developed. Fabrication of the Holographic Photonic Structures section summarizes the

TABLE 7.2
Holographic Sensors and Their Target Analytes.

Category	Analytes
Ions	pH [24] (Na^+, K^+) [25] Divalent metal ions (Ni^{2+}, Zn^{2+}, Co^{2+}, Ca^{2+}, Mg^{2+}) [26,27] Pb^{2+} Co^{3+} [28] Cu^{2+}, Fe^{2+} [29]
Metabolites	Glucose [30] Lactate [31] Urea [32] Acetylcholine [33] Testosterone [34]
Enzymes	Trypsin [35] Amylase [36]
Drugs	Penicillin G [37], cocaine [38]
Microorganisms	Bacteria, calcium dipicolinate [39]
Gasses	O_2 [40], N_2 [41], hydrocarbons [42,43], ammonia [44]
Mechanical stimuli	Pressure [10,27,45]
Environmental stimuli	Temperature [27,46], humidity [11,15]
Other analytes	Water activity in solvents [18], organic solvents [29], alcohol, ionic strength [47], antraquinone-2-carboxylate [40] magnetic field [48], light [49]

different methods for fabrication of holographic sensors, starting with the optical recording geometries and the mechanism of formation of the holographic photonic structures. Approaches to Functionalisation of Holographic Structures section outlines the different approaches to their functionalization. A number of recent reviews [9,10,23] focus on the materials for holographic sensors in great detail. Here, we briefly describe the basics and update with some recently reported findings. Challenges in Holographic Sensors Research section looks into the main challenges in the research field and possible approaches for their solution.

SENSOR PLATFORMS

The large variety of analytes that have been studied (Table 7.2) is evidence that sensor platforms using holograms are very flexible and relatively easy to tune to the specific task. Another important advantage of these sensor platforms is that they can be miniaturized and thus incorporated in mobile devices, for example, for point-of-care medical applications or for environmental monitoring. It has been demonstrated that these platforms are also compatible with modern mobile phone technologies [50], which makes them accessible to and useable by nonspecialists. They also allow for multiplexing and thus for development of sensor devices that can detect multiple analytes simultaneously. One approach is to deliver the analyte-sensing component to the layer by digital printing. Initial studies demonstrate that this is achievable [43].

The following subsections describe sensor platforms categorized by different criteria and consider their main characteristics, advantages, and challenges. Practically, all holographic sensor platforms described in the literature are built on phase holograms, which change the phase of the propagating wave while its amplitude is considered to remain constant. This happens when the holographic recording takes the form of either thickness variation of the photosensitive layer or as a spatial modulation of the refractive index of the layer. In amplitude holograms, the holographic recording results in a spatial modulation of the absorption coefficient of the layer. Amplitude holograms have limited diffraction efficiency (less than 8%) [1], and thus they are not currently explored due to the low dynamic range they can offer. In the following discussion, the holograms are assumed to be phase holograms, unless otherwise stated. For a discussion on the efficiency of different types of holograms, see Chapter 1: Thick Gratings Characteristics and Thin Gratings Characteristics.

The first criterion that will be considered is whether the diffractive photonic structure, the hologram, is recorded in the volume or at the surface the analyte sensitive layer.

Sensors Based on Structures in the Volume or at the Surface of the Analyte Sensitive Layer

Most of the studies reported in the literature utilize holograms that are recorded in the volume of the layer, and only recently sensors using surface diffractive structures created by holographic lithography have been reported [25c,26c,27,34]. Fig. 7.2 presents schematically the two types of sensing platform. The surface relief structure (Fig. 7.2A) can be a simple sinusoidal diffraction grating, created by interfering two plane waves. The flexibility in this approach is that the spatial frequency of the structure can be easily modified by simply varying the angle between the two recording beams [25c,26c]. Surface relief structures can also be fabricated by exposure to more than two beams to create more complex patterns such as Aztec gratings Fig. 7.2B [51].

Photonic structures in the volume of the layer are typically recorded using holographic lithography creating regions of varying refractive index in the material. In the case of two plane recording waves, the structure consists of parallel planes of varying refractive index as shown in Fig. 7.2C and D. Another fabrication approach utilizes ablation by exposure to a single laser pulse [52]. This approach is discussed in Fabrication of the Holographic Photonic Structures section. Table 7.3 compares the advantages and challenges related to surface and volume sensors.

Dynamic range

In terms of dynamic range, the holograms recorded in the volume of the layer offer the advantage of achieving nearly 100% diffraction efficiency, whereas the surface structures typically operate in the regime of thin holograms and the maximum diffraction efficiency is less than 33% for sinusoidal modulation [1]. Utilizing the volume of the layer also allows for recording of high diffraction efficiency reflection holograms, which is difficult to achieve when the diffractive structure is confined to the surface of the layer.

Selectivity

On the other hand, functionalization of surface holograms is more flexible, because the same surface structure can be used to detect different analytes by simply using a different coating layer (Fig. 7.2B). Functionalization of the volume sensors is carried out before

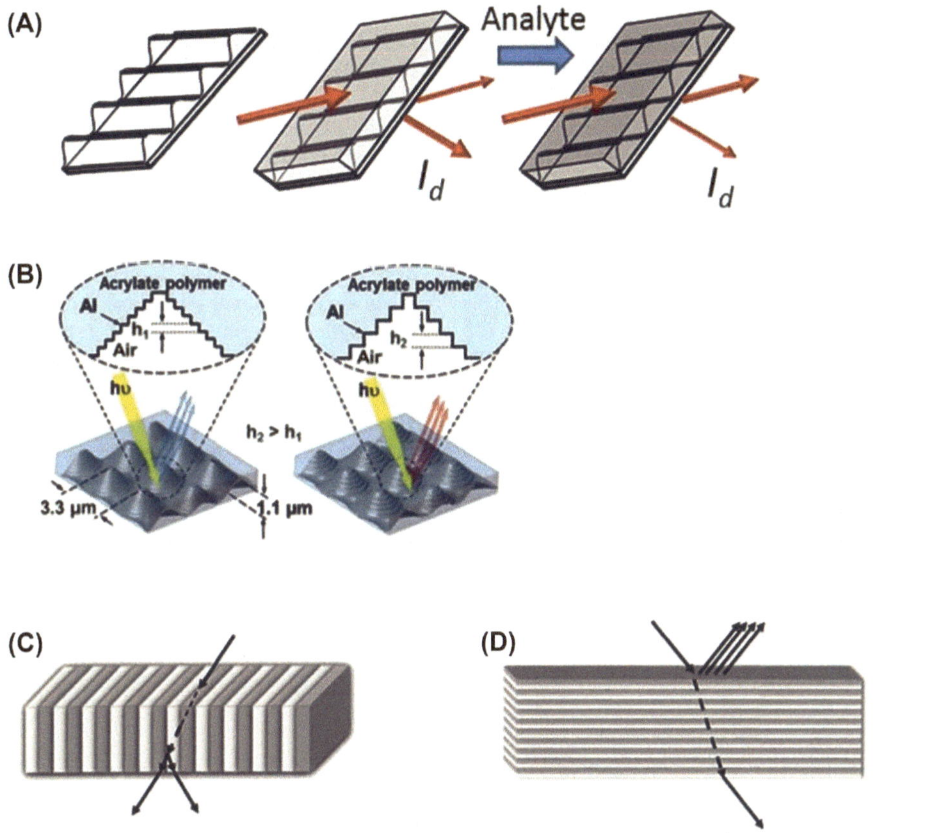

FIG. 7.2 Holograms recorded at the surface and in the volume of the layer: **(A)** surface relief structure before functionalization, after functionalization, and after exposure to the analyte; **(B)** Aztec gratings; volume structures; **(C)** transmission and **(D)** reflection hologram.

recording [11,46] or after recording by impregnation of the layer with the chemical component that can recognize the analyte [43]. In the first case, care must be taken to avoid any deterioration of the capability of the layer to record a hologram. Postrecording functionalization in both surface and volume sensors requires careful consideration of the effect this has on the original structure and its diffraction efficiency.

Response time of the sensor

One important parameter of the sensor is its response time. Sensors based on surface structures are inherently faster than those based on volume structures owing to the shorter diffusion lengths required for the analytes to travel; hundreds of nanometers versus tens of micrometers. Successful examples have been reported in the literature using both approaches. Theoretical studies of the time response properties of the sensors are rare. Recently, an insightful study of the dependence of the dynamics of the sensors depending on the chemical structure and size of the analyte has been reported [53].

Depending on the regime of operation of the hologram, holographic sensor platforms can be divided into two groups comprising thin or thick holograms.

Regime of Operation: Thin and Thick (Volume) Holograms

This criterion takes into account the optical properties of the created photonic structure and its interaction with the propagating wave. The division of holograms into thick and thin types is a well-established convention and is related to two different regimes of operation of the hologram. Thin holograms have maximum diffraction efficiency of 33% for sinusoidal refractive index modulation and produce multiple diffracted waves (Fig. 7.3A), at almost any angle of incidence of the probe beam. Thick holograms are usually called "volume" holograms in the literature, and this is related to their regime of operation and not just the fact that they are recorded in the volume of the layer. These holograms can theoretically achieve 100% diffraction efficiency, and they produce a single diffracted wave only at the Bragg angle of incidence (Fig. 7.3B).

Considering in which of the two regimes the sensor is going to operate is important because it determines the dynamic range of the sensor and its sensitivity [54], as well as the accuracy of the angle at which probe beam must be incident at the hologram.

TABLE 7.3
Comparison of Sensor Structures Recorded at the Surface and in the Volume of the Layer.

	Volume of the Layer	Surface of the Layer
Advantages	Easier to achieve large dynamic range, up to 100% diffraction efficiency is possible. High diffraction efficiency reflection holograms can be recorded	Shorter diffusion times of the analyte, faster response time More flexible in terms of functionalization
Disadvantages	Longer diffusion times Less flexible functionalization	Typically the recorded holograms operate in the regime of thin holograms, and the maximum theoretical diffraction efficiency is less than 33%

Two different approaches are used to evaluate whether a hologram operates as a thick or thin. Both are derived by evaluating the solutions of the wave equation describing the propagation of the probe beam in the layer containing the hologram and estimating the amount of light transferred to the higher orders of diffraction. The Klein-Cook Q parameter [55] takes into account the physical thickness of the grating d, as well as the grating period Λ, the average refractive index n, and the probe wavelength λ_p and is described by Eq. (7.1a):

$$Q = \frac{2\pi\lambda_p d}{n\Lambda^2} \tag{7.1a}$$

Q values > 10 correspond to thick holograms, whereas holograms with a Q value < 1 are considered thin. Another approach proposed by Moharam and Young [56] takes into account the refractive index modulation Δn instead of the thickness d when determining the regime of operation of the hologram. The ρ parameter is described by Eq. (7.1b):

$$\rho = \frac{\lambda_p^2}{\Lambda^2 n \Delta n} \tag{7.1b}$$

According to the second criteria, holograms characterized by $\rho \leq 1$ operate as thin holograms, whereas for $\rho > 10$, the gratings can be considered thick. As seen from Eqs. 7.1a and 7.1b, the regime of operation of the hologram depends on the wavelength of the probe beam, its spatial frequency/frequencies, and the effective refractive index of the medium.

Generally speaking, although the two criteria classify the holograms in the same category, deviations do

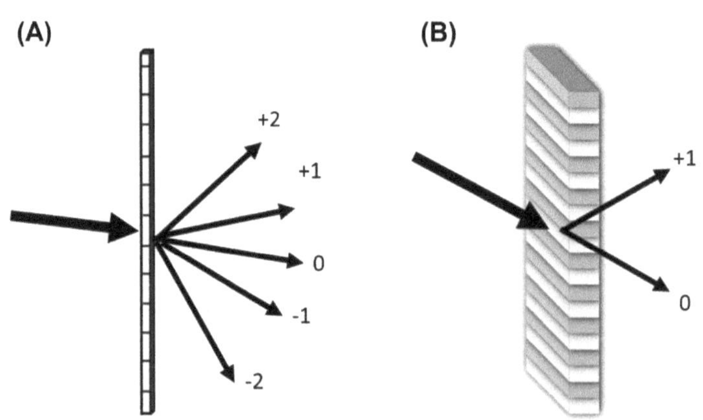

FIG. 7.3 Transmission holograms in regime of operation as thin hologram **(A)** and **(B)** thick (volume) hologram.

occur, for example, for very large refractive index modulations where gratings operating typically as thick gratings (characterized by large Q factor) could produce strong multiple orders of diffraction (behavior characteristic for a thin hologram predicted by the small ρ factor).

Mathematically, two theories are mostly utilized to predict the diffractive properties of thin and thick gratings depending on their physical properties and the probe beam wavelength. They are widely accepted within the optics community based on their prevalence in the literature—Raman-Nath Theory [57] is used to describe thin gratings. Kogenik's coupled-wave theory [5] is satisfactorily used to describe thick gratings, but for more accurate description, the rigorous coupled-wave theory [58] is utilized. In the following subsection, we will use their outcomes to define the diffraction efficiency of the simplest holograms—holographic gratings recorded in the two main regimes of recording—transmission and reflection. Based on these, we will also derive expressions for the sensitivity of sensors using these holograms. The main implications for the design of holographic sensors are presented in Table 7.4.

Dynamic range

In terms of dynamic range, the thick holograms offer the advantage of theoretically achievable 100% diffraction efficiency of the hologram, whereas the maximum achievable diffraction efficiency in thin holograms is 33% for thin phase holograms.

Sensitivity

The high angular selectivity in thick gratings can be utilized to achieve greater sensitivity of the sensor device. An example how this can be achieved is presented in Fig. 7.4. A grating undergoes a dimensional change, and this leads to a shift in the Bragg selectivity curve. If the probe beam angle is kept constant, the measured change in diffraction efficiency $\Delta\eta_1$ is much higher than that measured if the maximum diffraction efficiencies of the two curves (before and after the dimensional change) are compared. Thus, a significantly smaller change in the dimensions of the layer can be readily detected.

Signal to noise ratio

Owing to the fact that there is only one diffraction order in thick gratings, the signal to noise ratio is potentially higher.

Compactness

The condition for operation as a thin holographic grating is typically satisfied for spatial frequencies below a few hundreds of lines per millimeter [54], which means that the angle between the zero and the first diffraction order beams is typically a few degrees (Fig. 7.3A). Thus, to separate the two beams, the diffraction and the detection planes need to be quite far apart. This makes the sensors based on holograms operating in the thin regime less compact. Because of the larger angle between the zero and first diffraction order

TABLE 7.4
Two Main Regime of Operation of the Diffractive Gratings and the Implications for Their Use as Holographic Sensors.

Regime of Operation	Thin Grating	Thick Grating
Number of diffractive orders	Multiple orders	One diffractive order
Diffraction efficiency	Maximum diffraction efficiency of sinusoidal grating is 33%	Up to 100% theoretically possible
Angular selectivity	Low angular selectivity Diffraction orders are produced practically for every incident angle	Observed only when incident beam is at Bragg/near Bragg angle
Theoretical description	Described by Raman-Nath theory [57]	Rigorous coupled-wave theory or Kogelnik's coupled-wave theory [5,58]
Features in sensing	Smaller dynamic range Typically, the spatial frequency of the recorded gratings is smaller. Thus, to separate the diffracted beams from the zero order requires a longer distance between the diffraction and the detection planes	Larger dynamic range High angular selectivity can be utilized to achieve high sensitivity Potentially higher signal to noise ratio Easier to fabricate compact sensor devices

beam, holograms operating in the thick regime offer the possibility to fabricate much more compact sensor devices.

Depending on the geometry of recording and reconstruction, holograms can be divided in two groups: reflection and transmission holograms. The main characteristics of the two types and the implications for the design of sensor devices are discussed in substantial detail in Refs. [9,10,54]. Here, we summarize the main findings.

Reflection and Transmission Holograms

Recording of and reconstruction from reflection and transmission holographic gratings is presented schematically in Fig. 7.5. The main difference in the two geometries of recording is that in transmission mode of

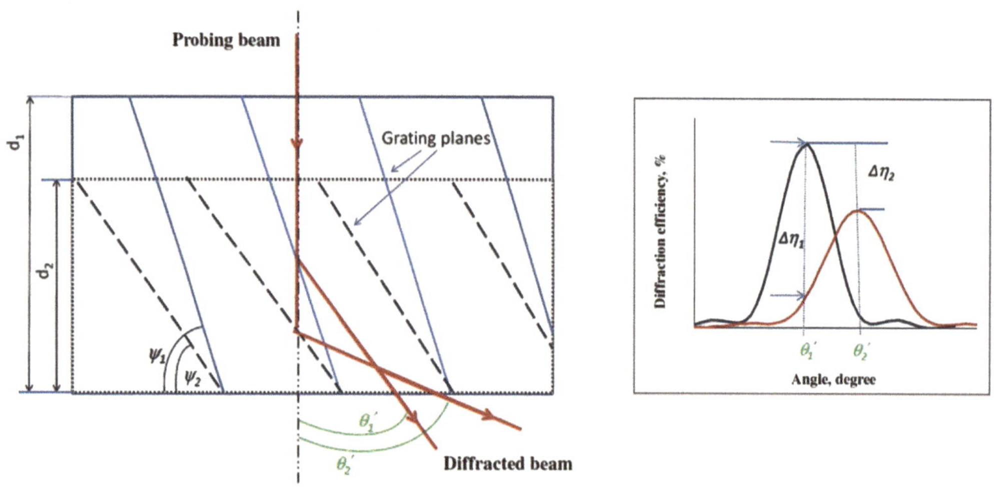

FIG. 7.4 Increasing the sensitivity of the sensor by utilizing slanted holographic gratings operating as thick (volume) gratings. (Reprinted with permission from T. Mikulchyk, Development of Holographic Sensors for Monitoring Relative Humidity and Temperature (Ph.D. thesis), 2016.)

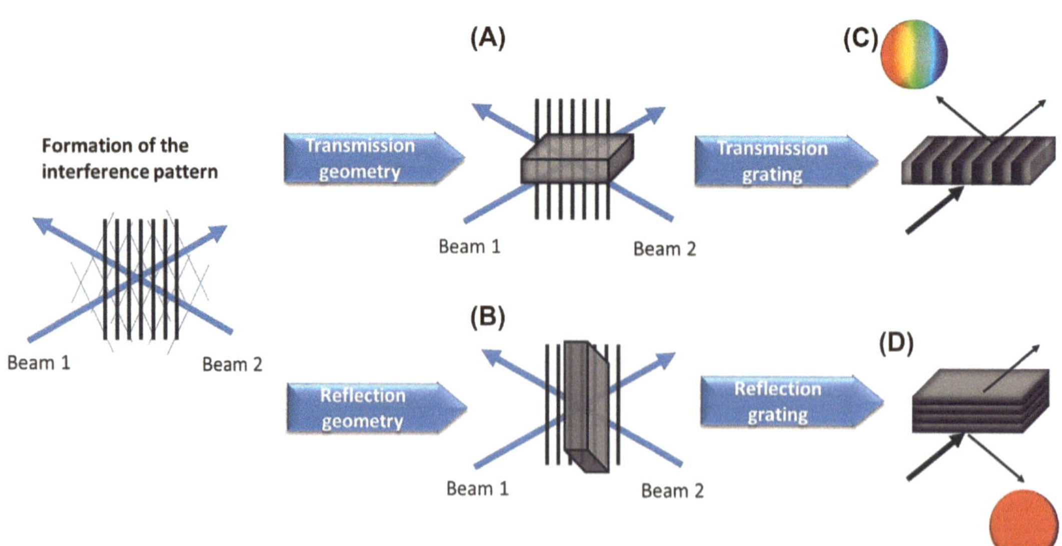

FIG. 7.5 Formation **(A)** and readout **(C)** of a transmission hologram; formation **(B)** and readout **(D)** of a reflection hologram.

TABLE 7.5
Diffraction Efficiency and Sensitivity of Holographic Sensors.

	Diffraction Efficiency	Sensitivity
Transmission	Thin $\eta = \frac{I_d}{I_o} = J_m^2\left(\frac{\varphi}{2}\right)$	$\frac{\partial(\eta)}{\partial(\Delta n)} = \frac{\pi d}{\lambda_p \cos\theta_B} J_1\left(\frac{\pi d}{\lambda_p \cos\theta_B}\Delta n\right)\left[J_0\left(\frac{\pi d}{\lambda_p \cos\theta_B}\Delta n\right) - J_2\left(\frac{\pi d}{\lambda_p \cos\theta_B}\Delta n\right)\right]$
$m\lambda_p = \Lambda \sin\theta_B$	$\varphi = \frac{2\pi\Delta n d}{\lambda_p \cos\theta_B}$	$\frac{\partial(\eta)}{\partial(\Delta d)} = \frac{\pi\Delta n}{\lambda_p \cos\theta_B} J_1\left(\frac{\pi\Delta n}{\lambda_p \cos\theta_B} d\right)\left[J_0\left(\frac{\pi\Delta n}{\lambda_p \cos\theta_B} d\right) - J_2\left(\frac{\pi\Delta n}{\lambda_p \cos\theta_B} d\right)\right]$
$\lambda_p = 2n\Lambda \sin\theta_B$	Thick $\eta = \sin^2\left(\frac{\varphi}{2}\right)$	$\frac{\partial(\eta)}{\partial(\Delta n)} = \frac{\pi d}{\lambda_p \cos\theta_B}\left[\sin\left(2\frac{\pi\Delta n d}{\lambda_p \cos\theta_B}\right)\right]$
	$\varphi = \frac{2\pi\Delta n d}{\lambda_p \cos\theta_B}$	$\frac{\partial(\eta)}{\partial(\Delta d)} = \frac{\pi\Delta n}{\lambda_p \cos\theta_B}\left[\sin\left(2\frac{\pi\Delta n d}{\lambda_p \cos\theta_B}\right)\right]$
Reflection $\lambda_p = 2n\Lambda \sin\theta_B$	$\eta = \tanh^2\left(\frac{\varphi}{2}\right)$	$\frac{\partial(\lambda_p)}{\partial(\Delta n)} = \frac{\pi d}{\tanh^{-1}(\sqrt{n})\cos\theta_B}$
	$\eta = \tanh^2\left(\frac{\pi\Delta n d}{\lambda_r \cos\theta_B}\right)$	$\frac{\partial(\lambda_p)}{\partial(d)} = \frac{\pi\Delta n}{\tanh^{-1}(\sqrt{n})\cos\theta_B}$

recording the photosensitive material is oriented so that the two recording beams are incident from the same side of the layer (Fig. 7.5A), whereas in reflection mode of recording the layer is oriented so that the recording beams are incident from opposite sides of the layer (Fig. 7.5B). The consequence is that the interference fringes recorded in the material produce in the first case a transmitted diffracted wave (Fig. 7.5C), whereas in the second case the diffracted wave is produced in reflection (Fig. 7.5D). If the two recording beams are incident to the recording layer at same angles, the recorded interference fringes are parallel (for a reflection grating) and perpendicular (for a transmission grating) to the surface of the layer. If the angles of incidence of the two beams are different, then the fringes are slanted with respect to the surface of the layer. See Chapter 1: Efficiency of Thick Gratings, for a description of the optical properties of reflection versus transmission diffraction gratings.

The analysis of the propagation of the waves in the layer containing the holographic grating allows to correlate the amount of energy that is transferred to the diffracted wave, to the properties of the structure—grating period, thickness of the structure, and the refractive index modulation and the properties of the probe wave—wavelength and wave vector. A theoretical description of the properties of different types of sensors based on unslanted holographic gratings has been reported previously [9,11d] and further studied in more detail recently in Refs. [53,54]. In Ref. [54] detailed theoretical studies were carried out in the context of developing gas sensors utilizing zeolite-doped holograms. Although Ref. [54] focuses on a particular material, namely zeolite-doped photopolymers, the described approach is applicable to sensors fabricated in other materials.

The mathematical equations used in Ref. [54] are summarized in Table 7.5. The equations presented in the table allow for estimation of the sensitivity of the sensor device when the expected variations in refractive index modulation Δn, thickness variation Δd, effective refractive index n, spatial period of the structure Λ, probe wavelength λ_p (outside the holographic layer), and Bragg angle θ_B (inside the layer for thick holograms) are known. It is important to note that because of the inherent lower dispersion and lower angular selectivity of transmission gratings, they are not typically used for development of sensors providing visual readout. A small deviation from the angle of observation will present a different color to the observer; thus, special care needs to be taken to ensure that the angle of incidence of the probe beam and the angle of observation stay unchanged.

Also at the typical spatial frequency of recording ($1000\ \mathrm{L\,mm^{-1}}$) the diffraction pattern constitutes not a single color, but a spectral range (Fig. 7.5C). This property introduces some ambiguity when observation of the change in color due to the presence of the target analyte is sought. For that reason, the response of transmission holograms to the external stimuli is usually recorded by measuring the change in their diffraction efficiency [10,27]. Accordingly, such sensors are used with electronic readout systems.

Reflection gratings are typically characterized by much higher spatial frequency and thus higher dispersion. This is particularly easy to observe for holograms recorded in the Denisyuk mode. For that reason, only a single color is seen at a particular angle of observation. The grating is practically an interference filter with very high spatial frequency; thus, the reflected part of the spectral range of the probe beam is a few nanometers in width. A hologram can also display an image if one of the recording beams has been spatially modulated in phase and amplitude by some object. These types of sensors are suitable for development of visual indicators that change color of the image under exposure to the analyte.

Readout of Sensor Response

The holographic sensors are photonic structures and thus require illumination with light to detect the change introduced by the external stimuli. Thus the performance of the sensor will depend on the probe light—laser light, LED light, normal daylight, white light sources, such as halogen lamps and on the detection system—spectrophotometers, digital cameras, and visual detection.

Light sources

Laser light. Laser light allows interrogation of the response of the hologram at a single wavelength. The laser output can be very stable, and thus this type of probe light is very suitable for detection of small changes in the diffraction efficiency of the diffractive sensing structure. As shown in Fig. 7.4, minor change in the dimensions of the hologram can be detected by measuring the diffraction efficiency in a predetermined direction. Laser light is collimated, and this allows for a good matching of the probe beam with one of the recording beams, thus achieving high diffraction efficiency. In sensors based on reflection holograms that are able to reconstruct a holographic image, the spatial coherence of the source also has a significant importance. The small size of the laser beam means high spatial coherence of the light source and enhances the sharpness of the reconstructed image.

***LED* light.** The use of LED light as a probe beam has a number of advantages and challenges. The main advantage probably is the fact that it has a relatively low cost, and thus is suitable for use in low-cost sensing systems. The challenges are related to three characteristics of the LED light source—it produces diverging light, the wavelength spectral range is broader, and it cannot be treated as a point source (at shorter distances from the hologram). The LED light is diverging, and unless a close match of the recording beam wavefront and the LED light wavefront is achieved, there will be always some substantial portion of light propagating without being diffracted. In some cases, it may be necessary to collimate the LED light for the diffractive optical sensor to perform with high efficiency. In this case, an additional challenge is the fact that the LED source is not a point source and its dimensions have to be considered. This is particularly a challenge when the distance between the LED source and the collimating element is small. The spatial coherence of LED sources is lower than that of laser light, and thus, blurring of the reconstructed holographic image may occur.

Halogen lamps. Halogen lamps are very often used in combination with spectrophotometers. Their intensity output can be stabilized and can provide a controlled and continuous light in a wavelength range fully covering the visible and NIR part of the spectrum. There are a number of suppliers on the market offering halogen light sources coupled to an optical fiber or fiber bundle. Thus the probe light can be easily guided to the location of the sensor, which in some cases may be in a hazardous environment. The spatial coherence of this light source is low (typically the source is 1−2 cm in size). Thus when the detection is based on image reconstruction, to minimize source-size blurring, the reconstruction source size must be less than 1 mm in diameter for every 1 m of distance that separates the illumination source and any point on the hologram [60].

Ambient day light. In some sensor systems, the readout is performed by illumination with day light. As the quality of light can differ significantly from place to place, this is probably the most challenging condition of operation. Most often, this arrangement is used with holograms recorded in Denisyuk mode, and the sensors are used as colorimetric indicators [11c,11d].

Signal detection

Spectrophotometers. Spectrophotometers are often used as detection systems in holographic sensors because they can provide information about both the spectral characteristics and the intensity of the zero order and/or of the diffracted beam. The measurement of the characteristics of the diffracted beam provides better signal to noise ratio. Fiber-based spectrophotometers are particularly suitable for this task. The signal is collected by a collimating lens from a predetermined direction and then coupled to an optical fiber and sent to the spectrophotometer. When a double channel spectrophotometer is used, both the zero- and first-order diffracted beams can be measured.

Photodiodes. Photodiodes are a low-cost alternative for detection of the intensity of the zero-order and/or intensity of the first-order diffracted beam. They are particularly useful in sensor systems where the change in the diffraction efficiency of the hologram is followed as a result of the presence of the analyte.

Digital cameras. A digital photographic recording of the spatial distribution of light diffracted by the hologram when it is illuminated by white light at its Bragg angle can be used in analytical measurements [61]. With a properly calibrated camera, the wavelength measurement accuracy can be better than 1 nm. In digital photography, each pixel of a colored image corresponds to a specific color vector in the RGB (red, green, and blue) system. The average wavelength of the diffracted light could be determined from the photograph of a hologram. The main limitations of using a commercial digital camera to measure the spectrum of light diffracted by the hologram are that (1) the diffracted light must be narrow band and (2) measurements must be performed in the spectral range in which at least two sensor types of the detecting array are simultaneously sensitive [61–63].

Direct visual observation. Direct readout is intended to be used when resources are limited, where electronic detection equipment is not readily available and mainly for semiquantitative detection, or a yes-no outcome of a simple test procedure. Direct visual observation poses some challenges—the angular range over which the hologram is visible is typically narrow—just a few degrees. The brightness of the hologram has to be high for the observer to be able to easily find the correct relative positions of light source and sensor for achieving optimum detection of the image.

FABRICATION OF THE HOLOGRAPHIC PHOTONIC STRUCTURES

There have been different approaches implemented in the fabrication of holographic sensors. They all utilize holographic lithography to achieve the required refractive index modulation in the volume of the layer or surface amplitude profile on its surface. Detailed descriptions of the main optical arrangements are presented below.

Optical Recording Geometries

Split beam geometry

In this arrangement, the beam produced by a laser is split in two beams, which are recombined in the space where the photosensitive layer is located (Fig. 7.6A and B). Depending on the orientation of the photosensitive layer with respect to the two interfering beams, a transmission (Fig. 7.6A) or a reflection (Fig. 7.6B) hologram is recorded. In the simplest arrangement, two collimated beams are used to record a holographic diffraction grating [27]. In more complex arrangements, a spatial light modulator (SLM) is positioned in the path of the one of the recording beams. This allows for the recording of diffraction patterns with increased complexity [64], which under illumination with the reference beam produces an image capable of changing properties in the presence of the target analyte.

The main advantage of the split beam geometry for holographic recording is that it allows for equalization of the intensities of the two recording beams, and thus, maximum visibility of the resulting interference fringes can be achieved. The equalization is typically achieved with the use of a combination of components—a polarizing beam splitter and two half-wave plates (Fig. 7.6A,B). The first half-wave plate is used to determine the ratio of the intensities of the transmitted and the reflected light, the second half-wave plate is used to rotate the state of polarization of one of the beams produced by the polarizing beam splitter, and thus, both beams have the same state of polarization. This is a necessary condition for achievement of maximum visibility of the resulting interference fringes. In most holographic setups used for recording in optically isotropic materials, the state of polarization of the recording beams is s-linearly polarized. Thus, any change in the spatial frequency of the recording grating will not result in deterioration of the visibility of the interference pattern.

In some studies reported in the literature, surface relief gratings are produced in photoanisotropic materials. In this case, the two recording beams have two orthogonal states of polarization [65]. It has been theoretically predicted and experimentally observed that the resulting surface profile amplitude is maximum when the two recording beams have orthogonal circular polarizations [4].

The photonic structure can also be fabricated by interfering more than two recording beams. An example of such structures is Aztec grating. Detailed description of their fabrication process is provided in Refs. [66,67]. The most interesting characteristic of these structures is that they have properties of reflection gratings and yet they can be easily manufactured by the proposed technology, which relies on the simultaneous recording of a transmission and a reflection hologram. After a number of steps including wet processing

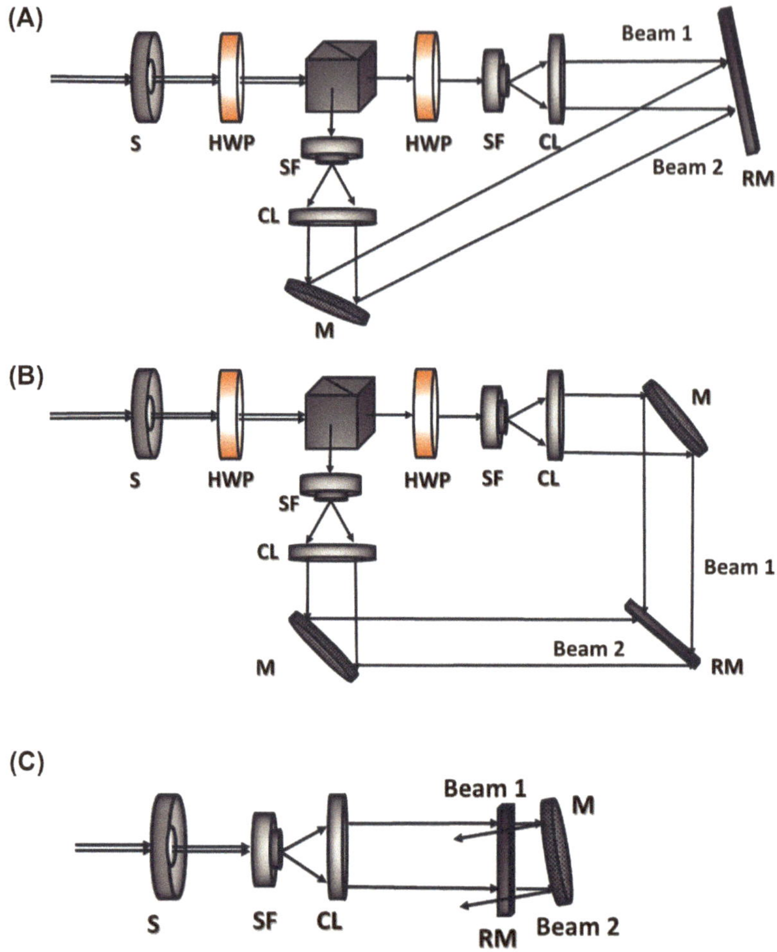

FIG. 7.6 Optical setup for recording a transmission **(A)** and reflection **(B)** hologram; Denisyuk recording geometry.

(etching) and production of master holograms, the pyramid structures can be mass produced by stamping [51]. For the technique of manufacturing embossed holograms, see Chapter 2: Photoresists and Embossed Holograms.

Denisyuk recording geometry

In this optical arrangement, first proposed by Denisyuk [68], the laser beam is transmitted through the recording material and reflected from a mirror positioned behind the object (Fig. 7.6C). In absence of an object, a reflection grating is recorded. The main advantage of the holograms recorded using this geometry is that they are highly selective with respect to reconstructing wavelength and angle, owing to the fact that the two recording beams are propagating with and angle of nearly 180 degrees between them. Such holograms are readily visible in white light and thus suitable for fabrication of holographic indicators that change color in the presence of the target analyte.

There are a number of drawbacks associated with this recording geometry. Firstly, equalization of the intensities of the two recording beams is not possible. Some of the light is absorbed when the beam is transmitted through the layer; thus the reflected beam always has lower intensity. A careful balancing of the absorption of the recording layer and achieving good visibility of the interference fringes is required. Also, it has been demonstrated that in addition to the reflection grating, a transmission volume and sometimes surface gratings are recorded. Their relative strengths can be balanced by consideration of the layer thickness and sensitizer concentration, as well as intensity of recording [69].

Despite the drawbacks associated with this geometry of recording, it must be emphasized that the largest number of reports in the literature utilize Denisyuk recording geometry. This is because the sensors produce a colorimetric response that can be easily interpreted even without extensive training of the user. In addition, with the help of modern mobile technologies, the color can be analyzed with the help of colorimetric apps [50] or the spectrum of the reflected light can be measured with the help of mobile spectrometers.

Mechanism of Formation of the Hologram in the Photosensitive Layer

For a discussion on holographic recording materials in a broader perspective than holographic sensor, see Chapter 2.

Mechanism of hologram formation in self-processing materials

Although in the initial literature reports, holographic sensors utilize mainly silver halide emulsions, the use of self-processing materials for fabrication of holographic sensors has continuously increased in recent years.

Photopolymer materials. Self-processing photopolymers allow for real-time control over the final properties of the recorded hologram, and because of the lack of multiple processing steps are more suitable for commercial applications [67].

The principle of holographic recording in photopolymer materials has been studied theoretically [70–78] and experimentally [79–85], and a good understanding of their behavior has been developed, allowing modification of their chemical composition to provide both good holographic recording properties and sensitivity to the target analyte.

Photopolymers are photosensitive materials, which undergo photopolymerization on exposure to light with sufficient energy to create a free radical and start the polymerization process. When the light is spatially patterned, this causes spatially varying change of the optical properties of the material. Both the absorption and the refractive index change. Both variations can produce diffractive patterns. It is known from theoretical modeling of these structures that the amplitude holograms, based on optically induced variation of the absorption coefficient, are much less effective than the phase holograms, based on refractive index variation. In photopolymers, primarily it is the refractive index modulation that is responsible for the efficiency of the hologram. Typically, the dye is bleached during the recording process or the layers are bleached after recording by exposure to UV light. Thus, there is no amplitude hologram to contribute to the diffraction.

The hologram formation in photopolymers involves three main processes—(1) spatially varying change of the refractivity of the molecules, as single monomer molecules combine to form polymer molecules, (2) spatially varying change of the density of the polymer layer due to concentration-driven diffusion of monomer molecules from dark to illuminated areas, and (3) change of the density of the polymer layer due to diffusion of mobile polymer chains (Fig. 7.7). In the case of hybrid materials containing nanoparticles dispersed in the host polymer matrix, there is a fourth process involving redistribution of the nanoparticles, which leads to spatial variation of the refractive index of the layer .

Recording in self-processing photopolymer layers is carried out either by sensitizing the photopolymer mixture with dyes absorbing in the visible wavelength range of the electromagnetic spectrum or by direct recording with UV light. Some self-processing materials successfully used in holographic sensing are pAAm-PVA [27], poly(2-hydroxyethyl methacrylate) (pHEMA) [86], and poly(dimethylsiloxane) (PDMS) [87]. A

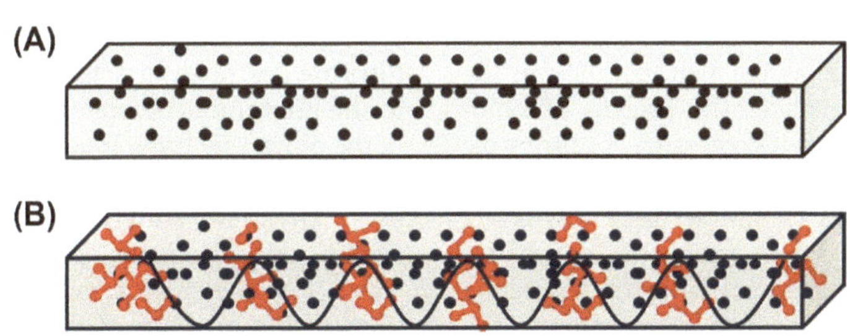

FIG. 7.7 Holographic recording in photopolymers: **(A)** unexposed layer; **(B)** after exposure to spatially varying intensity.

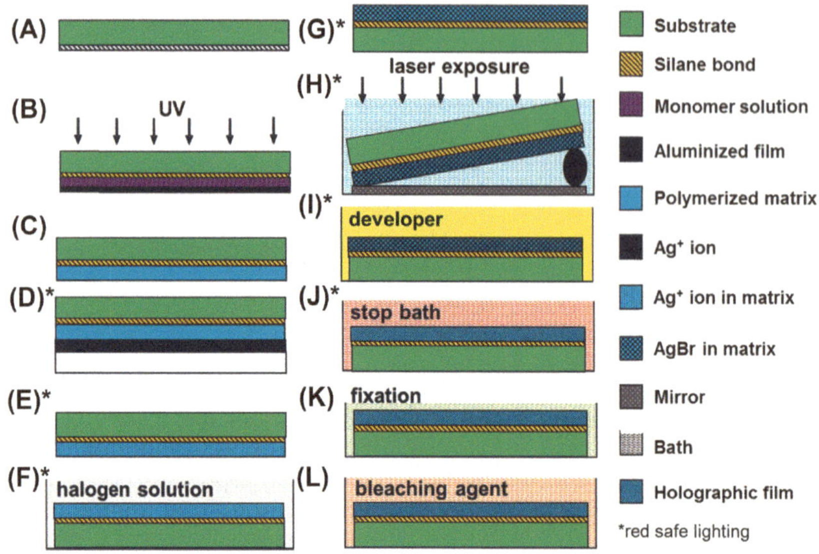

FIG. 7.8 Holographic image formation through the silver halide system. **(A)** The surface of a substrate is functionalized to allow the polymer layer to attach to the substrate. **(B)** The monomer solution is polymerized on the substrate. **(C)** The polymer–substrate system after polymerization. **(D)** The diffusion of silver salt solution into the polymer matrix. **(E)** The excessive silver salt solution is wiped off the surface, and the polymer layer is dried under a tepid air current. **(F)** The polymer layer-substrate system is placed in a halide ion solution to form a silver halide (AgBr). **(G)** The polymer layer-substrate system after silver halide formation. **(H)** Exposure to laser light with an object at the rear of the polymer layer. **(I)** Nanoparticle formation. **(J)** Stop bath. **(K)** Fixation. **(L)** Bleaching. (Reprinted with permission from A. K. Yetisen, I. Naydenova, F. Da Cruz Vasconcellos, J. Blyth, C. R. Lowe, Holographic sensors: three-dimensional analyte sensitive nanostructures and their applications, Chem. Rev. 114 (2014) 10654–10696, American Chemical Society.)

detailed description of the hologram formation process in each particular case is provided in Ref. [9].

Optically anisotropic materials. Optically anisotropic materials, which have not yet been fully explored for holographic sensing, have the potential to attract significant interest in the future, as they offer some important advantages. The main property of the holographic structures recorded in such materials is that in addition to changing the direction of propagation, they can also change the state of polarization of the propagating probe beam [4]. This introduces another parameter in the sensing system that can vary in the presence of the target analyte. The process of holographic recording in self-processing anisotropic materials involves trans-cis-trans isomerization of the absorbing photosensitive molecules, typically characterized by large dipole moments, which are either dispersed in a polymer matrix or covalently attached to the main chain or as side chains to the polymer backbone [4].

Azobenzene polymers are by far the most investigated optically anisotropic polymeric materials because of their high sensitivity and thermal stability, suitability for preparation of layers with high optical quality, and the possibility to perform multiple recordings. When the material is illuminated with a spatially varying light field, the structure of the layer can change as a result of the spatially varying orientation of the photosensitive molecules and the polymer chains to which they are attached. Thus, the birefringence and the dichroism (linear and/or circular) in the layer can vary in space. The diffractive properties of these holographic structures are described in detail in Refs. [4,88]. In addition to the photoinduced optical anisotropy in the volume of the polymeric layer, it has been observed that surface relief structures can be created in these materials [89–92]. These surface relief structures have been used in optical sensing, after coating with thin metal layers [65]. The spectral shift of surface plasmon polaritons in the presence of the target analyte have been analyzed.

Mechanism of hologram formation in materials requiring wet processing

Silver halides were the first materials utilized in development of holographic sensors and proved to be a flexible and successful material platform [9]. They are typically more photosensitive than photopolymer materials. A typical silver halide system involves (1) diffusion of silver ions into the analyte-sensitive polymer matrix, (2) formation of silver halide, (3) laser exposure to form the latent image, (4) amplification of the latent image to form a visible image through development, and (5) postprocessing steps to improve the diffraction efficiency. Each of these steps is shown in Fig. 7.8, which also shows generic photochemical processing used in the silver halide system. Detailed descriptions of each of the steps are provided in Refs. [9,10]. It should be kept in mind that they also have some disadvantages when commercially viable sensors are considered. Their important limitations include the cost of silver-based chemicals and the requirement for harsh postrecording chemical treatment to develop and stabilize the recorded hologram.

Formation of the hologram by laser ablation

This is a relatively recent approach to the recording of holograms and was first used in the context of holographic sensors research by the group of Chris Lowe at Cambridge University [52]. It was utilized to reduce the complexity of the photochemical processing and allow the formation of holograms in hydrophobic materials. The method consists of in situ particle size reduction through pulsed laser ablation occurring in the volume of the layer or surface ablation caused by dye absorption. To achieve this, metal particles or absorbing dye molecules are dispersed in the polymer layer. When illuminated with an interference pattern, the absorption of light in the layer is spatially modulated. This leads to spatially varying ablation in the volume (size of nanoparticles) [52] or at the surface (absorbing dye) [93].

The effectiveness of the process depends on multiple factors among which are laser power and wavelength, pulse duration, surface plasmon resonance of the nanoparticles, and their size. All these influence the subsequent absorption of light by the nanoparticles [94] Holograms produced by this method are reported in polymers doped with silver, gold, copper, tin, iron, nickel, and lead nanoparticles and dyes [95]. Typically the diffraction efficiency of the gratings utilizing laser ablation is relatively low. Recently, improvement of the diffraction efficiency of the ablated grating by 200% was reported at the expense of introduction of additional steps requiring wet chemical processing [96].

APPROACHES TO FUNCTIONALIZATION OF HOLOGRAPHIC STRUCTURES

To develop holographic sensors, different approaches have been adopted with respect to the incorporation of the target analyte/external stimulus–sensitive component into the layer containing the hologram. In most cases, chemical components—single molecules, polymers, or nanoparticles that can interact with the target analyte—have been added to the layer before recording the hologram. In sensors utilizing molecularly imprinted polymers, the functionalization of the layer occurs during the recording process. The third approach is to functionalize the hologram after recording, either by coating its surface or by introducing the analyte-sensitive component by ink jet printing or immersing the layer in it. In the following subsection, each one of the different approaches will be described in more details with examples.

Addition of the Functionalizing Material to the Photopolymer Mixture

This approach is encountered most often in the literature. The functionalizing component is added during the formation of the photosensitive layer. After the end of the recording, the functionalizing component is either redistributed or uniformly dispersed throughout the volume of the layer. Evidence that redistribution of the zeolite nanoparticles has been achieved is reported in Refs. [97,98].

Functionalization by selection of appropriate monomer in photopolymers

N-isopropylacrylamide (NIPA) has been used to develop holographic sensors responsive to temperature changes [27,46d,e]. During holographic recording, the monomer is converted to spatially patterned poly (N-isopropylacrylamide) (PNIPA). The temperature sensitivity of poly(N-isopropylacrylamide) (PNIPA) is derived from its ability to undergo phase transition at a critical temperature. PNIPA can be classified as a negative temperature-sensitive system and has a critical solution temperature (LCST) of about 32°C. This temperature can be modified by copolymerization of NIPA with other appropriate monomers [99].

NIPA-based photopolymer compositions have been developed for holographic recording in both transmission and reflection modes [46d,e]. Changes in the characteristics of transmission (Fig. 7.9E) and reflection (Fig. 7.9B, C and D) holograms recorded in NIPA-based photopolymer reveal the potential of this material for development of holographic temperature sensors.

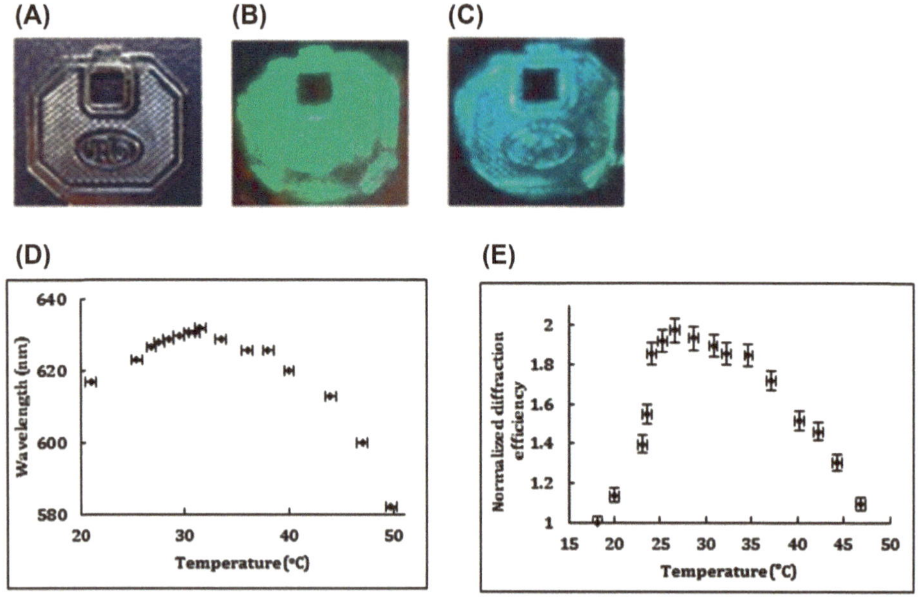

FIG. 7.9 Photographs of the object **(A)** and its Denisyuk hologram recorded in the NIPA-based photopolymer before temperature exposure **(B)** and during exposure to temperature of 50°C **(C)**. Bragg peak wavelength versus temperature **(D)**. Temperature dependence of normalized diffraction efficiency of a PNIPA-based transmission grating **(E)**. (Reprinted with remission from T. Mikulchyk, S. Martin, I. Naydenova, N-isopropylacrylamide-based photopolymer for holographic recording of thermosensitive transmission and reflection gratings, Applied Optics 56 (22) (2017) 6348–6356, https://doi.org/10.1364/AO.56.006348, Optical Society of America.)

Diacetone acrylamide has been used for development of pressure-sensitive holograms [27]. Reflection holograms that can change their color under application of pressure have been fabricated. Detailed characterization of the response of the sensors is described in Ref. [27]. Fig. 7.10 summarizes some of the results. Pressure was applied to the reflection gratings using an Instron Series 5569 Tensile Tester. A ball bearing of 3 mm diameter was attached to the tip of the Instron (Fig. 7.10A). The blueshift in the spectral characteristics of the hologram (Fig. 7.10B) clearly shows that the grating period decreases in result of the applied pressure. The dependence of the relative change in grating's period as a function of the applied pressure is presented in Fig. 7.10C. The hologram can be used to obtain a visual (Fig. 7.10D) and quantitative (Fig. 7.10E) map of the pressure.

There are many physical parameters that influence the sensitivity of the pressure sensor. Elasticity is one of the most important parameters, as it defines the ability of the material to deform under pressure. Fig. 7.10F shows a characterization of the influence of hologram thickness on the elastic modulus. By increasing the hologram thickness from 110 to 195 μm, the elasticity is seen to decrease. This demonstrates the possibility of tuning the response of the pressure sensor by modifying the grating thickness.

Boronic acids and their aggregates have been added to different hydrogels for development of sensors for saccharides [30,100]. Boronic acid (BA)–containing (co) polymers have attracted researchers attention for decades, because of their unique responsiveness toward 1,2- and 1,3-diols, including saccharides and nucleotides.

The biggest challenge for in vivo applications lies in the investigation of the potential interference and competing effects of all present biomolecules that can interact with BA-containing polymers and the fine-tuning of the BA materials to tailor the response to one single component, such as glucose.

Functionalization of volume phase holograms by addition of zeolite nanoparticles

The application of nanosized zeolites in chemical sensing and biomedical and biological analyses has shown significant potential [101–105]. Zeolite nanocrystals have highly tuneable size- and shape-dependent chemical and physical properties. Their

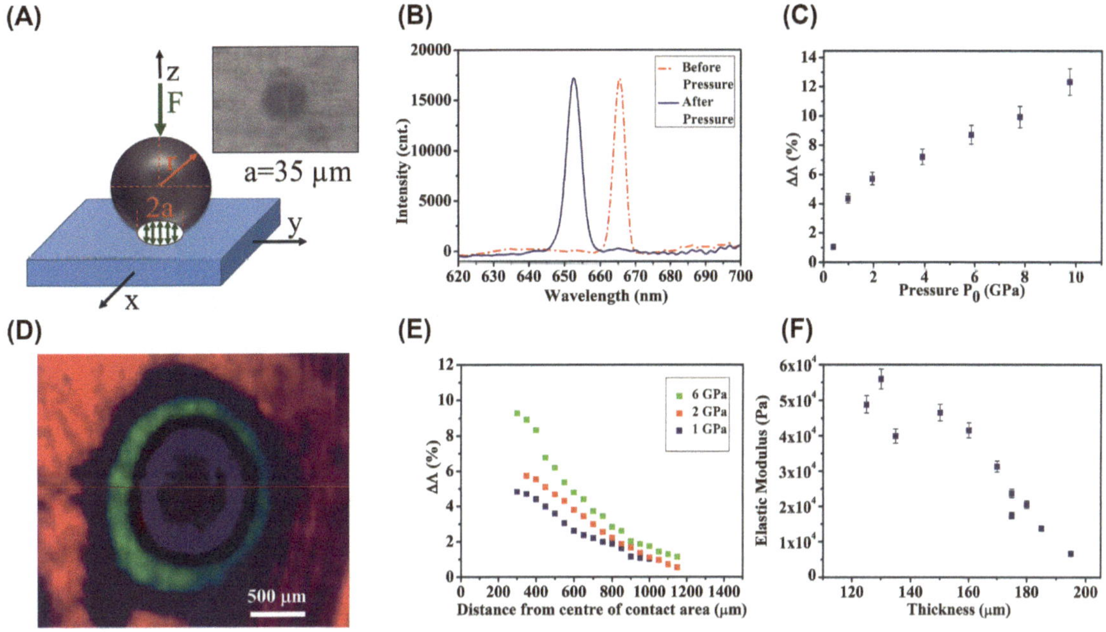

FIG. 7.10 **(A)** Schematic of pressure testing with ball bearing $r = 1.5$ mm, $a = 35$ μm; **(B)** spectra of light reconstructed from a reflection grating, illuminated by white light; **(C)** % change in hologram fringe spacing ($\Delta\Lambda$) versus pressure (GPa); **(D)** 2D color map of $\Delta\Lambda$ for $P_o = 1$ GPa; **(E)** $\Delta\Lambda$ as a function of distance from center of contact area (μm) for $P_o = 1$, 2 and 6 GPa; **(F)** elastic modulus (Pa) versus reflection grating thickness (μm). (Reprinted with permission from D. Cody, S.-E Gul, T. Mikulchyk, M. Irfan, A. Kharchenko, K. Goldyn, S. Martin, S. Mintova, J. Cassidy, I. Naydenova, Self-processing photopolymer materials for versatile design and fabrication of holographic sensors and interactive holograms, Applied Optics 57 22 (2018), pp. E173-E183. https://doi.org/10.1364/AO.57.00E173, Optical Society of America.)

unique surface chemistry, high degree of thermal stability, large surface area, and large micropore volume are important for designing of optical sensing devices. Photosensitive layers of self-processing photopolymers have been doped with zeolite nanoparticles to act as holographic sensors [43,106,107]. After recording, the zeolite nanoparticles are spatially redistributed as a result of the holographic recording [97,98]. Selective adsorption of the analyte by the zeolite nanoparticle leads to spatially varying change in the refractive index of the layer and thus a change in its diffraction efficiency. Fig. 7.11 outlines two approaches for addition of the zeolite nanoparticles to the photopolymer layer. The nanoparticles are either added to the photopolymer solution before formation of the dry layer or are printed with the help of an inkjet printer. The second approach allows for precise patterning with current resolution of 10 μm.

The ink jet deposition of zeolite nanoparticles on photopolymer layer was used to fabricate irreversible humidity sensor (Fig. 7.12) [43].

The response of zeolite-doped layers to VOC [106], humidity [43,107], and alcohols [43] has been characterized. The results reveal the potential of zeolite-doped holographic sensors based on volume phase holograms.

Molecularly Imprinted Polymer Layers

This approach is used, for example, in holographic molecularly imprinted polymer (MIP) films, which are structured by using interference photolithography. MIPs are prepared by templating at the molecular level by photopolymerization. The polymerization is performed in the presence of a template molecule (target), and the cross-linked polymer matrix consists of cavities that are complementary to the template molecule in terms of shape, size, and position of functional groups. Hence, MIPs have the capability to bind to target molecules with high affinity and specificity. An example of such a sensor for detection of testosterone is reported in Ref. [34].

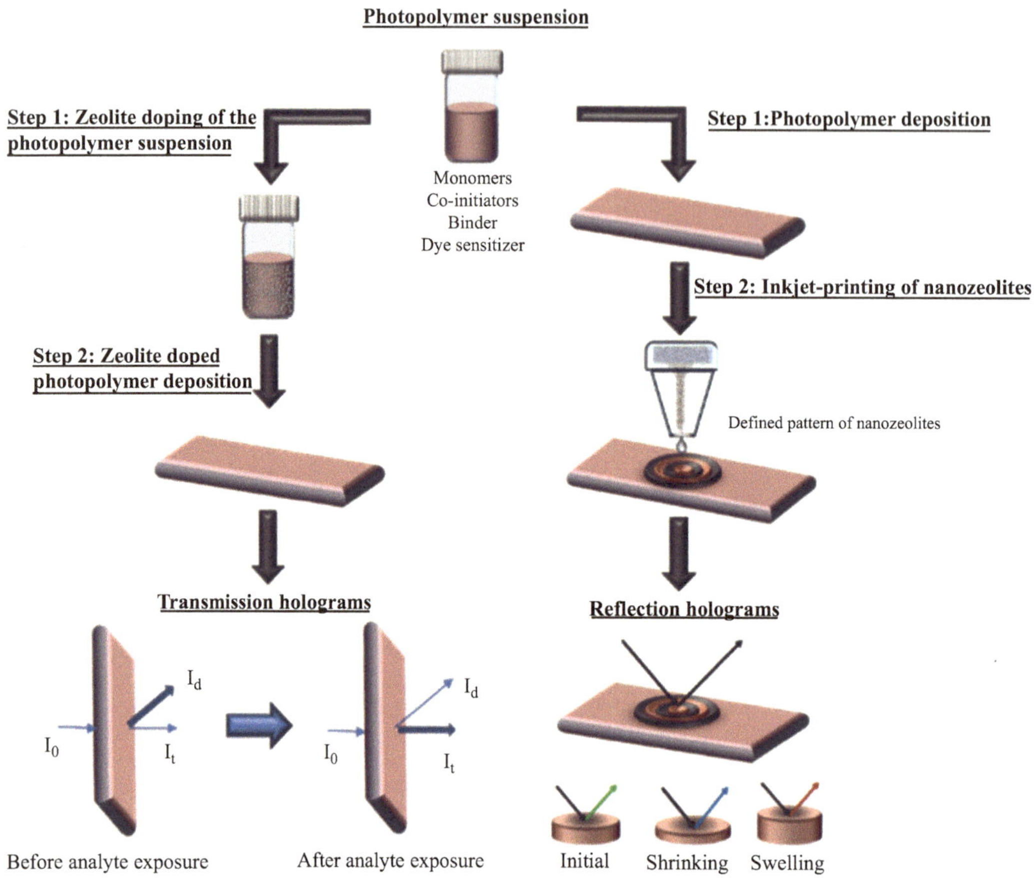

FIG. 7.11 Fabrication of the holographic sensors doped with zeolite nanoparticles: (left) direct addition of zeolite nanocrystals to the photopolymer solution for preparation of transmission holographic sensors, and (right) inkjet printing of zeolite nanocrystals for preparation of reflection holographic sensors. I_0, I_d, and I_t correspond to the intensity of the incident light, diffracted light, and transmitted light, respectively. (Reprinted with permission from I. Naydenova, J. Grand, T. Mikulchyk, S. Martin, V. Toal, V. Georgieva, S. Thomas, S. Mintova, Hybrid sensors fabricated by inkjet printing and holographic patterning, Chemistry of Materials 27 (17) (2015), 6097–6101, American Chemical Society.)

Functionalization after Recording of the Holographic Structure

Surface relief gratings have been created by holographic lithography [27]. The structure is modified by coating it with a sol-gel enriched with LTL zeolite nanoparticles (Fig. 7.13A). AFM characterization of the surface reveals that the groves in the surface are almost completely filled after coating (Fig. 7.13B). The response of the structure to different metal cations in water was then characterized by measuring its diffraction efficiency (Fig. 7.13C) after exposure to the analyte for different time periods. The sensitivity of the structure was determined from these measurements (Fig. 7.13D) and compared for Cu^{+2}, Ca^{+2}, Na^{+1}, and Pb^{2+} (Fig. 7.13E). The results reveal that there is significant difference between the response to mono- and dications.

CHALLENGES IN HOLOGRAPHIC SENSORS RESEARCH

The challenges in holographic sensors research can be divided in two: (1) challenges that are common for all sensors and these are related to the *selectivity, sensitivity, stability, response time, and reversibility* of the sensors; (2) challenges that can be attributed to the specific properties of holographic sensors—*sensitivity, readout, signal to noise ratio*.

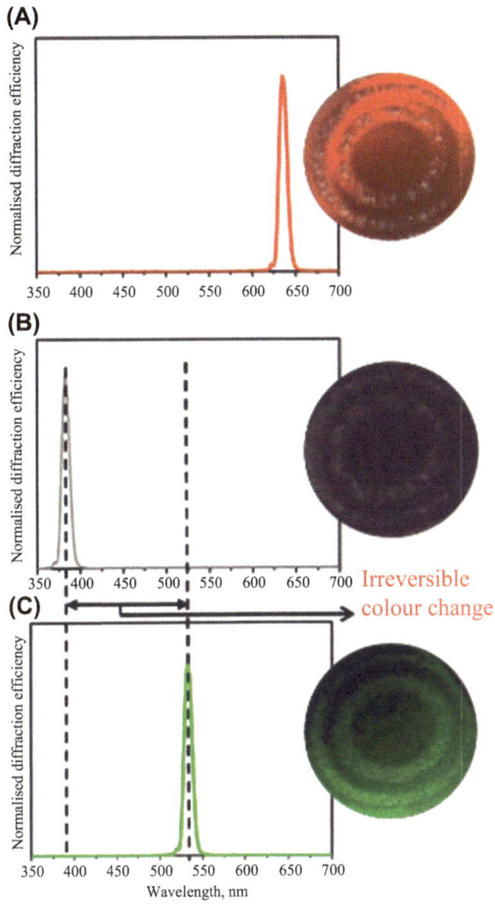

FIG. 7.12 Holographic sensors based on EMT-zeolite-doped layers exposed to relative humidity (RH) **(A)** 60% (as prepared), **(B)** 20% (sensor annealed at 120°C for 20 min), and **(C)** 20% (relative humidity in the chamber was brought back to 20% following exposure to 80% RH for 10 min). (Reprinted with permission from I. Naydenova, J. Grand, T. Mikulchyk, S. Martin, V. Toal, V. Georgieva, S. Thomas, S. Mintova, Hybrid sensors fabricated by inkjet printing and holographic patterning, Chemistry of Materials 27 (17) (2015), 6097–6101, American Chemical Society.)

The first group of challenges are addressed by development of novel materials capable of capturing the target analyte and of changing their properties—dimensions by swelling/shrinkage or by changing their refractive index. Research in development of other sensors can be directly transferred in holographic sensors design, providing that the materials that are used are optically clear. This typically means that the materials are characterized by low scattering and low absorption at the probe wavelength. Additional challenge for sensors designed for operation in liquid environment is that the holograms must be recorded in materials resistant to the solvent *(stability)*. For example, hydrogels modified to achieve *selectivity* toward different analytes have been widely used for development of holographic sensors operating in water environment [23]. Hydrogels have been functionalized to be sensitive to a wide range of analytes by creating covalently bound pendant recognition groups or molecularly imprinted sites [34].

The biggest challenge for in vivo applications, as for nonholographic sensors, remains the potential interference and competitive effects of all present molecules and ions that can interact with the analyte-sensitive component in the layer. Zeolite nanoparticles are also a contender to achieve high *selectivity* to target analytes, but further work is needed in relation to their synthesis to explore their full potential [27,104,106,107].

One approach to achieving faster *response time* is by controlling the porosity if the layer containing the hologram. The development of disposable, point-of-care sensors must take into account impact on environment. The use of biocompatible and biodegradable materials becomes an important factor in disposable sensors as justified in Ref. [29].

The second group of challenges are specific to holographic sensors and concern improvement of their sensitivity, signal to noise ratio, and readout. Theoretical modeling [54] has demonstrated that by using sensors based on dimensional changes of the layer, higher *sensitivity* is easier to achieve. Higher *sensitivity* has also been achieved by using self-supporting layers [108], which facilitates the penetration of the analyte in the layer. Higher *sensitivity* is also achieved by utilizing volume phase holograms, but this can impose challenges with the readout of the sensor. Holographic sensors utilizing volume phase holograms are known for their high angular selectivity. In sensor devices with electronic readout, this fact can be utilized to achieve increased sensitivity. When visual, semiquantitative detection is envisaged, the high angular selectivity is a challenge. The narrow angle of view can be a significant obstacle in their use by nonspecialists. Incorporation of diffusers [30], lenses, or other optical devices in holograms can increase the angle of view [109]. However, such additional devices inevitably decrease the intensity of the diffracted light in any given direction. To readily see the diffracted light, it is necessary to use holograms with high diffraction efficiency.

Another obstacle is the fact that when illuminated with white light, the hologram can be seen in different colors when viewed from different angles. Color changes due to variation in the viewing angle may be

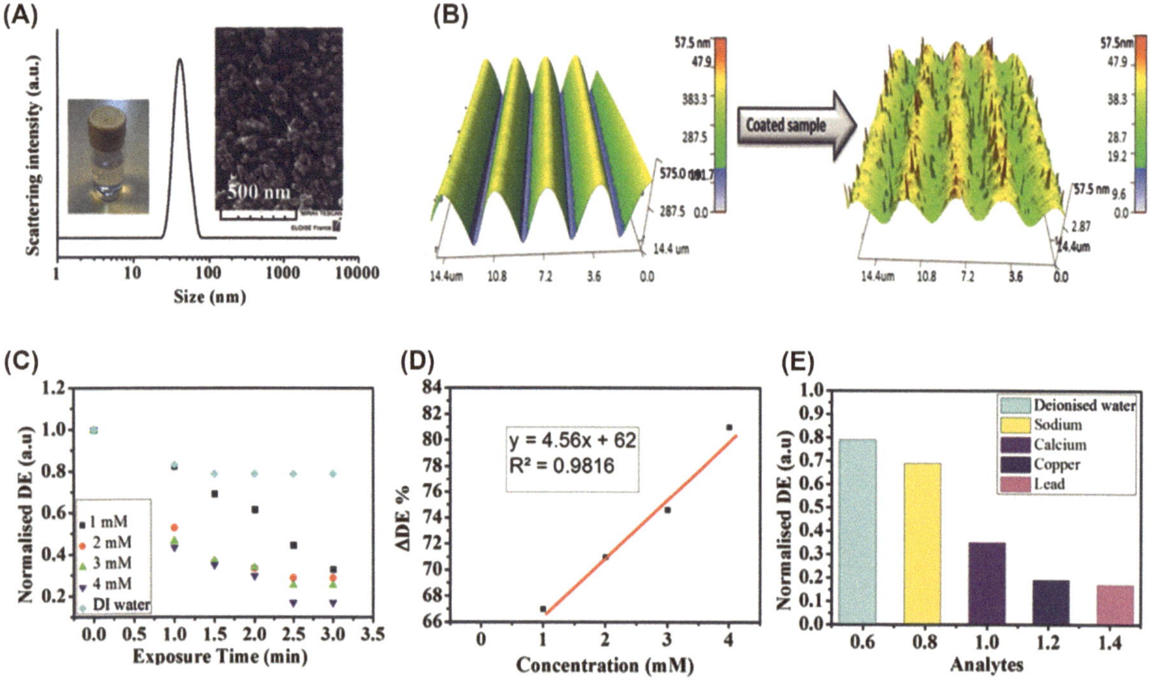

FIG. 7.13 **(A)** LTL-zeolite size distribution measured using DLS; **(B)** SEM image of the LTL zeolites **(C)** WLI images coated and uncoated with LTL zeolites; **(D)** normalized DE change response photonic structures in sol-gel matrix exposure to (1–4 mM) Pb^{+2}; **(E)** sensitivity of the response at 3 min for lead; **(F)** comparison of the change in DE to different analytes. (Reprinted with permission from D. Cody, S.-E Gul, T. Mikulchyk, M. Irfan, A. Kharchenko, K. Goldyn, S. Martin, S. Mintova, J. Cassidy, I. Naydenova, Self-processing photopolymer materials for versatile design and fabrication of holographic sensors and interactive holograms, Applied Optics 57 22 (2018), E173–E183. https://doi.org/10.1364/AO.57.00E173, Optical Society of America.)

misinterpreted as color change, caused by swelling/shrinkage of the hologram or change of its refractive index due to the presence of the analyte or other external stimulus. Thus for a precise visual detection, innovative optical arrangements are needed to avoid any ambiguity at the readout stage. An example of such arrangement is combination of a hologram and a retroreflector [110]. To address these challenges, novel, more complex diffractive structures recorded by holographic lithography have been proposed [111–113]. Some of them are inspired from structural color designs in nature [114].

The holographic sensor technology has been brought another step closer to real-life application by development of novel smartphone technologies as a platform to quantify the sensor readouts [50]. One significant advantage of these technologies is that they offer real-time reporting. Development of software applications for smartphones can be readily adapted to other electronic devices, such as tablets, computers, smartwatches, or other wearable technologies.

Although initial clinical testing of holographic sensors have been reported [115,116], further improvements, particularly of the readout technologies, are required for them to be widely used in real-life applications.

CONCLUSIONS

Holographic sensors are photonic devices that utilize diffraction of light to detect a target analyte or other external stimuli. The flexibility in their design is a result of a variety of different holographic recording geometries and the variety of diffractive structures that can be produced. Structures producing either discrete diffracted beams of light or images have been successfully used in sensing. The effect of the type of diffractive structure on the dynamic range, the readout mode, the sensitivity, and the response time has been studied theoretically and experimentally. In addition, the sensitivity, response time, and reversibility of the sensor are also affected by the functionalizing material. Selectivity of the sensors is determined solely by the material used to functionalize the layer containing the hologram. It has been demonstrated that both changing hologram and creation of a hologram can be utilized in sensing. The second approach is characterized by high signal to noise ratio because the signal is detected in direction where it previously did not exist.

The main challenges specific to holographic sensors as optical devices are related to improvement of their sensitivity and readout. Innovative diffractive structures, some inspired by nature, have been proposed. Recent studies reveal that improved readout properties can be achieved when the hologram is combined with other optical components such as retroreflectors. It has been demonstrated that holographic sensors are compatible with smartphone technologies. These features make holographic sensors promising candidates for development of point-of-care applications.

REFERENCES

[1] R.J. Collier, C.B. Burckhardt, L.H. Lin, Optical Holography, Academic press, 1971.

[2] G. Saxby, S. Zacharovas, Practical Holography, CRC Press, 2015.

[3] V. Toal, Introduction to Holography, CRC Press, Boca Raton, FL, 2011.

[4] L. Nikolova, P. S. Ramanujam, Polarisation Holography, Cambridge University Press.

[5] H. Kogelnik, Bell Syst. Technol. J. 48 (1969) 2909.

[6] P. Hariharan, Optical Holography. Principles, Techniques and Applications, Cambridge studies of modern optics, 1996.

[7] a) G. Andersen, F. Ghebremichael, K. Gurley, Proceedings of SPIE 6488 (2007) 64880I, https://doi.org/10.1117/12.701231;
b) F. Ghebremichael, G.P. Andersen, K.S. Gurley, Applied Optics 47 (2008) A62–A69, https://doi.org/10.1364/AO.47.000A62;
c) F. Kong, A. Lambert, Applied Optics 55 (2016) 3615–3625, https://doi.org/10.1364/AO.55.003615.

[8] a) M.A.A. Neil, M.J. Booth, T. Wilson, New modal wavefront sensor: a theoretical analysis, Journal of the Optical Society of America A 17 (2000) 1098–1107, https://doi.org/10.1364/JOSAA.17.001098;
b) M.A.A. Neil, M.J. Booth, T. Wilson, Closed-loop aberration correction by use of a modal Zernike wave-front sensor, Optics Letters 25 (2000) 1083–1085, https://doi.org/10.1364/OL.25.001083.

[9] A.K. Yetisen, I. Naydenova, F. Da Cruz Vasconcellos, J. Blyth, C.R. Lowe, Holographic sensors: three-dimensional analyte sensitive nanostructures and their applications, Chemical Reviews 114 (2014) 10654–10696.

[10] M. Zawadzka, T. Mikulchyk, D. Cody, S. Martin, A.K. Yetisen, H.B.J.L. Martinez-Hurtado, E. Mihaylova, H. Awala, S. Mintova, S.H. Yun, I. Naydenova, Photonic Materials for Sensing, Biosensing and Display Devices, vol. 229, 2016.

[11] a) J. Blyth, R.B. Millington, A.G. Mayes, E.R. Frears, C.R. Lowe, Analytical Chemistry 68 (1996) 1089;
b) R.B. Millington, A.G. Mayes, J. Blyth, C.R. Lowe, Sensors and Actuators, A: Chemical 33 (1996) 55;
c) I. Naydenova, R. Jallapuram, S. Martin, V. Toal, Humidity sensors: types, in: C.T. Okada (Ed.), Nanomaterials and Environmental Monitoring, Nova Science Publishers, Hauppauge, 2011, p. 117;
d) I. Naydenova, R. Jallapuram, V. Toal, S. Martin, Sensors and Actuators, A: Chemical 139 (2009) 35;
e) I. Naydenova, R. Jallapuram, V. Toal, S. Martin, Applied Physics Letters 92 (2008) 031109;
f) T. Mikulchyk, S. Martin, I. Naydenova, Journal of Optics 15 (2013) 105301;
g) D. Yu, H. Liu, D. Mao, Y. Geng, W. Wang, L. Sun, J. Lv, Holographic humidity response of slanted gratings in moisture-absorbing acrylamide photopolymer, Applied Optics Research 54 (22) (August 1, 2015) 6804–6912;
(h) D.A. Ilatovskii, V. Milichko, A.V. Vinogradov, V.V. Vinogradov, Holographic sol–gel monoliths: optical properties and application for humidity sensing, Royal Society Open Science 5 (2018) 172465. https://doi.org/10.1098/rsos.172465;
(i) T. Mikulchyk, S. Martin, I. Naydenova, Investigation of the sensitivity to humidity of an acrylamide-based photopolymer containing N-phenylglycine as a photoinitiator, Optical Materials 37 (2014) 810–815.

[12] a) I. Naydenova, S. Martin, V. Toal, Journal of European Optical Society-Rapid Publications 4 (2009) 09042.b) V. Farrelly, S. Martin, I. Naydenova, V. Toal, A Holographic Method and Product. EP Patent Application 2054779 A2, May 6, 2009.
c) S. Martin, I. Naydenova, V. Toal, Selective Hologram Formation. U.S. Patent 8,383,294 B2, February 26, 2013.

[13] D. Wuest, R. Lakes, Color control in reflection holograms by humidity, Applied Optics 30 (1991) 2363–2367.

[14] G. Naik, A. Mathur, S. Pappu, Dichromated gelatin holograms: an investigation of their environmental stability, Applied Optics 29 (1990) 5292–5297.

[15] R.C. Spooncer, F.A. Al-Ramadhan, B.E. Jones, International Journal of Optoelectronics 7 (1992) 449.

[16] C.R. Lowe, R.B. Millington, J. Blyth, A.G. Mayes, Hologram Used as a Sensor, WO Patent Application 1995026499 A1, October 5, 1995.

[17] R.B. Millington, A.G. Mayes, J. Blyth, C.R. Lowe, A holographic sensor for proteases, Analytical Chemistry 67 (23) (1995) 4229–4233, https://doi.org/10.1021/ac00119a004.

[18] J. Blyth, R.B. Millington, A.G. Mayes, E.R. Frears, C.R. Lowe, Holographic sensor for water in solvents, Analytical Chemistry 68 (1996) 1089–1094.

[19] A.G. Mayes, J. Blyth, R.B. Millington, C.R. Lowe, A holographic sensor based on a rationally designed

synthetic polymer, Journal of Molecular Recognition 11 (1–6) (1998) 168–174.

[20] A.G. Mayes, J. Blyth, M. Kyrolainen-Reay, R.B. Millington, C.R. Lowe, A holographic alcohol sensor, Analytical Chemistry 71 (16) (1999) 3390–3396, https://doi.org/10.1021/ac990045m.

[21] El Joudi, EM; Dress, P; Podgorsek, RP; Franke, H, Moisture sensitivity of holographic gratings recorded in photosensitive polyimide, Japanese Journal Of Applied Physics Part 1-Regular Papers Brief Communications & Review Papers, Volume: 36 Issue: 2 Pages: 714-719.

[22] Ley, C; Calderara, I; Lougnot, DJ, Holographic gratings recorded in polymer hydrogels – an original application as a sensor in aqueous environment, Measurement Science and Technology,Volume: 8 Issue: 9 Pages: 997-1000, DOI: 10.1088/0957-0233/8/9/007.

[23] A.K. Yetisen, H. Butt, L.R. Volpatti Lisa, I. Pavlichenko, M. Humar, S.J.J. Kwok, H. Koo, K.S. Kim, I. Naydenova, A. Khademhosseini, S.K. Hahn, S.H. Yun, Photonic hydrogel sensors, Biotechnology Advances (2015), https://doi.org/10.1016/j.biotechadv.2015.10.005.

[24] a) C.P. Tsangarides, A.K. Yetisen, F.D. Vasconcellos, Y. Montelongo, M.M. Qasim, T.D. Wilkinson, C.R. Lowe, H. Butt, RSC Advances 4 (2014) 10454;
b) A.K. Yetisen, H. Butt, F.D. Vasconcellos, Y. Montelongo, C.A.B. Davidson, J. Blyth, L. Chan, J.B. Carmody, S. Vignolini, U. Steiner, J.J. Baumberg, T.D. Wilkinson, C.R. Lowe, Advanced Optical Materials 2 (2014) 250;
c) A.J. Marshall, J. Blyth, C.A. Davidson, C.R. Lowe, Analytical Chemistry 75 (2003) 4423;
d) L.L. Bell, A.A. Seshia, C.A.B. Davidson, C.R. Lowe, Procedia Engineering 5 (2010) 1352;
e) A.V. Kraiskii, V.A. Postnikov, T.T. Sultanov, A.V. Khamidulin, IEEE Journal of Quantum Electronics 40 (2010) 178.f) Liu, HP; Yu, D; Zhou, K; Wang, SC; Luo, SH; Li, L; Wang, WB; Song, QG, Novel pH-sensitive photopolymer hydrogel and its holographic sensing response for solution characterization, Optics & Laser Technology, Volume: 101 Pages: 257-267, DOI: 10.1016/j.optlastec.2017.11.025;

[25] a) J. Blyth, C.R. Lowe, A.G. Mayes, R.B. Millington, Holographic Sensors and Their Production, WO Patent Application 1999063408 A1, December 9, 1999;
b) A.G. Mayes, J. Blyth, R.B. Millington, C.R. Lowe, Analytical Chemistry 74 (2002) 3649;
(c) Sabad-e-Gul, S. Martin, J. Cassidy, I. Naydenova, Development of sensitive holographic devices for physiological metal ion detection, Proceedings of SPIE 10354 (2017) 103540C, https://doi.org/10.1117/12.2275734, 1.

[26] a) B.M. Gonzalez, G. Christie, C.A.B. Davidson, J. Blyth, C.R. Lowe, Analytica Chimica Acta 528 (2005) 219;
b) D. Bhatta, G. Christie, B. Madrigal-Gonzalez, J. Blyth, C.R. Lowe, Biosensors and Bioelectronics 23 (2007) 520;
c) S.e Gul, D. Cody, A. Kharchenko, S. Martin, S. Mintova, J. Cassidy, I. Naydenova, LTL type nanozeolites utilized in surface photonics structures for environmental sensors, Microporous and Mesoporous Materials 261 (2018) 268–274.d) Yetisen, AK; Montelongo, Y; Qasim, MM; Butt, H; Wilkinson, TD; Monteiro, MJ; Yun, SH, Analytical Chemistry Photonic nanosensor for colorimetric detection of metal ions, Volume: 87 Issue: 10 Pages: 5101-5108.

[27] D. Cody, S.-E. Gul, T. Mikulchyk, M. Irfan, A. Kharchenko, K. Goldyn, S. Martin, S. Mintova, J. Cassidy, I. Naydenova, Self-processing photopolymer materials for versatile design and fabrication of holographic sensors and interactive holograms, Applied Optics 57 (22) (2018) E173–E183. https://doi.org/10.1364/AO.57.00E173.

[28] A.V. Kraiskii, V.A. Postnikov, T.T. Sultanov, A.V. Khamidulin, IEEE Journal of Quantum Electronics 40 (2010) 178.

[29] A.K. Yetisen, M.M. Qasim, S. Nosheen, T.D. Wilkinson, C.R. Lowe, Journal of Materials Chemistry C 2 (2014) 3569.

[30] a) S. Kabilan, A.J. Marshall, F.K. Sartain, M.C. Lee, A. Hussain, X.P. Yang, J. Blyth, N. Karangu, K. James, J. Zeng, D. Smith, A. Domschke, C.R. Lowe, Biosensors and Bioelectronics 20 (2005) 1602;
b) G.J. Worsley, G.A. Tourniaire, K.E. Medlock, F.K. Sartain, H.E. Harmer, M. Thatcher, A.M. Horgan, J. Pritchard, Clinical Chemistry 53 (2007) 1820;
c) A.K. Yetisen, Y. Montelongo, F. da Cruz Vasconcellos, J.L. Martinez-Hurtado, S. Neupane, H. Butt, M.M. Qasim, J. Blyth, K. Burling, J.B. Carmody, M. Evans, T.D. Wilkinson, L.T. Kubota, M.J. Monteiro, C.R. Lowe, Nano Letters 14 (2014) 3587;
d) C.R. Lowe, A.P. James, E. Rayne, S. Kabilan, A.J. Marshall, Holographic Sensor Having Heterogeneous Properties, WO Patent Application 2006008524 A1, January 26, 2006;
e) C.R. Lowe, C.A.B. Davidson, J. Blyth, S. Kabilan, A.J. Marshall, B. Madrigal Gonzalez, A.P. James, Method of Detecting an Analyte in a Fluid, WO Patent Application 2003087899 A1, October 23, 2003;
f) S. Kabilan, J. Blyth, M.C. Lee, A.J. Marshall, A. Hussain, X.P. Yang, C.R. Lowe, Journal of Molecular Recognition 17 (2004) 162;
g) M.C. Lee, S. Kabilan, A. Hussain, X. Yang, J. Blyth, C.R. Lowe, Analytical Chemistry 76 (2004) 5748;
h) C.R. Lowe, J. Blyth, S. Kabilan, A. Hussain, X.P. Yang, F.K. Sartain, M.C. Lee, Holographic Sensor, WO Patent Application 2004081624 A1, September 23, 2004;
i) A. Domschke, S. Kabilan, R. Anand, M. Caines, D. Fetter, P. Griffith, K. James, N. Karangu, D. Smith, M. Vargas, J. Zeng, A. Hussain, Y. Xiaoping, J. Blyth, A. Mueller, P. Herbrechtsmeier, C.R. Lowe, Sensors Proceedings of IEEE (2004) 1320;
j) C.R. Lowe, S. Kabilan, J. Blyth, A. Domschke, D. Smith, N. Karangu, Ophthalmic Device Comprising

a Holographic Sensor, WO Patent Application 2005031442 A1, April 7, 2005;
k) A.M. Horgan, A.J. Marshall, S.J. Kew, K.E.S. Dean, C.D. Creasey, S. Kabilan, Biosensors and Bioelectronics 21 (2006) 1838;
l) X. Yang, M.C. Lee, F. Sartain, X. Pan, C.R. Lowe, Chemistry 12 (2006) 8491;
m) S. Kabilan, M.C. Lee, A.M. Horgan, K.E.S. Medlock, Novel Boronate Complex and its use in a Glucose Sensor, WO Patent Application 2007054689 A1, May 18, 2007;
n) A. Domschke, W.F. March, S. Kabilan, C. Lowe, Diabetes Technology & Therapeutics 8 (2006) 89;
o) X.P. Yang, X.H. Pan, J. Blyth, C.R. Lowe, Biosensors and Bioelectronics 23 (2008) 899;
p) G.J. Worsley, G.A. Tourniaire, K.E. Medlock, F.K. Sartain, H.E. Harmer, M. Thatcher, A.M. Horgan, J.J. Pritchard, Diabetes Science and Technology 2 (2008) 213;
q) B. Burles, R.B. Millington, C.R. Lowe, S. Kabilan, J. Blyth, Ophthalmic Device Comprising a Holographic Sensor, US Patent 8,241,574 B2, August 14, 2012;
r) E. Vezouviou, C.R. Lowe, Biosensors and Bioelectronics 68 (2015) 371–381;
s) G. Vancoillie, R. Hoogenboom, From glucose sensors to autonomous drug delivery, Sensors 16 (2016) 1736, https://doi.org/10.3390/s16101736;
t) E. Mohamed, M.U. Hassan, A.K. Yetisen, H. Butt, Glucose sensing with phenylboronic acid functionalized hydrogel-based optical diffusers, ACS Nano 12 (2018) 2283–2291;
u) G. Khalili Moghaddam, H. Margerison, J. Suzuki, J. Blyth, C.R. Lowe, A transparent glucose-sensitive double polymerised holographic sensor, Sensors and Actuators B 267 (2018) 1–4.

[31] a) F.K. Sartain, X.P. Yang, C.R. Lowe, Analytical Chemistry 78 (2006) 5664;
b) F.K. Sartain, X.P. Yang, C.R. Lowe, Chemistry – A European Journal 14 (2008) 4060.

[32] A.J. Marshall, D.S. Young, J. Blyth, S. Kabilan, C.R. Lowe, Analytical Chemistry 76 (2004) 1518.

[33] a) E.V. Tan, C.R. Lowe, Analytical Chemistry 81 (2009) 7579;
b) C.R. Lowe, C.A. Davidson, E.V. Tan, Holographic Sensor and its use in the Detection of Enzyme Inhibition, WO Patent Application 2010041078 A1, April 15, 2010.

[34] a) Y. Fuchs, S. Kunath, O. Soppera, K. Haupt, A.G. Mayes, Advanced Functional Materials 24 (2014) 688;
b) Y. Fuchs, O. Soppera, A.G. Mayes, K. Haupt, Advanced Materials 25 (2013) 566.

[35] a) R.B. Millington, A.G. Mayes, J. Blyth, C.R. Lowe, Sensors and Actuators, A: Chemical 33 (1996) 55;
b) A.G. Mayes, J. Blyth, R.B. Millington, C.R. Lowe, Journal of Molecular Recognition 11 (1998) 168.

[36] a) J. Blyth, C.R. Lowe, A.G. Mayes, R.B. Millington, Holographic Sensors and Their Production, WO Patent Application 1999063408 A1, December 9, 1999;
b) C.A. Davidson, J. Blyth, C.R. Lowe, Method of Production of a Holographic Sensor, WO Patent Application 2010041079 A1, April 15, 2010.

[37] a) A.J. Marshall, D.S. Young, J. Blyth, S. Kabilan, C.R. Lowe, Analytical Chemistry 76 (2004) 1518;
b) C.R. Lowe, A.J. Marshall, Use of Holographic Sensor, WO Patent Application 2007023282 A1, March 1, 2007.

[38] N.C.L. Oliveira, G. El Khoury, J.M. Versnel, G.K. Moghaddam, L.S. Leite, J.L. Lima, C.R. Lowe, A holographic sensor based on a biomimetic affinity ligand for the detection of cocaine, Sensors and Actuators B: Chemical 270 (2018) 216–222.

[39] a) D. Bhatta, G. Christie, J. Blyth, C.R. Lowe, Sensors and Actuators, A: Chemical 134 (2008) 356;
b) G. Christie, C.R. Lowe, Sensor for Spores, WO Patent Application 2008087407 A1, July 24, 2008.

[40] C.R. Lowe, C.A.B. Davidson, J. Blyth, S. Kabilan, A.J. Marshall, B. Madrigal Gonzalez, A.P. James, Method of Detecting an Analyte in a Fluid, WO Patent Application 2003087899 A1, October 23, 2003.

[41] B.D. Lawrence, M. Cronin-Golomb, I. Georgakoudi, D.L. Kaplan, F.G. Omenetto, Biomacromolecules 9 (2008) 1214.

[42] a) J.L. Martinez-Hurtado, C.A. Davidson, J. Blyth, C.R. Lowe, Langmuir 26 (2010) 15694;
b) J.L. Martinez-Hurtado, C.A.B. Davidson, C.R. Lowe, in: Proc. SPIE; Orlando, FL, 2011, p. 80240Y.

[43] I. Naydenova, J. Grand, T. Mikulchyk, S. Martin, V. Toal, V. Georgieva, S. Thomas, S. Mintova, Hybrid sensors fabricated by inkjet printing and holographic patterning, Chemistry of Materials 27 (17) (2015) 6097–6101.

[44] J.L. Hurtado, C.R. Lowe, ACS Applied Materials and Interfaces 6 (2014) 8903–8908.

[45] a) C.R. Lowe, J. Blyth, A.P. James, Interrogation of a Sensor, WO Patent Application 2006008531 A1, January 26, 2006;
b) Ermold, ML; Rai, K; Fontecchio, AK, Hydrostatic pressure response of polymer-dispersed liquid crystal gratings, Journal of Applied Physics Volume: 97 Issue: 10 Part: 1, Article Number: 104905 c) K. Rai, A.K. Fontecchio, Optimization of pressure response in HPDLC gratings based on polymer composition, Molecular Crystals and Liquid Crystals 450 (2006) 383–390, https://doi.org/10.1080/15421400600588207.

[46] a) L. Hongpeng, Y. Dan, K. Zhou, M. Dongyao, L. Liu, W. Hui, W. Wang, S. Qinggong, Temperature-induced spectrum response of volume grating as an effective strategy for holographic sensing in acrylamide polymer part I: sensing, Applied Optics 55 (35) (December 10, 2016) 9907–9916;
(b) L. Hongpeng, Y. Dan, K. Zhou, M. Dongyao, L. Liu, W. Hui, W. Wang, S. Qinggong, Temperature-induced spectrum response of a volume grating as an effective strategy for holographic sensing in an acrylamide polymer part II: physical mechanism, Applied Optics 55 (35) (December 10, 2016) 9917–9924;

(c) S. Jung, K.I. MacConaghy, J.L. Kaar, M.P. Stoykovich, Enhanced optical sensitivity in thermoresponsive photonic crystal hydrogels by operating near the phase transition, ACS Applied Materials and Interfaces 9 (2017) 27927–27935;
(d) T. Mikulchyk, S. Martin, I. Naydenova, N-isopropylacrylamide-based photopolymer for holographic recording of thermosensitive transmission and reflection gratings, Applied Optics 56 (22) (2017) 6348–6356. https://doi.org/10.1364/AO.56.006348;
(e) T. Mikulchyk, W. James, D. Cody, S. Martin, I. Naydenova, Humidity and temperature induced changes in the diffraction efficiency and the Bragg angle of slanted photopolymer-based holographic gratings, Sensors and Actuators B (Chemical) 239 (2017) 776–785. https://doi.org/10.1016/j.snb.2016.08.052.

[47] A.J. Marshall, D.S. Young, S. Kabilan, A. Hussain, J. Blyth, C.R. Lowe, Analytica Chimica Acta 527 (2004) 13.

[48] C.R. Lowe, J. Blyth, A.P. James, Interrogation of a Sensor, WO Patent Application 2006008531 A1, January 26, 2006.

[49] S. Kabilan, A.J. Marshall, Interrogation of a Light-Sensitive Sensor, WO Patent Application 2007010244 A1, January 25, 2007.

[50] a) A.K. Yetisen, J.L. Martinez-Hurtado, A. Garcia-Melendrez, F. Cruz Vasconcellos, C.R. Lowe, A smartphone algorithm with inter-phone repeatability for the analysis of colorimetric tests, Sensors and Actuators B: Chemical 196 (June 2014) 156–160. https://doi.org/10.1016/j.snb.2014.01.077;
b) G.K. Moghaddam, C.R. Lowe, Smartphone-based quantitative measurements on holographic sensors, PLoS One 12 (11) (2017) e0187467. https://doi.org/10.1371/journal.pone.0187467.

[51] A.K. Yetisen, H. Butt, T. Mikulchyk, R. Ahmed, Y. Montelongo, M. Humar, N. Jiang, S. Martin, I. Naydenova, S.H. Yun, Color-selective 2.5D holograms on large-area flexible substrates for sensing and multilevel security, Advanced Optical Materials (2016), https://doi.org/10.1002/adom.201600162.

[52] a) J.L. Martinez-Hurtado, C.A. Davidson, J. Blyth, C.R. Lowe, Langmuir 26 (2010) 15694;
b) K. Kaneko, H.B. Sun, X.M. Duan, S. Kawata, Applied Physics Letters 83 (2003) 1426;
c) T.N. Smirnova, L.M. Kokhtych, A.S. Kutsenko, O.V. Sakhno, J. Stumpe, Nanotechnology 20 (2009) 405301;
d) A.O. Rybaltovskii, V.I. Gerasimova, N.V. Minaev, V.I. Sokolov, P.S. Timashev, E.A. Troitskaya, V.V. Firsov, V.I. Yusupov, V.N. Bagratashvili, Nanotechnology in Russia 5 (2010) 435;
e) V.N. Bagratashvili, A.O. Rybaltovsky, N.V. Minaev, P.S. Timashev, V.V. Firsov, V.I. Yusupov, Laser Physics Letters 7 (2010) 401;
f) B.Al Qattan, H. Butt, A. Sabouri, A.K. Yetisen, R. Ahmeda, N. Mahmoodia, Holographic direct pulsed laser writing of two dimensional nanostructures, RSC Advances 6 (2016) 111269;
g) R. Ahmed, A.K. Yetisen, A. El Khourya, H. Butt, Printable ink lenses, diffusers, and 2D gratings, Nanoscale 9 (2017) 266–276.

[53] J.L. Martinez Hurtado, C.R. Lowe, An integrated photonic-diffusion model for holographic sensors in Polymeric matrices, Journal of Membrane Science 495 (2015) 14.

[54] a) D. Cody, I. Naydenova, Theoretical modeling and design of photonic structures in zeolite nanocomposites for gas sensing. Part II: volume gratings, JOSA A 35 (1) (2018) 12–19. https://doi.org/10.1364/JOSAA.35.000012;
b) D. Cody, I. Naydenova, Theoretical modeling and design of photonic structures in zeolite nanocomposites for gas sensing. Part I: surface relief gratings, JOSA A 34 (12) (2017) 2110–2119. https://doi.org/10.1364/JOSAA.34.002110;
c) H. Liu, Y. Dan, M. Dongyao, Y. Geng, W. Wang, Modeling swelling and absorption dynamics for holographic sensing in analytes sensitive photopolymer, Optics Communications 368 (2016) 95–104.

[55] W.R. Klein, B.D. Cook, Unified approach to ultrasonic light diffraction, IEEE Transactions on Sonics and Ultrasonics 14 (1967) 123–134.

[56] M.G. Moharam, L. Young, Criterion for Bragg and Raman-Nath diffraction regimes, Applied Optics 17 (1978) 1757–1759.

[57] C.V. Raman, N.S. Nagendra Nath, The diffraction of light by high frequency sound waves: Part 1, Proceedings of the Indian Academy of Sciences – Section A 32 (1935) 406–412.

[58] M.G. Moharam, T.K. Gaylord, Rigorous coupled-wave analysis of planar-grating diffraction, JOSA 71 (7) (1981) 811–818. https://doi.org/10.1364/JOSA.71.000811.

[59] T. Mikulchyk, Development of Holographic Sensors for Monitoring Relative Humidity and Temperature, PhD thesis, 2016.

[60] H. Bjelkhagen, D. Brotherton-Ratcliffe, Ultra-Realistic Imaging: Advanced Techniques in Analogue and Digital Colour Holography, Taylor & Francis, Boca Raton, FL, 2013, p. 89.

[61] A.V. Kraiskii, T.V. Mironova, T.T. Sultanov, IEEE Journal of Quantum Electronics 40 (2010) 652.

[62] A.V. Kraiskii, T.V. Mironova, T.T. Sultanov, Quantum Electronics 42 (2012) 1137.

[63] A.K. Yetisen, J.L. Martinez-Hurtado, F. da Cruz Vasconcellos, M.C. Simsekler, M.S. Akram, C.R. Lowe, Lab on a Chip 14 (2014) 833.

[64] Z. Zhou, Z. Shi, X. Cai, S. Zhang, S.G. Corder, X. Li, Y. Zhang, G. Zhang, L. Chen, M. Liu, D.L. Kaplan, F.G. Omenetto, Y. Mao, Z. Tao, T.H. Tao, The use of functionalized silk fibroin films as a platform for optical diffraction-based sensing applications, Advanced Materials 29 (2017), 1605471.

[65] D. Nazarova, L. Nedelchev, P. Sharlandjiev, Surface plasmon-polariton resonances in metal-coated

holographic azopolymer gratings, Bulgarian Chemical Communciations 45 (Special Issue B) (2013) 115–118.
[66] a) J.J. Cowan, Optical Engineering 24 (1985) 245796;
b) J.J. Cowan, W. Slafer, The Imaging Science Journal 31 (1987) 100.
[67] J.J. Cowan, Journal of the Optical Society of America A 7 (1990) 1529.
[68] Y.N. Denisyuk, Doklady Akademii Nauk SSSR 144 (1962) 1275.
[69] A.K. Yetisen, Y. Montelongo, N.M. Farandos, I. Naydenova, C.R. Lowe, S.H. Yun, Mechanism of multiple grating formation in high-energy recording of holographic sensors, Applied Physics Letters 105 (2014) 261106. https://doi.org/10.1063/1.4905352.
[70] a) G. Zhao, P. Mourolis, Diffusion model of hologram formation in dry photopolymer materials, Journal of Modern Optics 41 (1994) 1929–1939;
b) G.H. Zhao, P. Mouroulis, Extension of a diffusion-model for holographic photopolymers, Journal of Modern Optics 42 (1995) 2571–2573.
[71] D.J. Lougnot, P. Jost, L. Lavielle, Polymers for holographic recording: VI. Some basic ideas for modelling the kinetics of the recording process, Pure Applied Optics 6 (1997) 225–245.
[72] T. Babeva, I. Naydenova, D. Mackey, S. Martin, V. Toal, Journal of the Optical Society of America B 27 (2010) 197.
[73] J.T. Sheridan, J.R. Lawrence, Journal of the Optical Society of America A 17 (2000) 1108.
[74] J.T. Sheridan, M. Downey, F.T. O'Neill, Journal of Optics 3 (2001) 477.
[75] J. Lawrence, F. O'Neill, J. Sheridan, Adjusted intensity nonlocal diffusion model of photopolymer grating formation, Journal of the Optical Society of America B 19 (2002) 621–629.
[76] S. Liu, M.R. Gleeson, D. Sabol, J.T. Sheridan, Extended model of the photo initiation mechanisms in photopolymer materials, Journal of Applied Physics 106 (2009) 104911.
[77] D. Mackey, P. O'Reilly, I. Naydenova, Theoretical modeling of the effect of polymer chain immobilization rates on holographic recording in photopolymers, JOSA A 33 (5) (2016) 920–929.
[78] S. Piazzolla, B. Jenkins, First-harmonic diffusion model for holographic grating formation in photopolymers, Journal of the Optical Society of America B 17 (2000) 1147–1157.
[79] S. Martin, C.A. Feely, V. Toal, Holographic recording characteristics of an acrylamide-based photopolymer, Applied Optics 36 (1997) 5757–5768.
[80] V. Moreau, Y. Renotte, Y. Lion, Characterisation of DuPont photopolymer: determination of kinetic parameters in a diffusion model, Applied Optics 41 (2002) 3427–3435.
[81] V. Colvin, R. Larson, A. Harris, M. Schilling, Quantitative model of volume hologram formation in photopolymers, Journal of Applied Physics 81 (1997) 5913–5923.
[82] J.H. Kwon, H.C. Hwang, K.C. Woo, Analysis of temporal behavior of beams diffracted by volume gratings formed in photopolymers, Journal of the Optical Society of America B 16 (1999) 1651–1657.
[83] C. Carre, D. Lougnot, Photopolymers for holographic recording—from standard to self-processing materials, Journal of Physics 3 (1993) 1445–1460.
[84] I. Naydenova, R. Jallapuram, S. Martin, R. Howard, V. Toal, Investigations of the diffusion processes in self-processing acrylamide-based photopolymer system, Applied Optics 43 (14) (2004) 2900. https://doi.org/10.1364/AO.43.002900.
[85] S. Gallego, M. Ortuño, C. Neipp, A. Márquez, A. Beléndez, I. Pascual, J.V. Kelly, J.T. Sheridan, Physical and effective optical thickness of holographic diffraction gratings recorded in photopolymers, Optics Express 13 (2005) 1939–1947.
[86] A.J. Marshall, J. Blyth, C.A. Davidson, C.R. Lowe, Analytical Chemistry 75 (2003) 4423.
[87] J.L. Martinez-Hurtado, C.A. Davidson, J. Blyth, C.R. Lowe, Langmuir 26 (2010) 15694.
[88] T. Todorov, L. Nikolova, N. Tomova, Applied Optics 23 (1984) 4309.
[89] P. Rochon, E. Batalla, A. Natansohn, Optically induced surface gratings on azoaromatic polymer films, Applied Physics Letters 66 (1995) 136–138.
[90] D.Y. Kim, S.K. Tripathy, L. Li, J. Kumar, Laser induced holographic surface relief gratings on nonlinear optical polymer films, Applied Physics Letters 66 (1995) 1166–1168.
[91] P.S. Ramanujam, N.C.R. Holme, S. Hvilsted, Atomic force and optical near-field microscopic investigation of polarization holographic gratings in a liquid crystalline azobenzene side-chain polyester, Applied Physics Letters 68 (1996) 1329–1331.
[92] I. Naydenova, L. Nikolova, T. Todorov, N.C.R. Holme, P.S. Ramanujam, S. Hvilsted, Diffraction from polarization holographic gratings in side – chain azobenzene polyesters, Journal of the Optical Society of America B 15 (1998) 1257. https://doi.org/10.1364/JOSAB.15.001257.
[93] M. Zawadzka, I. Naydenova, Surface holograms for sensing application, in: IOP Conf. Series: Journal of Physics: Conf. Series, vol. 961, 2017, p. 012002.
[94] a) M. Dell'Aglio, R. Gaudiuso, R. ElRashedy, O. De Pascale, G. Palazzo, A. De Giacomo, Physical Chemistry Chemical Physics 15 (2013) 20868;
b) P. Wagener, S. Ibrahimkutty, A. Menzel, A. Plech, S. Barcikowski, Physical Chemistry Chemical Physics 15 (2013) 3068;
c) B. Toftmann, B. Doggett, C.B.J. Rgensen, J. Schou, J.G. Lunney, Journal of Applied Physics 113 (2013) 083304;
d) Y. Montelongo, J.O. Tenorio-Pearl, W.I. Milne, T.D. Wilkinson, Nano Letters 14 (2014) 294.
[95] C.A. Davidson, J. Blyth, C.R. Lowe, Method of Production of a Holographic Sensor, WO Patent Application 2010041079 A1, April 15, 2010.

[96] J.M. Versnel, C. Williams, C.A.B. Davidson, T.D. Wilkinson, C.R. Lowe, Signal intensity enhancement of laser ablated volume holograms, Optical Materials 73 (2017) 400e–407.

[97] A. Ostrowski, I. Naydenova, V. Toal, Light induced redistribution of Si - MFI zeolite nanoparticles in acrylamide-based photopolymer holographic gratings, Journal of Optics A: Pure and Applied Optics 11 (2009) 034004, https://doi.org/10.1088/1464-4258/11/3/034004, 4pp.

[98] I. Naydenova, E. Leite, T. Babeva, N. Pandey, T. Baron, T. Yovcheva, S. Sainov, S. Martin, S. Mintova, V. Toal, Optical properties of photopolymerisable nanocomposites containing nanosized molecular sieves, Journal of Optics (10pp) (2011) 044019, https://doi.org/10.1088/2040-8978/13/4/044019.

[99] M.V. Deshmukh, A.A. Vaidya, M.G. Kulkarni, P.R. Rajamohanan, S. Ganapathy, LCST in poly(N-isopropylacrylamide) copolymers: high resolution proton NMR investigations, Polymer 41 (2000) 7951–7960.

[100] T. Elshaarani, H.J. Yu, L. Wang, Zain-ul-Abdin, R.S. Ullah, M. Haroon, R.U. Khan, S. Fahad, A. Khan, A. Nazir, M. Usman, K.U.R. Naveed, Synthesis of hydrogel-bearing phenylboronic acid moieties and their applications in glucose sensing and insulin delivery, Journal of Materials Chemistry B 6 (23) (2018) 3831–3854, https://doi.org/10.1039/c7tb03332j.

[101] S. Mintova, J.-P. Gilson, V. Valtchev, Nanoscale 5 (2013) 6693.

[102] X. Xu, J. Wang, Y. Long, Sensors 6 (2006) 1751.

[103] D. Banerjee, Z. Hu, J. Li, Dalton Transactions 43 (2014) 10668.

[104] M. Zaarour, B. Dong, I. Naydenova, R. Retoux, S. Mintova, Microporous and Mesoporous Materials 189 (2014) 11.

[105] A. Dubbe, Solid State Ionics 179 (2008) 1645.

[106] a) D. Yu, H. Liu, D. Mao, Y. Geng, W. Wang, L. Sun, and Jiang L., Enhancement of spectrum strength in holographic sensing in nanozeolites dispersed acrylamide photopolymer , Vol. 23, No. 22 | DOI:10.1364/OE.23.029113 | Optics Express 29113, #249433.; b) D. Mao, Y. Geng, H. Liu, K. Zhou, L. Xian, D. Yu, Two-way shift of wavelength in holographic sensing of organic vapor in nanozeolites dispersed acrylamide photopolymer, Applied Optics 55 (23) (August 10, 2016) 6212–6221.

[107] E. Leite, T. Babeva, E.-P. Ng, V. Toal, S. Mintova, I. Naydenova, Optical properties of photopolymer layers doped with aluminophosphate nanocrystals, Journal of Physical Chemistry C (2010), https://doi.org/10.1021/jp1060073.

[108] K. Zhou, Y. Geng, H. Liu, S. Wang, D. Mao, D. Yu, Improvement of Sensitivity Substrate free structures -Improvement of holographic sensing response in substrate-free acrylamide photopolymer, Applied Optics 56 (13) (2017) 3714–3724.

[109] Q. Zhao, A.K. Yetisen, A. Sabouri, S.H. Yun, H. Butt, Printable nanophotonic devices via holographic laser ablation, ACS Nano 9 (2015 b) 9062–9069.

[110] R. Ahmed, A.K. Yetisen, S.H. Yun, H. Butt, Color-selective holographic retroreflector array for sensing applications, light, Science & Applications 6 (2017) e16214.

[111] H. Butt, A.K. Yetisen, D. Mistry, S.A. Khan, M.U. Hassan, S.H. Yun, Morpho butterfly-inspired nanostructures, Advanced Optical Materials 4 (4) (2016) 497–504.

[112] N. Jiang, H. Butt, Y. Montelongo, F. Liu, S. Afewerki, G.L. Ying, Q. Dai, S.H. Yun, A.K. Yetisen, Laser interference lithography for the nanofabrication of stimuli-responsive Bragg stacks, Advanced Functional Materials 28 (24) (2018) 1702715.

[113] M. Bajgrowicz-Cieslak, Y. Alqurashi, M. Ismail Elshereif, A.K. Yetisen, M.U. Hassan, H. Butt, Optical glucose sensors based on hexagonally packed 2.5-dimensional photonic concavities imprinted in phenylboronic acid functionalized hydrogel films, RSC Advances 7 (2017) 53916.

[114] J.-W. Oh, W.-J. Chung, K. Heo, H.-E. Jin, B.Y. Lee, E. Wang, et al., Biomimetic virus-based colourimetric sensors, Nature Communications 5 (2014).

[115] A. Domschke, W.F. March, S. Kabilan, C. Lowe, Initial clinical testing of a holographic non-invasive contact lens glucose sensor, Diabetes Technology & Therapeutics 8 (1) (2006) 89–93.

[116] G.V. Zaitseva, A.R. Zaritskii, M.N. Kirichenko, M.A. Krasnova, A.V. Kraiskii, V.A. Postnikov, M.A. Shevchenko, Effect of decreasing glucose concentration in blood plasma after blood saturation with oxygen, 2017, Bulletin of the Lebedev Physics Institute 44 (6) (2017) 155–158.

CHAPTER 8

Holographic Security

VINCENT TOAL, BSC, MSC, PHD

INTRODUCTION

The wavelength and spatial distribution of amplitude, phase, and polarization state of a light wavefront can all be holographically recorded. Optical components including lenses and free space itself with their spatial transformation properties, as well as diffusers and spatial light modulators (SLMs) of phase, amplitude, and polarization state, all enable controlled manipulation of light wavefront characteristics. The properties of light combined with the means for their control and recording are valuable tools for advanced security applications.

THE COUNTERFEIT PROBLEM

Counterfeit goods constituted a global market of US$461 billion and accounted for around 2.5% of global imports in 2013 [1], up from an estimated 1.9% in 2008 [2]. The threats posed by counterfeit goods are not only economic but also physical in the form of dangerous medicines, equipment, spare parts, domestic appliances, and toys.

It is widely accepted that in addressing such problems it is wise to use overt, covert, and tamper evident devices and forensic techniques. The choice of devices and methods depends on the scale of the threat and the consequences if it is not addressed. In this context, it is apparent that some manufacturers appear to tolerate some volume of counterfeit product, particularly in cases where the end user is aware that the product is probably fake. In some markets, however, such compromise is clearly unacceptable where the health and safety of persons is at stake. Examples are pharmaceuticals, medicines, diagnostic instruments, food, animal feed and pharmaceuticals, spare parts for motor vehicles, ships, and aircraft.

Regionally based pricing strategies give rise to authentic goods manufactured in low-cost economies being made available in higher-cost economies. Although not in itself a threat to health or safety, such fraudulent activity implies weakness in the supply chain management system and may indicate the possibility of worse to follow. Only careful checks at each link in the chain by cross-reference to a secure data base will help to alleviate the problem.

Verification is required for product authentication, brand protection, and supply chain security as well as for documentation such as export/import licenses, end user certificates, passports, driving licenses, credit and debit cards, person identification, admission tickets, and legal tender.

The growth in fake certification [3] demands thorough scrutiny of academic qualifications and specialized skills.

Contraband, narcotics, weapons, and explosives are also of worldwide concern, and rapid, effective means of detection are essential.

HOLOGRAPHIC SECURITY AND AUTHENTICATION

Since the mid-1960s, when fully three-dimensional imaging by off-axis holographic methods was introduced [4], the idea of using recorded optical wavefronts for authentication has received much attention because of the large amount of information that can be carried. Wavelength, spatial distribution of amplitude and phase, and polarization states may all be exploited for authentication purposes. The choice of which of the information bearing properties of light to use depends on the level of security required and cost.

MASS-PRODUCED HOLOGRAMS

Mass-produced holograms are usually made either in thin metallic foil or foil backed plastic or in a photopolymer, preferably one requiring no physical or chemical processing following the recording step. Other materials such as silver halide or dichromated gelatin may be used, but these require chemical processing,

Optical Holography-Materials, Theory and Applications. https://doi.org/10.1016/B978-0-12-815467-0.00008-6

with attendant financial and environmental costs. Holographic recording materials are extensively explained in Chapter 2.

Mass-produced holograms are widely used on debit and credit cards and product packaging.

Rainbow Holograms

The rainbow holography technique [5] enables one to make a hologram, which can be viewed with ordinary, as opposed to laser light, which is an advantage for its use as a security tag.

Recording a rainbow hologram involves two steps. First a transmission hologram (H1) is made in the usual way (Fig. 8.1A) by recording the pattern of interference between two mutually coherent light beams, one of which is reflected or scattered from some object.

During reconstruction of the real image using the conjugate reference beam, a horizontal slit aperture, S, is placed in contact with H1. If the original reference beam is collimated, it becomes its own conjugate by simply rotating the hologram in-plane through an angle of 180°. A second hologram, H2, is recorded (Fig. 8.1B), and using the original reference beam, a real orthoscopic image is obtained, which can only be viewed through the reconstructed, conjugate image of the slit aperture, S* (Fig. 8.1C). In white light illumination of the hologram, the image is seen in different colors due to the dispersion effect of the slit (Fig. 8.1D). Horizontal perspective is maintained, but vertical perspective is lost due to the slit aperture.

One can arrange recording of the H2 hologram so that the real image reconstructed from H1 is intersected by the recording medium for H2 (Fig. 8.2A). Using this configuration, a so-called image-plane hologram is then recorded.

If H2 is recorded in photoresist, a surface relief hologram is formed, which can then be electroplated with nickel to form a durable surface relief master hologram. This is used to make multiple copies by stamping in thin, high reflectivity metal foil or in plastic backed by reflective foil so that the resulting copied holograms may be viewed in reflected light.

Such holograms are now ubiquitous but of questionable value for anticounterfeit purposes, as all those made from a single master are identical and fake holograms are easily produced. Even very crude versions of such holograms escape detection simply because they are not examined thoroughly, if at all. Considerable

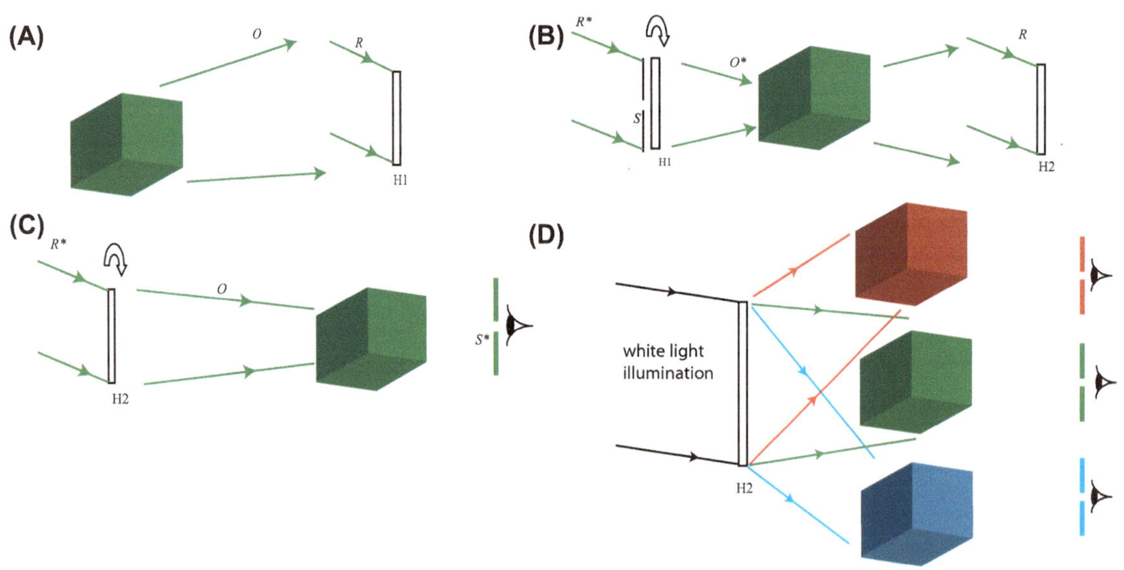

FIG. 8.1 Rainbow holography. A transmission hologram (H1) is recorded in the usual way **(A)**. A horizontal slit aperture, S, is placed in contact with H1 during reconstruction of the real image by the conjugate reference beam and a second hologram, H2, is recorded. **(B)** Using the original reference beam, a real orthoscopic image is obtained, which can only be viewed through the reconstructed, conjugate image of the slit aperture **(C)**. When white light is used in the final reconstruction, multiple real images of the slit aperture are produced in different spectral colors as shown in **(D)** along with real orthoscopic images of the original object in spectral colors.

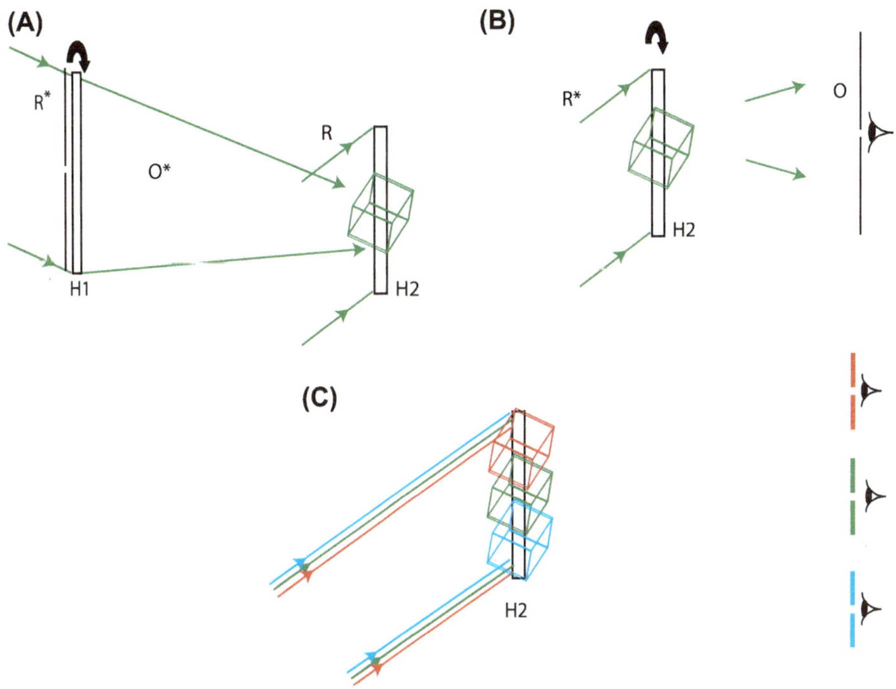

FIG. 8.2 Image plane rainbow holography.

effort has been made to produce foil holograms, which display increasingly complicated and interesting images, including images that change with viewpoint, to invite inspection. In addition, serial numbers have been ink printed or laser engraved underneath, beside or on the surface of the holograms.

Moiré techniques can provide additional security [6]. After recording the H2 hologram in the photoresist, a grating pattern may also be recorded. The pattern in the form of a transparency is placed in contact with the photoresist and illuminated normally by an expanded laser (typically of wavelength 488 nm). The pattern is copied in surface relief in the developed photoresist. To authenticate a foil copy hologram made from the master, a second decoder transparency is placed on top of the copy and the moiré pattern observed in reflected light. A counterfeit hologram shows the wrong pattern or no pattern at all. A recorded surface relief pattern could, however, be characterized by phase contrast or optical scanning microscopy and subsequently incorporated in counterfeit holograms. To overcome this difficulty, one may record the pattern as a hologram using an encrypted reference beam [7], see phase coded reference beam encryption section later in this chapter.

Volume Reflection Holograms

Volume reflection holograms [8] made by recording the interference pattern produced by overlapping, counter-propagating mutually coherent light beams (Fig. 8.3), can be viewed in white light, as they also function as multilayer, dielectric passband filters (see Chapter 1 Inline Reflection Hologram (Denisyuk) section). The recorded fringe pattern is oriented approximately parallel to the hologram plane with spacing of $\lambda/(2n)$, where λ is the laser wavelength and n is the average refractive index of the recording material. Unlike rainbow holograms, there is no loss of vertical perspective.

Increased image complexity can be achieved by using additional beams to illuminate the object from a range of angles of incidence to obtain a wider viewing perspective. Multiplexing using reference beams at different angles of incidence enables the display of quite different images when the hologram is illuminated from different directions.

Mass production of reflection holograms by contact copying from a high-quality master hologram, using low coherence light, means that the same reconstructed image is visible in all of them. However, volume reflection holograms are not so easy to replicate, as the information stored in them is distributed throughout the

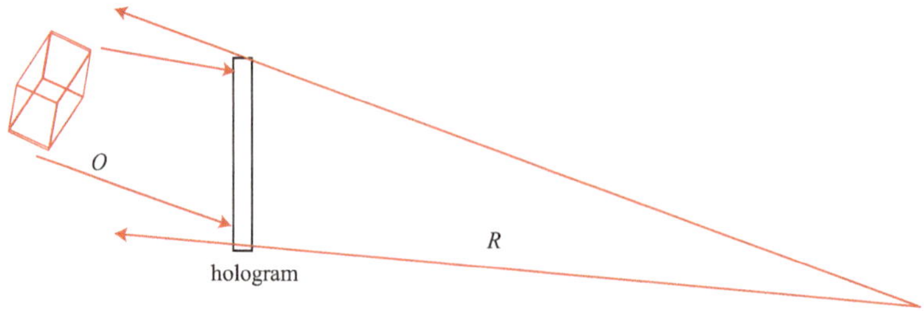

FIG. 8.3 Reflection (Denisyuk) holography.

depth of the recording layer. For example, in a layer of thickness 30 μm, there are typically 150 fringes of maximum brightness in an interference pattern recorded at 633 nm.

SERIALIZED HOLOGRAPHY

The ideal hologram for anticounterfeiting applications is one that is unique to each individual product item or package, known as a serialized hologram. The hologram need not be particularly eye catching; what is most important that it should display the data needed for authentication purposes. At the same time, it must be secure against the possibility of copying. If a hologram contains all of the information that is needed about the product that it accompanies, then any data obtained from it must match that held in the manufacturer's database.

The information provided by a hologram can be as minimal as the product manufacturer's name and logo but could also include a serial number, lot number, sell-by date if applicable, and place of manufacture. All this information can be encoded in a holographic quick response code if required.

One way to create a serialized hologram is to make use of the data to spatially modulate a beam of laser light so that it becomes the object beam in a hologram recording system. The data can be written onto a transmissive liquid crystal spatial light modulator (SLM), which can modulate both the phase and amplitude of a light wave front, or onto a liquid crystal on silicon (LCOS) phase only reflective SLM or a digital micromirror (DMM) array. The data are updated for each new serialized hologram that is to be recorded.

Production volumes of security holograms for packaging applications may run to many millions per year so that rapid recording without physical or chemical processing is essential, implying the need for a self-processing photopolymer. Production of serialized holograms of this type has been pioneered by Optrace [9], an Irish start-up company, using machines that record up to 10,000 holograms per hour in a photopolymer developed by researchers in the Centre for Industrial and Engineering Optics at the Technological University, Dublin [10].

Surface ablation by single 6 ns laser pulses has been used to record holographic surface gratings in well-ordered printed ink on a substrate, a process requiring a few minutes to complete [11] and which may also enable mass production of serialized holograms.

FOURIER METHODS FOR IMAGE COMPARISON

The techniques discussed so far are all aimed at making holograms more difficult or even impossible to replicate by ensuring the uniqueness of the hologram itself. Holographic methods may also be used to compare images for authentication purposes by making use of the Fourier transform properties of lenses.

Optical Matched Filtering

Optical matched filtering by holographic means, also known as Vander Lugt filtering [12] or optical character recognition, has been extensively researched for security applications.

Starting with the definition of a 2-D spatial Fourier transform $F(x_2, y_2)$ of a function $f(x_1, y_1)$

$$F(x_2, y_2) = \iint f(x_1, y_1) e^{\frac{jk(x_1 x_2 + y_1 y_2)}{f}} dx_1 dy_1 \quad (8.1)$$

A point source at $y_1 = b$ of unit amplitude in the front focal plane (x_1, y_1) of a lens is specified by the delta function $\delta(y_1 - b)$ and has its Fourier transform $\exp(-jkby_2/f)$ with $k = 2\pi/\lambda$ and λ the wavelength, in the back focal plane (x_2, y_2) (Fig. 8.4A).

An optical character is written on transparency or an SLM and illuminated by laser light, coherent with the point source, to give complex amplitude $f(x_1, y_1)$. The

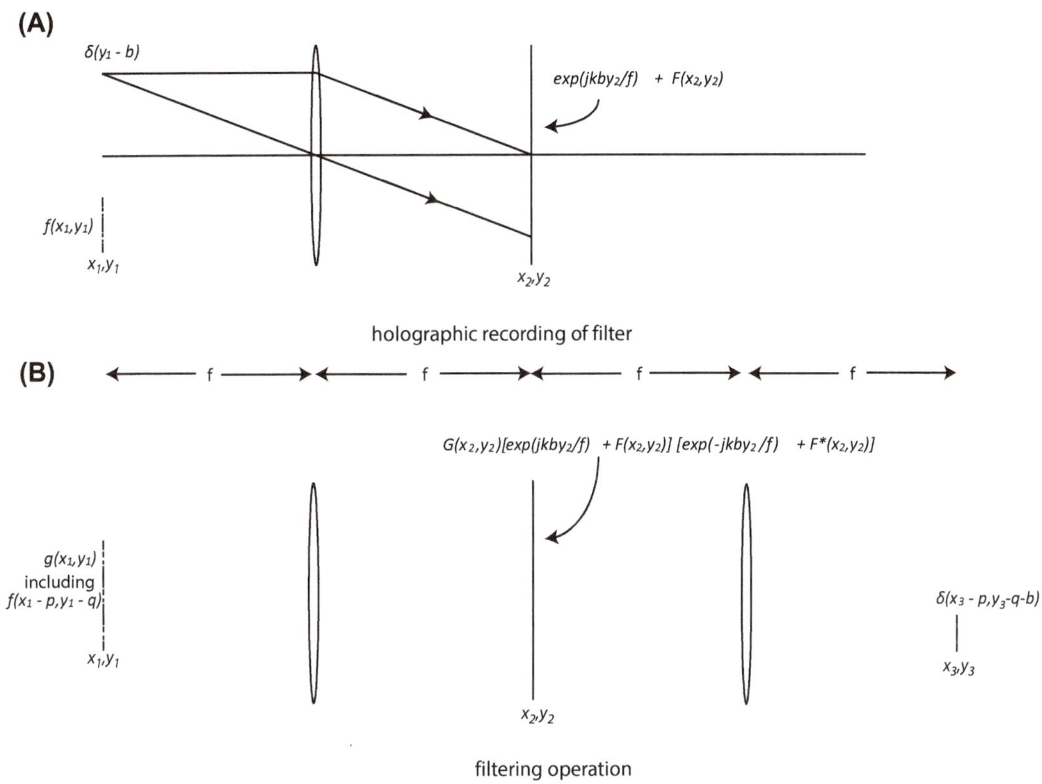

FIG. 8.4 Optical matched filtering.

combined intensity distribution in the back focal plane of the lens is

$$\left[\exp\left(-\frac{jkby_2}{f}\right)+F(x_2,y_2)\right]\left[\exp\left(\frac{jkby_2}{f}\right)+F^*(x_2,y_2)\right]$$

where $F^*(x_2, y_2)$ is the complex conjugate of $F(x_2, y_2)$.

This intensity distribution is recorded and subsequently illuminated by the Fourier transform of $g(x_1, y_1)$ (Fig. 8.4B) to produce the output

$$G(x_2,y_2)\left[\exp\left(-\frac{jkby_2}{f}\right)+F(x_2,y_2)\right]\left[\exp\left(\frac{jkby_2}{f}\right)+F^*(x_2,y_2)\right]$$

If $g(x_1, y_1)$ includes the character $f(x_1, y_1)$ at coordinates (p, q), i.e., $f(x_1-p, y_1-q)$, then $G(x_2, y_2)$ includes $F(x_2, y_2)\exp[-jk(px_2+qy_2)]$

and $$G(x_2,y_2)\left[\exp\left(-\frac{jkby_2}{f}\right)+F(x_2,y_2)\right]\left[\exp\left(\frac{jkby_2}{f}\right)+F^*(x_2,y_2)\right]$$

includes $|F(x_2,y_2)|^2\exp[-jk(px_2+(q+b)y_2)/f]$

which is a plane wavefront that can be transformed by the second lens in Fig. 8.4B into a focused point of light in its back focal plane at $x_3=-p$, $y_3=-(q+b)$, i.e., at the location of the character in the original input $g(x_1, y_1)$. This means that the procedure can determine the presence and the location of a character or group of characters in a set of characters.

Each spatial frequency point must be located within an area determined by the resolution limit of the Fourier transform lens, i.e., within $2.44\lambda f/D$ where D is the aperture diameter.

The procedure can be used for example to determine whether a given optical pattern such as a human fingerprint already exists in an optical database, to protect against fraud or to verify the identities of persons.

The technique is invariant with respect to position, but changes in scale or a rotation of the character constitute a different filter since the Fourier transform $F(x_2, y_2)$ is also rescaled and rotated. To overcome this difficulty, the Fourier transform $F(x_2, y_2)$ can first be converted to polar coordinates $F(r,\theta)$ with $\theta=\arctan(y_2/x_2)$ and $r=\sqrt{x_2^2+y_2^2}$. Scale change then affects only r, which can be handled by a scale invariant 1-D Mellin transform, which is a 1-D Fourier transform in ln(r). Rotational invariance is effected by a 1-D Fourier transform in θ, as this converts the rotations into phase factors, which have no effect on the magnitude of the transformed output [13,14]. To put it more simply, a point $x_2 + jy_2$ in the Fourier plane is rewritten as

$$w = re^{j\theta} \tag{8.2}$$

$$\ln(w) = \ln(r) + j\theta \tag{8.3}$$

and change in scale by a factor β accompanied by a rotation through angle α in θ space gives

$$\ln(w) = \ln(r) + \ln\beta + j\theta + j\alpha \tag{8.4}$$

Eq. (8.4) shows that separate 1-D Fourier transforms with respect to ln (r) and θ are invariant with respect to the now spatial shifts $ln\beta$ and $j\alpha$, and the system is immune to rotations and scale changes of $f(x_1, y_1)$.

Joint Transform Correlation

In many security applications, optical patterns can be directly compared for authentication purposes by joint transform correlation. Two patterns represented by complex amplitudes $f\left(x_1 - \frac{p}{2}, y_1 - \frac{q}{2}\right)$ and $g\left(x_1 + \frac{p}{2}, y_1 + \frac{q}{2}\right)$ in the front focal plane of a lens are illuminated by a collimated beam of laser light. Their Fourier transforms

$$F(x_2, y_2)\exp\left[-\frac{jk(x_2p + y_2q)}{2f}\right] \quad and$$
$$G(x_2, y_2)\exp\left[\frac{jk(x_2p + y_2q)}{2f}\right]$$

interfere in the back focal plane, and a recording is made of the interference pattern, having the form

$$I(x_2, y_2) = FG^* \exp\left[-\frac{jk(x_2p + y_2q)}{f}\right] + F^*G\exp\left[\frac{jk(x_2p + y_2q)}{f}\right] + |F|^2 + |G|^2 \tag{8.5}$$

When the recording is illuminated, the first term in Eq. (8.5) is transformed by the second lens in the 4f system of Fig. 8.4B and the correlation integral

$$\iint f(x_3, y_3)g^*(u - x_3 + p, v - y_3 + q)dudv$$

is obtained in the back focal plane (x_3, y_3), whereas the second term gives

$$\iint f^*(u - x_3 - p, v - y_3 - q)g(x_3, y_3)dudv$$

The term $|F|^2 + |G|^2$ gives rise to a point of light at the focus of the second lens.

The correlation integral has maximum value if $f(x_1, y_1)$ and $g(x_1, y_1)$ are identical.

For real-time applications, four-wave mixing in a photorefractive medium [15] is used. Fourier transforms F and G of coherent spatial distributions of horizontally plane-polarized light f and g are allowed to interfere with H, the Fourier transform of a third, vertically polarized distribution, h, in the photorefractive medium (Fig. 8.5). An output light signal S is obtained with

$$S \sim (FG^*)H$$

If F is a plane wavefront produced by a point source at the focus of a lens, then

$$S \sim G^*H$$

which provides the correlation of h and g in the front focal (Fourier) plane of lens L_1 where a CCD sensor array is located.

Alternatively, if G is a plane wavefront, joint transform correlation between f and h is obtained in the CCD sensor plane. since

$$S \sim FH^*$$

ENCRYPTION METHODS IN HOLOGRAPHY

In the holographic reconstruction process, the object beam is reproduced by diffraction of the reference beam by the hologram. If the object beam is to be reconstructed accurately, the correct reference beam must be used. This means that the wavelength and spatial distribution of phase in the reference beam must be the same as those used during the recording step. This requirement makes it possible to implement several hologram encryption methods.

Phase-Coded Reference Beam Encryption

Random phase encoding

To implement random phase encoding, a diffuser in the path of the reference beam in holographic recording imposes a random spatial phase distribution on the reference beam [16], creating a speckled reference beam. The minimum size, σ, of the speckles at the hologram recording plane is given by

$$\sigma \cong \frac{\lambda l}{d} \tag{8.6}$$

where d is the diameter of the laser beam of wavelength λ at the diffuser and l is the distance between diffuser and recording plane [17].

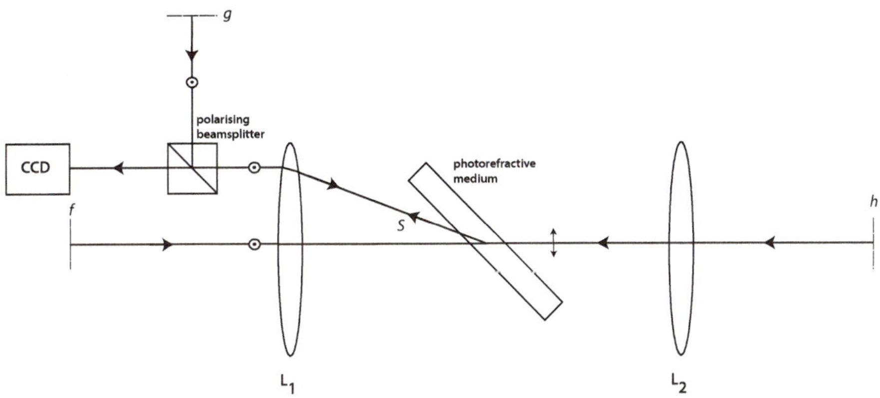

FIG. 8.5 Real-time matched filtering/joint transform correlation.

Successful reconstruction requires the diffuser to be precisely relocated in its original position. An error in positioning resulting in speckle displacement of greater than σ results in speckle decorrelation, and the corresponding holographic image will not be reconstructed.

Conventional ground glass diffusers can significantly reduce the optical power of the reference beam, but holographic diffusers with greater than 85% transmissivity are available. Alternatively an LCOS phase-only SLM may be used. Such devices with up to 4 million pixels and frame rate of tens of Hz allow rapid switching of the reference beam phase mask between hologram recordings. Again, reconstruction of the holographic image can only be successful if the correct phase mask is used as a decryption key.

An additional encryption key can be added by specifying a virtual wavelength. It can be shown that even with the correct phase mask in place at reconstruction, a deviation of 0.00002 nm from the specified wavelength of 623 nm means that the image cannot be decrypted [18].

In an extension of the random phase encoding technique [19], a key hologram, KH, is recorded using a diffuser as object. Then the conjugate image is reconstructed to serve as a reference beam with two converging object beams to record a security hologram (SH) (see Fig. 8.6). If the security hologram is reconstructed using the wrong phase-coded reference, then the focused spots of light produced by lenses L1 and L2 do not appear. Security is further enhanced by recording two holograms of each converging object beam with a slight axial or transverse displacement of lenses L1 and L2 between exposures. On reconstruction, a concentric ring fringe pattern is seen in the case of axial translation or straight-line parallel fringes in the case of an in-plane translation, as verification of the key hologram.

Deterministic orthogonal phase coding

A form of deterministic phase coding, called orthogonal phase coding [20], offers significant advantages over the use of random diffusers, as there are no mechanical parts in the system and no need for mechanical repositioning. The technique may be used to ensure secure storage of information in the form of data pages, each data page being recorded as a hologram.

A set of laser beams, each having the same amplitude and phase φ, which is either 0 or π relative to the others, comprises the reference wave in a holographic recording setup and interferes in the recording medium, usually a photorefractive crystal, with a data page O_n, n being an integer. The complete wavefront of the phase-coded reference wave must interfere with the data page. To ensure that this is the case, the page is Fourier transformed and overlapped in the recording medium with the reference wave focused so that it is discretized into its angular spectrum.

If N pages are to be recorded, there will be N phase-coded reference beams, and for each fresh data page, the set of reference beam phases is altered, so that the complete holographic recording of the N data pages is given by

$$\left(\sum_{n=1}^{N} O_n\right)\left(\sum_{n=1}^{N}\sum_{m=1}^{N} exp(j\varphi_{n,m})\right)^* + \text{complex conjugate}, \quad n, m \text{ integers}$$

On reconstruction, the holographic memory is illuminated by the phase-coded reference beam

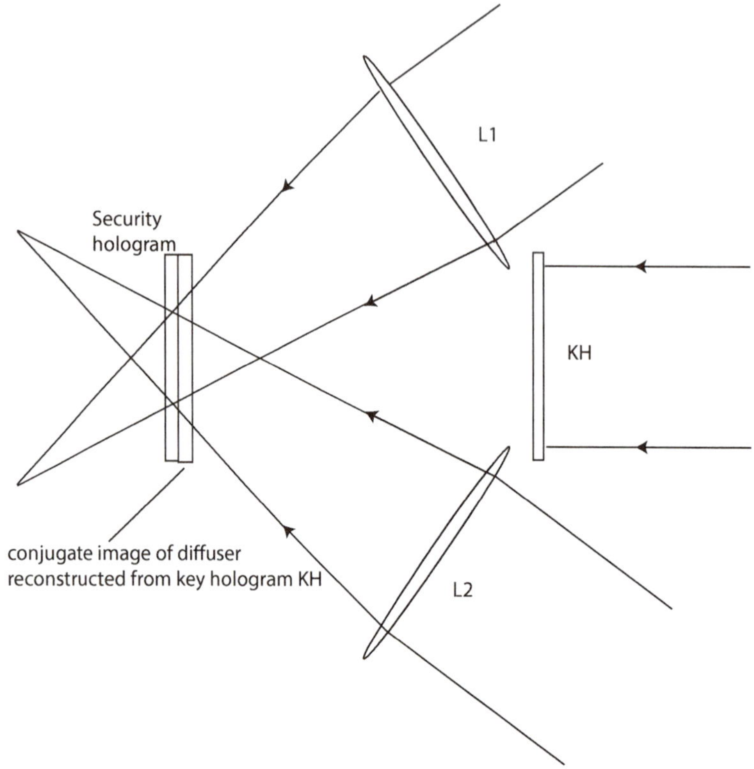

FIG. 8.6 Enhanced phase-coded reference beam security hologram.

$$\sum_{m=1}^{N} exp\left(j\varphi_{k,m}\right)$$

and the data page O_n is retrieved from the recorded set of N pages without cross-talk, provided that the phase codes are orthogonal.

$$\left(\sum_{m=1}^{N} exp\left(j\varphi_{k,m}\right)\right) O_n \left(\sum_{n=1}^{N}\sum_{m=1}^{N} exp\left(j\varphi_{n,m}\right)\right)^{*} = \begin{matrix} 0, k \neq n \\ 1 \end{matrix} \tag{8.7}$$

Eq. (8.7) means that to retrieve a data page, one must use the phase-coded reference beam with which it was recorded. The use of any other code produces zero output.

The appropriate orthogonal phase codes are obtained using the Walsh-Hadamard algorithm

$$H^{(0)} = 1, \quad H^{(1)} = \begin{bmatrix} 1 & 1 \\ 1 & -1 \end{bmatrix}, \quad H^{(p+1)} = \begin{bmatrix} H^{(p)} & H^{(p)} \\ H^{(p)} & -H^{(p)} \end{bmatrix} \tag{8.8}$$

where p is a positive integer. From Eq. (8.8), we obtain

$$H^{(2)} = \begin{bmatrix} 1 & 1 & 1 & 1 \\ 1 & -1 & 1 & -1 \\ 1 & 1 & -1 & -1 \\ 1 & -1 & -1 & 1 \end{bmatrix} \tag{8.9}$$

Zero phase shift is represented by 1 and a phase shift of π is represented by -1. For example, the second row of the matrix in Eq. (8.9) has phase code $0,\pi,0,\pi$. Then, if we record a data page using a reference wave with this phase code and attempt to retrieve it using the phase code in the third row, we obtain zero output.

The use of binary phase codes based on Hadamard matrices means that the number of phase shifts is either just 2 or an even integer power of 2 so the SLM is not used very efficiently. The use of unitary matrices to implement phase coding makes more efficient use of the SLM [21].

Double Random Encryption

Double random encryption techniques for image security applications have been extensively studied in the Fourier [22] and fractional Fourier domains (Fractional

Fourier Encryption section). The encrypted data can also be holographically recorded [23] using random or deterministic phase-coded reference beams.

An image $f(x_1, y_1)$ is multiplied by a phase mask in which phase is randomly distributed between *0* and 2π. The result

$$f(x_1, y_1)\exp[j2\pi n(x_1, y_1)]$$

(where $n = RND[0,1]$) is convolved with an optical transfer function $h(x_1,y_1)$ whose Fourier transform is a constant amplitude, random phase function exp $[j2\pi\nu(x_2,y_2)]$ giving encrypted output

$$\psi(x_1, y_1) = f(x_1, y_1)\exp[j2\pi n(x_1, y_1)] \otimes h(x_1, y_1) \qquad (8.10)$$

which is holographically recorded ($\otimes$ means convolution).

To decrypt the reconstructed image, $\psi(x_1,y_1)$ is Fourier transformed and then multiplied by exp $[-j2\pi\nu(x_2, y_2)]$, and the result is Fourier transformed again to give output

$$f(x_1, y_1)\exp[j2\pi n(x_1, y_1)]$$

A CMOS or other imaging device will provide output $|f(x_1, y_1)|^2$. A similar double encryption technique involves first converting the input $f(x_1, y_1)$ into a phase-only image [24] providing better performance.

Polarization Encryption

In double random polarization encryption [25], a binary image is converted to a pixel array of orthogonal linear polarization states using an SLM so that the image is rendered visible only by means of a linear polarizer. Double encryption is by means of polarization modulation masks using SLMs at the input and Fourier planes. Each SLM imposes a random phase shift between the orthogonal polarizations at each pixel so that the output light from the pixel is in a random elliptical polarization state. Polarization holograms are recorded in bacteriorhodopsin by photoinduced anisotropy and are optimally reconstructed using counterpropagating, orthogonal circularly polarized beams.

Fractional Fourier Encryption

It has been pointed out [26,27] that the propagation of light through a system of lenses is mathematically described by a continual process of fractional Fourier transformation. The amplitude and phase spatial distributions evolve through fractional transforms of increasing order.

A system of lenses placed between a planar, partially transparent object in the *x,y* plane and a hologram recording plane can act as a fractional Fourier encryption key [28] whose parameters, a, b, and α in the example shown in Fig. 8.7 are determined by the distances d_1, d_2, d_3, and d_4 and the focal lengths f_1 and f_2, all of which can be chosen arbitrarily.

The fractional Fourier transform calculated in the intermediate plane x',y' (with y' dependence omitted for simplicity) is

$$g(x') = K\int f(x)e^{j\pi\left(\frac{a^2x^2+b^2x'^2}{\tan\alpha}-\frac{2abx'x}{\sin\alpha}\right)}dx \qquad (8.11)$$

Its parameters are given by

$$a^2 = \frac{\sqrt{f_1 - d_2}}{\lambda\sqrt{f_1 - d_1}\left[f_1^2 - (f_1 - d_1)(f_1 - d_2)\right]^{\frac{1}{2}}}$$

$$b^2 = a^2\frac{f_1 - d_1}{f_1 - d_2}$$

$$\alpha = \cos^{-1}\sqrt{\frac{(f_1 - d_1)(f_1 - d_2)}{f_1}} \qquad (8.12)$$

The fractional Fourier transform at the hologram plane is calculated from Eq. (8.11) with the parameters a, b, and α obtained by substituting d_3 for d_1, d_4 for d_2 and f_2 for f_1 in Eq. (8.12).

Reconstruction by the conjugate reference wave leads to the formation of a decrypted real image at the original location of the object, only if the fractional Fourier decryption key is applied. This means that the light wave diffracted by the hologram at reconstruction must propagate through precisely the same optical system in reverse.

Additional security may be provided by double encryption (Double Random Encryption section), placing a random phase mask in contact with the input image and a second, statistically independent phase mask in the intermediate plane of the system shown in Fig. 8.7. Knowledge of either the phase masks alone or of the fractional Fourier encryption key alone will not suffice to decrypt the image [29].

Additional lenses may be included in the optical system to make the encryption key more secure.

A virtual axial translation may be given to each of the pixels in the input image that is to be encrypted. The translation distance is unique to each pixel, effectively converting the image into a 3-D object [30,31]. The

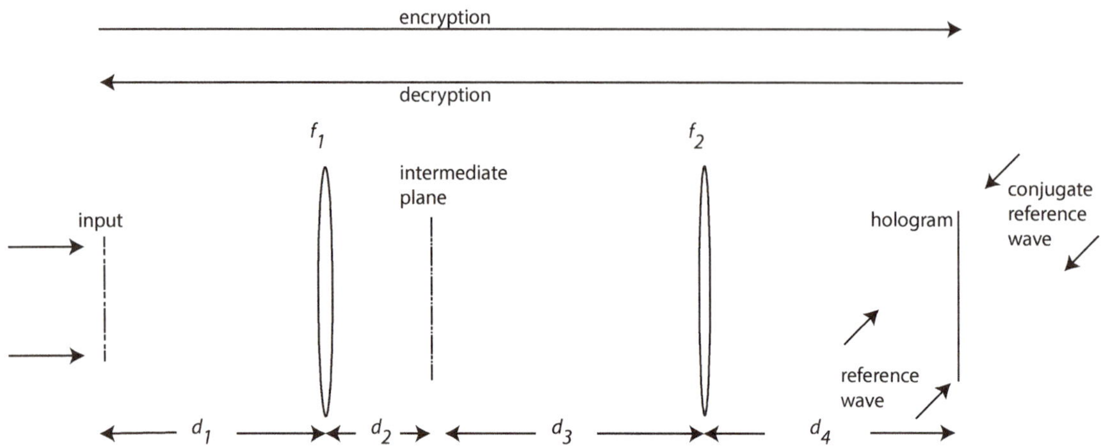

FIG. 8.7 Fractional Fourier holography.

spatial distribution of the translations, $d(x,y)$, is thus an additional encryption key.

Fresnel Domain Double Random Phase Encryption

Double random phase encryption may be implemented in the Fresnel domain for added security [32]. Only the two phase masks are required. Lenses are not needed.

In this case, the keys that can be used for encryption purpose are the distance d_1 between the two phase masks, the distance d_2 between the second phase mask and the output plane of the doubly encrypted image, as well as the wavelength.

The phase mask encryption keys alone are insufficient for decryption, as Fresnel transformation determines the complex amplitude distributions in the planes of the phase masks and errors in d_1, d_2, or wavelength will cause the decryption to fail. Decryption of the image requires reconstruction using the conjugate reference wave, or four-wave mixing, with the phase masks in their original positions.

One of the most important features of techniques such as double random, fractional Fourier and Fresnel encryption, is that the information may be secured and decrypted by computational methods.

DIGITAL HOLOGRAPHIC SECURITY

Digital holography using CCD or CMOS cameras as hologram recording devices has led to the development of holographic microscopy as well as applications in metrology and security. Here we describe some phase-shift methods for digital recording of encrypted holograms, as well as reconstruction of the encrypted images.

Phase Shift Digital Holography

It is not possible to retrieve the phase information from a single interference pattern without ambiguity. This is because the intensity modulation between two locations can be interpreted as having either a positive or a negative slope. Thus a difference in phase $\Delta\varphi$ between two locations in the pattern is given by $\Delta\varphi = \pm acos(\Delta I)$ with ΔI the difference in intensity and a the modulation amplitude of the pattern. Similarly, any abrupt reversal in the slope direction that cannot be resolved leads to a misinterpretation of the phase modulation.

To remove this ambiguity, phase shift methods are needed to obtain the encrypted image from the digital hologram. Phase shift digital holography normally requires the object and reference beams to be spatially separated in a Mach-Zehnder interferometer (Fig. 8.8) and the low spatial resolution of digital cameras ($\sim$100 lines mm^{-1}) requires that the beams are subsequently recombined onto a common propagation path before incidence on the CMOS sensor.

There are several algorithms involving two or more phase shifts for obtaining the map of phase difference, $\varnothing_O - \varnothing_R$ between the object and reference waves. A reliable algorithm should have a high degree of immunity to deviation of a phase shift from its specified value [33] and should not necessarily require that the phase shifts

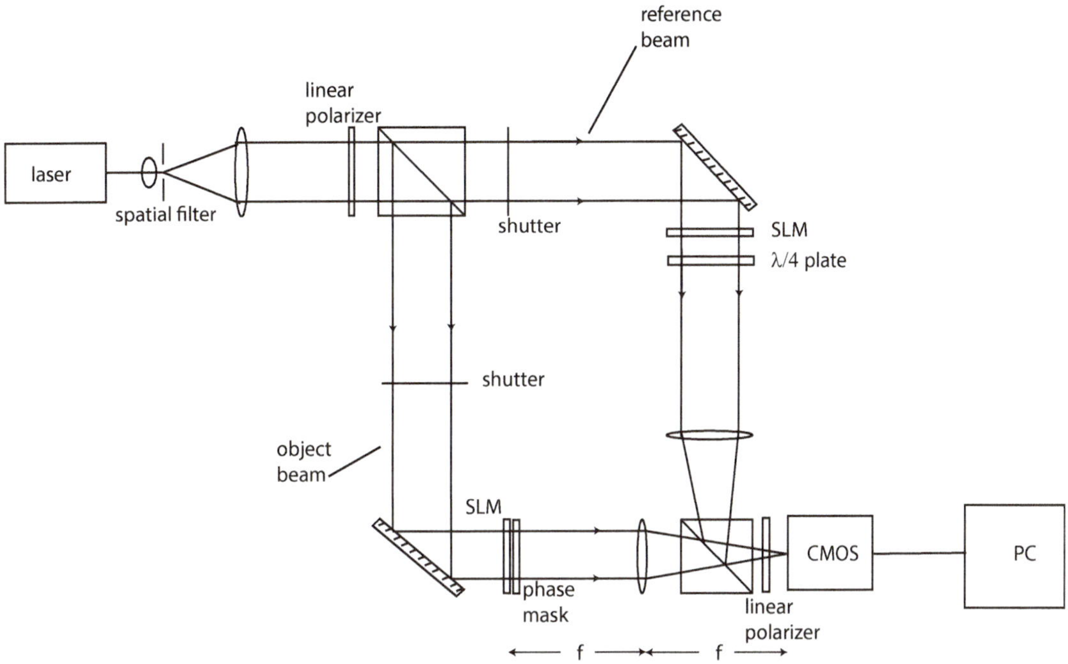

FIG. 8.8 Quadrature phase shift digital holographic encryption system.

be known. However, the greater the number of phase-shifted holograms, the greater the computational burden associated with the decryption and the greater the channel capacity required for transmission over a communication network.

Two phase shifts

A collimated light beam is spatially modulated by a 2-D image mask in input plane (x_1, y_1) to give an object wave $o(x_1, y_1)$, which is then randomly phase modulated by a phase-only SLM giving [34].

$$o(x_1, y_1)\exp[j\phi(x_1, y_1)]$$

This is Fourier transformed giving

$$O(x_2, y_2) = |O(x_2, y_2)|exp[j\varnothing_O(x_2, y_2)] \tag{8.13}$$

in the plane (x_2, y_2) of the camera sensor. A collimated reference wave spatially modulated by a phase-only SLM using random phase values of 0 and π is also Fourier transformed to become

$$R(x_2, y_2) = |R(x_2, y_2)|exp[j\varnothing_R(x_2, y_2)] \tag{8.14}$$

acting as an encryption key.

A hologram $I(x_2, y_2)$ is recorded with

$$I(x_2, y_2) = |O(x_2, y_2)|^2 + |R(x_2, y_2)|^2 + 2|O(x_2, y_2)||R(x_2, y_2)| \times |cos(\varnothing_O - \varnothing_R) \tag{8.15}$$

A second hologram recorded with a phase shift of $\pi/2$ between the beams is given by

$$I'(x_2, y_2) = |O(x_2, y_2)|^2 + |R(x_2, y_2)|^2 + 2|O(x_2, y_2)||R(x_2, y_2)| \times |sin(\varnothing_O - \varnothing_R) \tag{8.16}$$

The D.C. terms $|O(x_2, y_2)|^2$ and $|R(x_2, y_2)|^2$ can be obtained separately and removed from the expressions for I and I' by alternately blocking the reference and object beams, to obtain

$$J = |O(x_2, y_2)||R(x_2, y_2)|cos(\varnothing_O - \varnothing_R) \tag{8.17}$$

$$J' = |O(x_2, y_2)||R(x_2, y_2)|sin(\varnothing_O - \varnothing_R) \tag{8.18}$$

where

$$\varnothing_O - \varnothing_R = \tan^{-1}\left({J'}/{J}\right), \quad |O(x_2, y_2)||R(x_2, y_2)| = \sqrt{J^2 + J'^2}$$

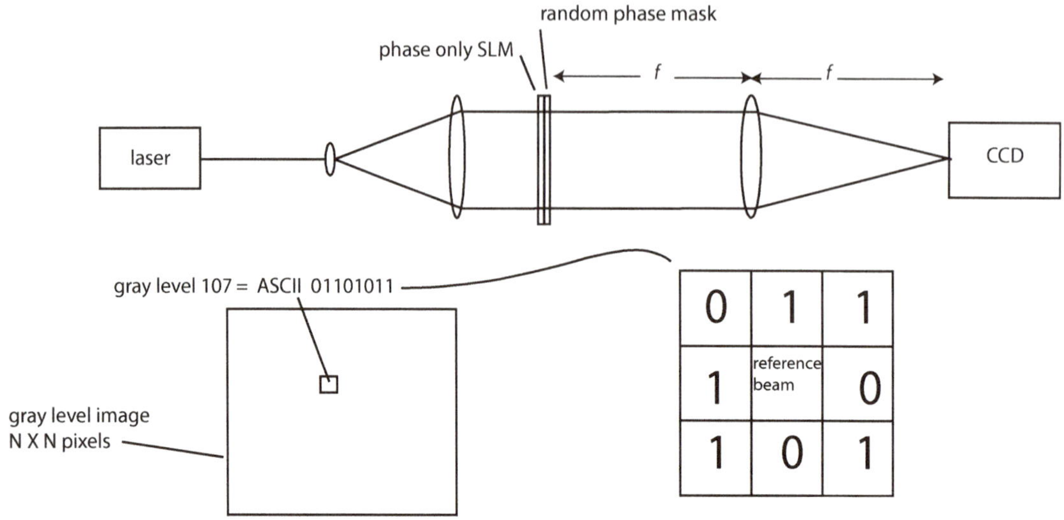

FIG. 8.9 Phase shift digital holography using a partitioned SLM.

from which we obtain

$$H(x_2, y_2) = |O(x_2, y_2)||R(x_2, y_2)|\exp[j(\varnothing_O - \varnothing_R)] \quad (8.19)$$

To decrypt the image, $H(x_2, y_2)$ is multiplied by the encryption key $R(x_2, y_2)$ and the result divided by $|R(x_2, y_2)|^2$ giving

$$|O(x_2, y_2)|\exp[j\varnothing_O(x_2, y_2)]$$

which is inverse Fourier transformed and the modulus of the result $|o(x_1, y_1)|$ is the original input image.

Phase shift can be implemented by a piezoelectrically translated plane mirror in the path of the reference beam.

Alternatively, as shown in Fig. 8.8, light that is plane polarized at 45 degrees to the plane of the Mach-Zehnder interferometer is split between the two arms, the reference beam passing through a quarter wave plate whose fast axis is set normal to the interferometer plane. The beams are recombined at a second beam splitter. A linear polarizer with transmission axis normal to the interferometer plane allows interference between vertically plane polarized components of the object and reference waves. In-plane rotation of the linear polarizer by 90 degrees enables interference of the horizontally plane-polarized components but with a phase shift of 90 degrees between them.

Partitioned spatial light modulator

In a simpler approach, a gray level image is first converted into ASCII form and multiplied by a random spatial phase mask in one part of a phase-only SLM (Fig. 8.9) located in the front focal plane of a lens and illuminated normally by a collimated laser beam [35]. The other part of the SLM is used to display a statistically independent, random phase pattern acting as a reference beam and encryption key. Two digital Fourier holograms with a phase shift of $\pi/2$ between them are recorded by a CCD or CMOS sensor in the back focal plane of a lens. To increase the dynamic range of the Fourier transform and thus make optimal use of the image sensor, a random phase mask is placed in contact with the SLM.

The procedure is essentially as described in the Two phase shifts section.

The SLM is partitioned so that gray levels are 8-bit encoded clockwise into 3×3 arrays of pixels starting at the upper left pixel. A 128×128 input image thus occupies 384×384 pixels, the middle pixel in each 3×3 array being used for the reference beam. A disadvantage is that the ratio of object beam to reference beam intensity is as high as 8:1, but the object beam intensity range can be reduced to obtain a more favorable ratio.

Quadrature phase shift modulation

In quadrature phase shift keying (QPSK) modulation [36], one starts by ASCII encoding the binary gray level at each pixel of the image. A random phase mask is added. Quadrature phase values are then assigned to each pair of binary digits starting with the most significant bit, for 8-bit gray level as follows.

Binary	Phase
00	0
01	$\pi/2$
10	π
11	$3\pi/2$

Four pixels are needed for each image pixel with the phases written clockwise to a phase-only SLM in the manner shown by the following examples.

Gray Level	ASCII	QPSK	
45	00101101	0	π
		$\pi/2$	$3\pi/2$
141	10001101	π	0
		$\pi/2$	$3\pi/2$

The phase image is recorded as a Fourier hologram. Reconstruction by the reference beam is followed by inverse Fourier transformation. Multiplication by the conjugate of the encryption phase map decrypts the original ASCII image.

HOLOGRAPHY FOR IMAGING OF CONCEALED OBJECTS

The movement of contraband, chemical and biological weapons, firearms, explosives, and narcotics within and across national boundaries is a matter of worldwide concern. Such items are usually concealed under optically opaque materials or in vehicle compartments lacking ease of rapid access, so the problem calls for the use of penetrating radiation usually in the millimeter and micrometer wavelength ranges. The competing demands for high spatial resolution and the ability to penetrate concealing materials favor the use of frequencies ranging from 300 to 1000 GHz [37].

The need for rapid acquisition of images suggests the use of a planar array of equally spaced detectors to record the interference pattern, but such systems have low resolution, with a small field of view and are expensive. A reasonable compromise may be achieved by rapid scanning in the hologram plane using a single detector or a linear array of detectors scanning in one direction.

Direct detection methods use a vector network analyzer (VNA) to measure both the amplitude and phase of the scattered radiation, relative to the amplitude and phase of the wave propagating directly from the transmitter and detected at a fixed location [38,39]. The scattered wave is detected at a number of points in the hologram plane. Thus, the method allows one to obtain directly the spatial distribution of amplitude and phase of the wavefront scattered by the object. VNAs are rather expensive, however.

Indirect recording of the complete interference pattern has been implemented by microwave heating of a cholesteric liquid crystal layer [40]. This produces changes in the color of light reflected from the layer, which can be photographed in monochrome to obtain a copy hologram, from which the image is reconstructed using a visible laser. The sensitivity of the liquid crystal layer is rather low, at about 10 mW cm^{-2}.

In another method [41], the illuminating radiation is phase locked to a local oscillator (LO). The receiver antenna is scanned over the hologram plane. The received signal is mixed with the LO to obtain an intermediate frequency, which is low-pass filtered. The resulting voltage is proportional to the local phase of the received wave and modulates the intensity of a cathode ray tube light spot whose motion pattern is the same as that of the receiver. A photographic recording of the phase front of the wave scattered by the object may then be obtained, that is, a phase-only hologram.

In indirect methods, the spatial pattern of intensity due to interference of the object and reference waves is scanned in orthogonal directions in the hologram plane by an antenna connected to a detector or along just one axis in that plane by a linear array of detectors. This method resembles the optical holographic method of Leith and Upatneiks (see Ref. [4] and Chapter 1 Off-Axis Transmission Hologram (Leith and Upatnieks) section) in that the phase difference between the radiation scattered by the object and that of the directly received reference radiation is encoded in the local spacing of an actual interference pattern.

Typically the spatial Fourier transform of the recorded hologram is calculated and spatially filtered to extract the angular spectrum of the object wave. Recall that in optical off-axis holography, one can use a plane reference wave incident at an angle θ to the normal at the hologram recording plane (x, y) so that the phase of the reference wave varies linearly along the x axis (say) of the hologram plane. However, at millimeter or micrometer wavelengths this is rather difficult.

One solution is to synthesize the reference beam [42]. The output from a microwave transmitter of wavelength λ is split into two parts by a directional coupler. One part illuminates the object lying in a plane parallel to the hologram plane, at $z = -z'$, whereas the other is passed through a phase shifter and a variable attenuator to provide a synthesized reference signal of constant amplitude. The amplitude of the reference signal is set

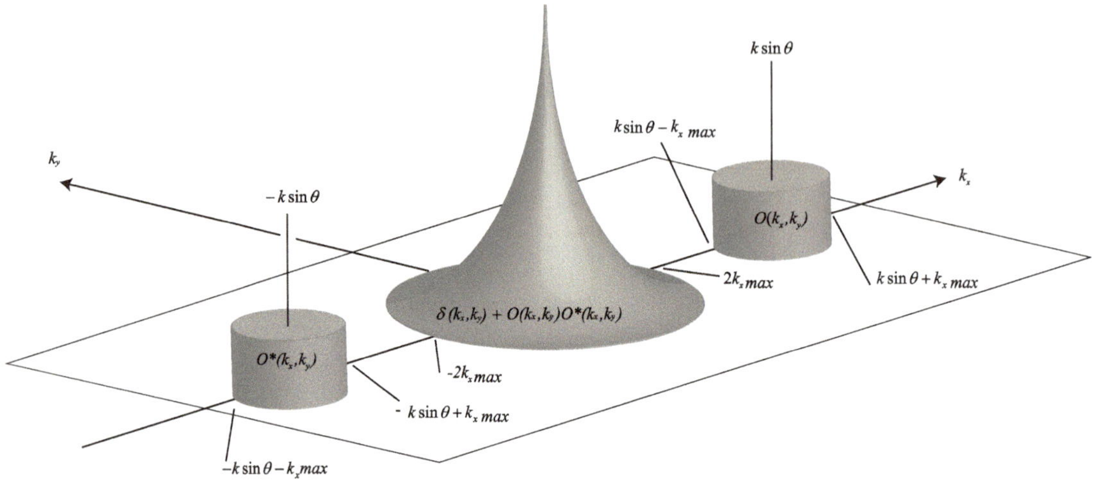

FIG. 8.10 Angular spatial frequency spectrum.

at unity, and its phase is made linearly dependant on position along the x axis of the hologram recording plane. Thus the reference signal is given by $exp(jkxsin\theta)$ where $k = 2\pi/\lambda$ and $\theta =$ constant, and its angular spatial frequency measured along the x axis is $2\pi sin\theta/\lambda$.

A receiver antenna scans the hologram plane (x, y). The wave scattered by the object is $o(x, y)$ in the hologram plane and is mixed with the reference signal in a hybrid tee to provide an output proportional to the intensity of interference at each point of the scan.

The spatially dependant intensity is given by

$$I(x,y) = |o(x,y)|^2 + 1 + o^*(x,y)\exp(jkxsin\theta) + o(x,y)\exp(-jkxsin\theta) \tag{8.20}$$

Spatial Fourier transformation (SFT) gives the angular spatial frequency spectrum of $I(x, y)$ and, using the convolution theorem, one obtains

$$SFT[I(x,y)] = [O(k_x,k_y)\otimes O^*(k_x,k_y) + \delta(k_x)] + O^*(k_x,k_y)\otimes\delta(k_x - ksin\theta) + O(k_x,k_y)\otimes\exp(k_x + ksin\theta) \tag{8.21}$$

The spectrum is shown in Fig. 8.10.

The value of θ is chosen so that $sin\theta \geq 3k_{x_{max}}$ where $k_{x_{max}}$ is the maximum angular spatial frequency in the object wave, to ensure that the three terms in the angular spatial frequency spectrum (Eq. 8.21) do not overlap. A band-pass filter centered on spatial frequency $(ksin\theta,0)$ is used to isolate $O(k_x, k_y)$, which is then spatial frequency shifted to center on (0,0). The object wave in the hologram plane is obtained from the spatial Fourier transform of $O(k_x, k_y)$.

$$o(x,y) = \int\int_{k_x,k_y} O(k_x,k_y)dk_x\, dk_y \tag{8.22}$$

The angular spectrum of the object wave consists of a set of plane waves with propagation vectors $k_x, k_y, \sqrt{1-k_x^2-k_y^2}$ where $1-k_x^2-k_y^2 = k_z^2$, and the object wave in a plane (x, y) at $z = -z'$ is related to that at $z = 0$ by

$$O_{z=-z'}(x,y) = O_{z=0}exp(-jk_zz') = \int\int_{k_x,k_y} O_{z=0}(k_x,k_y)\exp[-j(k_xx + k_yy + k_zz')]dk_x\, dk_y \tag{8.23}$$

from which by carrying out a spatial Fourier transform, a 2-D image of an object located at $z = -z'$ can be obtained. A 3-D image can be built up by calculating the object wave in a set of closely spaced planes normal to the z-axis, around the midplane of the object. The resolution along the z-axis is quite low unless a large area is scanned in the hologram plane. This method of image reconstruction is known as single-frequency backward wave projection.

Wideband illumination enables the reconstruction of 3-D images rather than 2-D image slices [38]. The reflective properties of a 3-D object are described by the function $f(x, y, z)$. The response of a detector at x', y', Z_1 and at temporal angular frequency ω is

$$s(x', y', \omega) = \iiint f(x, y, z) e^{-2jk\sqrt{(x-x')^2+(y-y')^2+(z-Z_1)^2}} dxdydz \tag{8.24}$$

where the amplitude of the transmitted wave is assumed constant and the receiver coincides with the transmitter.

The expression $e^{-2jk\sqrt{(x-x')^2+(y-y')^2+(z-Z_1)^2}}$ can be written

$$e^{-2jk\sqrt{(x-x')^2+(y-y')^2+(z-Z_1)^2}} = \iint e^{-j[k_{x'}(x-x')+k_{y'}(y-y')+k_z(z-Z')]} dk_{x'} dk_{y'} \tag{8.25}$$

The angular spatial frequencies $k_{x'}$ and $k_{y'}$ range from $-2k$ to $2k$.

From Eqs. 8.24 and 8.25, we obtain

$$s(x', y', \omega) = \iint F(k_{x'}, k_{y'}, k_z) e^{j(k_{x'}x'+k_{y'}y'+k_z Z_1)} dk_{x'} dk_{y'} \tag{8.26}$$

where

$$F(k_{x'}, k_{y'}, k_z) = \left[\iiint f(x, y, z) e^{-j(k_{x'}x+k_{y'}y+k_z z)} dxdydz\right]$$

is a 3-D spatial Fourier transform of $f(x,y,z)$.

Eq. (8.26) can be rewritten as

$$s(x', y', \omega) = IFT\{F(k_{x'}, k_{y'}, k_z) e^{jk_z Z_1}\} \tag{8.27}$$

Fourier transforming each side of Eq. (8.27) and removing the distinction between primed and unprimed coordinates, gives

$$F(k_x, k_y, k_z) = FT_{2D}[s(x, y, \omega)] e^{-jk_z Z_1} \tag{8.28}$$

We can eliminate k_z since $k_x^2 + k_y^2 + k_z^2 = 4k^2$ and carry out an inverse transformation to form the image, i.e.,

$$f(x, y, z) = IFT_{3D}\left\{FT_{2D}\{s(x, y, \omega)\} \exp\left(-j\sqrt{4k^2 - k_x^2 - k_y^2}\, Z_1\right\}\right. \tag{8.29}$$

where $s(x, y, \omega)$ is the receiver response at x, y in the hologram plane, at temporal angular frequency ω, and Z_1 is the distance between the transmitter/receiver and the x, y plane.

Because $s(x, y, \omega)$ is sampled at equal intervals in x, y and ω, resampling by linear interpolation is required at equal intervals of angular spatial frequency k_z.

Techniques that make use of both forward and back-scattered radiation have also been implemented [43], and 3-D images have been obtained using single-frequency radiation [44].

Finally an interesting application of the backpropagation method has been reported [45]. The coherence of the radiation from a 5 GHz Wi-Fi router is exploited to record holograms over an area of 1 m^2. The router bandwidth of 70 MHz implies a coherence length of 4.3 m. Reflections from walls create speckle, but summing several reconstructed images acquired at a number of frequencies within the band helps to reduce its effect. A type of dark ground illumination is implemented to enhance the images. The technique may be exploited in tracking radiofrequency identification tags, 3-D motion capture, and tracking for gaming and imaging through walls for security applications. It does, however, raise concerns about privacy and personal security.

REFERENCES

[1] OECD/EUIPO, Trade in Counterfeit and Pirated Goods: Mapping the Economic Impact, Paris, 2016.

[2] OECD, The Economic Impact of Counterfeiting and Piracy, 2008.

[3] H. Clifton, M. Chapman, S. Cox, 'Staggering' Trade in Fake Degrees Revealed – BBC News, 2018. Available: https://www.bbc.com/news/uk-42579634.

[4] E. Leith, J. Upatneiks, Wavefront reconstruction with diffuse illumination and three-dimensional objects, Journal of the Optical Society of America 54 (11) (1964) 1295–1301.

[5] S. Benton, Hologram Reconstructions with extended incoherent sources, Journal of the Optical Society of America 59 (1969) 1545–1546.

[6] X. Zhang, E. Dalsgaard, S. Liu, H. Lai, J. Chen, Concealed holographic coding for security applications by using a moiré technique, Applied Optics 36 (31) (1997), pp. 8096–7.

[7] A.K. Aggarwal, S.K. Kaura, D.P. Chhachhia, A.K. Sharma, Concealed moiré pattern encoded security holograms readable by a key hologram, Optics & Laser Technology 38 (2) (2006) 117–121.

[8] Y.N. Denisyuk, Photographic reconstruction of the optical properties of an object in its own scattered radiation field, Soviet Physics – Doklady 7 (1962) 543.

[9] D. Vather, et al., Serialised holography for brand protection and authentication, Applied Optics 57 (22) (2018) E131–E137.

[10] I. Naydenova, H. Sherif, S. Martin, R. Jallapuram, and V. Toal, US Patent 8,535,853 B2, (2013).

[11] F.D.C. Vasconcellos, et al., Printable surface holograms via laser ablation, ACS Photonics 1 (6) (2014) 489–495.
[12] A.B. Vander Lugt, Signal detection by complex spatial filtering, IEEE Transactions on Information Theory 10 (2) (1964) 139–145.
[13] D. Casasent, D. Psaltis, Position, rotation, and scale invariant optical correlation, Applied Optics 15 (7) (1976) 1795–1799.
[14] P. Bone, R. Young, C. Chatwin, Position-, rotation-, scale-, and orientation-invariant multiple object recognition from cluttered scenes, Optical Engineering 45 (7) (2006) 077203, 8.
[15] D. Vacar, A.J. Heeger, B. Volodin, B. Kippelen, N. Peyghambarian, Compact, low power polymer-based optical correlator, Review of Scientific Instruments 68 (2) (1997) 1119.
[16] A.M. Darskii, V. Markov, Shift selectivity of holograms with a reference speckle wave, Opt. Spectoskopy 65 (1988) 392–395.
[17] J.C. Dainty (Ed.), Laser Speckle and Related Phenomena, Springer, 1976.
[18] X. Peng, L. Yu, L. Cai, Double-lock for image encryption with virtual optical wavelength, Optics Express 10 (1) (2002) 41–45.
[19] A.K. Aggarwal, S.K. Kaura, D.P. Chhachhia, A.K. Sharma, Encoded reference wave security holograms with enhanced features, Journal of Optics A: Pure and Applied Optics 6 (2) (2004) 278–281.
[20] C. Denz, K.O. Müller, T. Heimann, T. Tschudi, Volume holographic storage demonstrator based on phase-coded multiplexing, IEEE Journal of Selected Topics in Quantum Electronics (1998).
[21] X. Zhang, Unitary matrices for phase-coded holographic memories, Optics Letters 31 (8) (2006) 1047–1049.
[22] P. Refregier, B. Javidi, Optical image encryption based on input plane and Fourier plane random encoding, Optics Letters 20 (7) (1995) 767.
[23] B. Javidi, G. Zhang, J. Li, Encrypted optical memory using double-random phase encoding, Applied Optics 36 (5) (1997) 1054–1058.
[24] N. Towghi, B. Javidi, Z. Luo, Fully phase encrypted image processor, Journal of the Optical Society of America A 16 (8) (1999) 1915.
[25] O. Matoba, B. Javidi, Secure holographic memory by double-random polarization encryption, Applied Optics 43 (14) (2004) 2915–2919.
[26] H.M. Ozaktas, D. Mendlovic, Fractional Fourier optics, Journal of the Optical Society of America A 12 (4) (1995) 743.
[27] H.M. Ozaktas, D. Mendlovic, Every Fourier optical system is equivalent to consecutive fractional-Fourier-domain filtering, Applied Optics 35 (17) (1996) 3167–3170.
[28] Y.S. Zeng, Y.K. Guo, F.H. Gao, J.H. Zhu, Principle and application of multiple fractional Fourier transform holography, Optics Communications 215 (1–3) (2003) 53–59.
[29] X. Wang, D. Zhao, L. Chen, Image encryption based on extended fractional Fourier transform and digital holography technique, Optics Communications 260 (2) (2006) 449–453.
[30] W. Chen, X. Chen, Space-based optical image encryption, Optics Express 18 (26) (2010) 1575–1577.
[31] C. Lin, X. Shen, B. Li, Four-dimensional key design in amplitude, phase, polarization and distance for optical encryption based on polarization digital holography and QR code, Optics Express 22 (17) (2014) 20727.
[32] G. Situ, J. Zhang, Double random-phase encoding in the Fresnel domain, Optics Letters 29 (14) (2004) 1584–1586.
[33] J. Li, et al., An advanced phase retrieval algorithm in N-step phase-shifting interferometry with unknown phase shifts, Scientific Reports 7 (March) (2017) 1–12.
[34] S.K. Gil, 2-Step quadrature phase-shifting digital holographic optical encryption using orthogonal polarization and error analysis, Journal of the Optical Society of Korea 16 (4) (2012) 354–364.
[35] S.H. Jeon, Y.G. Hwang, S.K. Gil, Optical encryption of gray-level image using on-axis and 2-f digital holography with two-step phase-shifting method, Optical Review 15 (4) (2008) 181–186.
[36] S.H. Jeon, S.K. Gil, QPSK modulation based optical image cryptosystem using phase-shifting digital holography, Journal of the Optical Society of Korea 14 (2) (2010) 97–103.
[37] D.M. Sheen, D.L. McMakin, J. Barber, T.E. Hall, R.H. Severtsen, Active imaging at 350 GHz for security applications, Proceedings of SPIE 6948 (March 2015) (2008), p. 69480M–69480M–10.
[38] D.M. Sheen, D.L. McMakin, T.E. Hall, Three-dimensional millimeter-wave imaging for concealed weapon detection, IEEE Transactions on Microwave Theory and Techniques 49 (9) (2001) 1581–1592.
[39] D.M. Sheen, D. McMakin, T.E. Hall, Near-field 3-dimensional radar imaging techniques and applications, Applied Optics 49 (19) (2010) E83–E93.
[40] C.F. Augustine, C. Deutsch, D. Fritzler, E. Marom, Microwave holography using liquid crystal detectors, Proceedings of the IEEE 57 (1969) 1333.
[41] N.H. Farhat, W.R. Guard, Millimeter wave hologrphic imaging of concealed weapons, in: Proc. IEEE, no. September, 1971, pp. 1383–1384.
[42] D. Smith, M. Elsdon, M. Leach, M. Fernando, S.J. Foti, 3D Microwave imaging for medical and security applications, 2006 Int. RF Microw. Conf. Proc. 00 (4) (2006) 233–237.
[43] R.K. Amineh, M. Ravan, A. Khalatour, N.K. Nikolova, 3-D Near-field microwave holography using reflected and transmitted signals, IEEE Tranactions Antennas Propag 59 (12) (2011) 4777–4789.
[44] R.K. Amineh, M. Ravan, R. Sharma, S. Baua, Three-dimensional holographic imaging using single frequency microwave data, International Journal of Antennas and Propagation (2018).
[45] P.M. Holl, F. Reinhard, Holography of Wi-fi radiation, Physical Review Letters 118 (2017), pp. 183901-01 183901-05.

Index

Note: Page numbers followed by "t" indicate tables, "f" indicate figures.

Kolibri 360 GmbH
Ferdinand-Jühlke-Straße 7
99095 Erfurt, Deutschland
produktsicherheit@kolibri360.de